Network Scheduling Techniques for Construction Project Management

Nonconvex Optimization and Its Applications

Volume 16

The titles published in this series are listed at the end of this volume.

Network Scheduling Techniques for Construction Project Management

by

Miklós Hajdu
Technical University of Budapest
Budapest, Hungary

SPRINGER-SCIENCE+BUSINESS MEDIA, B.V.

A C.I.P. Catalogue record for this book is available from the Library of Congress

ISBN 978-1-4419-4765-9 ISBN 978-1-4757-5951-8 (eBook)
DOI 10.1007/978-1-4757-5951-8

Printed on acid-free paper

CONTENTS

PREFACE

Industrial, financial, commercial or any kinds of project have at least one common feature: the better organized they are, the higher the profit or the lower the cost. Project management is the principle of planning different projects and keeping them on track within time, cost and resource constraints. The need for effective project management is ever-increasing. The complexity of the environment we live in requires more sophisticated methods than it did just a couple of decades ago. Project managers might face insurmountable obstacles in their work if they do not adapt themselves to the changing circumstances. On the other hand, better knowledge of project management can result in better plans, schedules and, last but not least, more contracts and more profit. This knowledge can help individuals and firms to stay alive in this competitive market and, in the global sense, utilize the finite resources of our planet in a more efficient way.

Project management is a multidiscipline which has emerged to answer practical problems. We address this book to those readers who aspire to a still higher degree of professionalism in management, especially in network techniques. Network techniques are the basic mathematical tools for planning and scheduling any kind of project. Dozens of different network techniques are known in the project management science, but few are used in practice. This book gives a comprehensive overview of deterministic network techniques, the most often used techniques in practice. Special emphasis was put on the Precedence Diagramming

Method, as more than ninety percent of the industrial applications are based on PDM. New theoretical improvements of PDM are also discussed in this book. The introduction of maximal type of precedence relationships and the discussion of PDM time-cost trade-off cannot be found in any other text.

A PC-based software packege called ProjectDirector, containing the theoretical improvements described in this book, is available from the author. It is an improvement over the commercial computer programs. For example, it can handle the maximal type of precedence relationships and shows the critical path characteristics and other improvements described in this book. More information can be found in Appendix B.

The author would appreciate any comments and suggestions for further improvements. Please send your comments to:

Dr. Miklos HAJDU
Internet:MHajdu@eporg.ekt.bme.hu
http://www.plansys.hu

Budapest , June 1996 Miklos Hajdu, Ph.D.

ACKNOWLEDGEMENTS

I started to write this book in the summer of 1995 in Holland, where I spent my sabbatical year at the Department of Civil Engineering Informatics at Delft University of Technology. I would like to thank my colleague, Prof. Peter van der Veer, chair of the department, for his continuous support. He also provided me with a quiet and peaceful working environment, which is the most important factor in completing this kind of work.

The author also wishes to express his sincere appreciation to Dr Jozsef Cser and Dr Reza Beheshti, faculty of the department, for their helpful comments during the work.

Finally, I wish to thank Mr. Arpad Horvath from the Department of Civil and Environmental Engineering at Carnegie Mellon University for all of his help, particularly in evaulating and editing the manuscript.

Chapter 1

INTRODUCTION.

1.1 WHAT IS A SCHEDULE? WHAT IS IT GOOD FOR?

Looking for the word "schedule" in Webster's Ninth New Collegiate Dictionary we find the following description:

schedule *(noun)*

1 a *obs.*:a written document **b:** a statement of supplementary details appended to legal or legislative document

2: written or printed list, catalog; or inventory; also TIMETABLE 1

3: PROGRAM; *esp.*: a procedural plan that indicates the time and sequence of each operation

4: a body of items to be dealt with: AGENDA

schedule *(verb)*

1 a: to place in a schedule **b:** to make a schedule of

2: to appoint, assign, or designate for a fixed time

Everyone makes a schedule to organize his or her own everyday life. We register the date of meetings, business lunches, when to go shopping, when to pick up the children from school, etc. As we, ourselves are the only human resource for such a plan, it has to reflect that we cannot be at two or more places at the same time. A failure of this kind of planning will result in a lot of confusion, last-minute rush in our work or sometimes problems in our private life.

Making a schedule for a construction or any kinds of project means the same: a plan with the sequence of operations and the list of resources, i.e., work force, materials, machines, as they correspond to the project. Of course making a schedule for a construction project is far more difficult than making a schedule of our everyday life.

Why do we say that? First, in a construction project we have to organize hundreds sometimes thousands of activities and the proper sequencing of so many activities demands a lot of knowledge and years of experience. Second, we can use hundreds of independent resources in a construction project, and these resources are very often limited both in time and size. The third reason is that during the preparation of a schedule we have to satisfy other requests: the deadline of the major phases of the project, the completion date of the construction, budget constraints, change orders, etc. To stay within the deadline and the budget is usually the most serious problem for a project manager. Finally, the reason why it is so difficult to make a schedule is the complexity and the computational work involved! Our mind is unfortunately limited and we are not always able to keep all the necessary data that we need for efficient decision making in our head.

A failure to make a schedule or devising a wrong schedule can result in delay of a deadline and can cost the various stakeholders in the project a lot of money. No doubt, a proper and well-established schedule can help a lot to avoid excess costs and unwanted delays. A well-organized construction project is finished quicker and cheaper than a badly organized. No wonder that we use more sophisticated scheduling techniques in the construction industry than in our everyday life.

In industry special tools have been developed to assist project managers to make their work more efficient. These tools are summarized in the next section.

1.2 TRADITIONAL SCHEDULING TECHNIQUES

Scheduling techniques can be classified in many ways. In this chapter we classify them as traditional or non-network and network-based scheduling techniques. Network-based scheduling techniques are used only since the second half of this century. What kinds of techniques were used before that? We know a little about the history of project scheduling. Were there project planning and scheduling methods in Egyptian times? They might have had scheduling problems however, they certainly did not have limitations in manpower and this could have been a solution for many problems.

The first scientific consideration of scheduling problems can be dated back to the turn of this century when Henry Gantt published his work. This work is considered as the basis of today's bar charts. A bar chart is a collection of activities listed vertically and the horizontal scale represents the time.

The bar chart is an easy-to-understand and efficient tool, which is widely used even in today's construction practice. A sample bar chart is reproduced in Fig. 1-1.

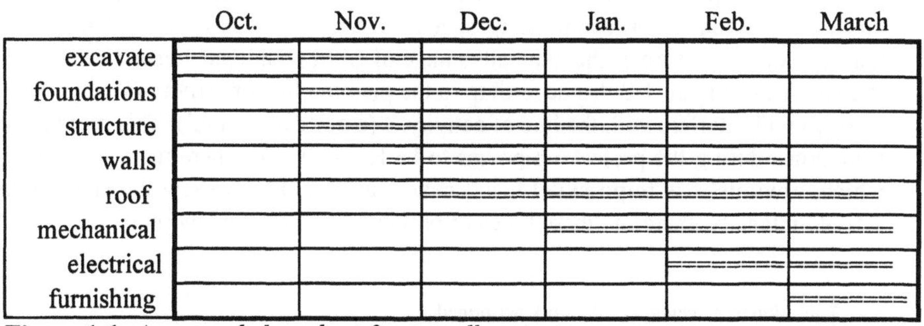

Figure 1-1 *An example bar chart for a small project*

There are lot of advantages in using a bar chart. The project manager can get information in a graphical way, and no special knowledge is necessary to understand a Gantt chart. The biggest disadvantage of using a Gantt chart manifests when the number of activities exceed a certain number. A Gantt chart cannot retain the logical relationship among activities; it only shows when activities have to start. Suppose that we have a project with thousands of activities and during the construction one activity has finished behind schedule. This delay has numerous impacts for an unknown number of activities. Unfortunately, the Gantt chart will not provide any information about the number of activities affected. Even the best designed and set up schedule has to be modified, updated or corrected during the construction and above a certain, relatively small number of activities a Gantt chart cannot help in this work. Maintaining a plan is simply too burdensome for large projects. Another problem arises during the preparation process of a bar chart: During the preparation the scheduler may be influenced by the desired completion dates and -- as the bar chart can be flexibly modified -- this desire can reflect in the plan without connection to the reality. The reason for this is that the planning and scheduling phases are considered simultaneously. The use of these two elements together often makes the plan a set of wishful thinking.

Summarizing the pros and cons of using bar charts in scheduling we can say that a bar chart is a very good tool in scheduling of small projects and can be the best visualization form of any size projects. However, it is not an adequate tool for planning and scheduling big projects.

An alternative traditional scheduling technique is the LSM (Linear Scheduling Method). The origin of this method is not quite clear. It has some relation to the line of balance (LOB) schedules. A basic format of an ordinary LSM schedule is shown in Figure 1-2. The horizontal axis plots time, the vertical axis plots location along the length of project or percentage of the work performed. Individual activities are plotted separately and can be broken at any time. This means that the completion of the activity in question has been suspended. The slope of activity line can change; this indicates that the accomplishing speed of the activity has changed for some reason. This feature gives the main advantage of LSM technique versus Gantt chart. We can see in Figure 1-2 that according to the plan the line of activity C (e.g., 'pipe layout') will interfere with the line of activity B (e.g., 'excavation'). This means that sometimes during the project a pipe has to be laid down to where there is no ditch which, **obviously, is impossible!** Gantt charts are not able to show the changes in activity intensity, therefore these kinds of problems can arise at any time.

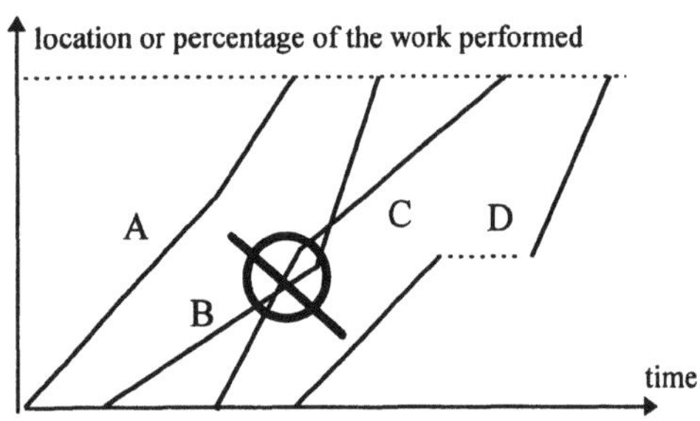

Figure 1-2 *Linear Scheduling Method Diagram*

This is really an important advantage of LSM compared to the bar chart.

Why do we not use LSM technique instead of bar charts? The LSM technique has a great disadvantage compared to the bar chart. Bar chart can be used for any type of construction projects. The LSM technique can be used for projects where activities usually follow each other, like activities in an assembly line, that is, where activity lines do not cross each other. A lot of criss-crossing makes the plan chaotic. Application of the LSM technique can be especially efficient in railway and road construction. In these cases project managers can benefit more from LSM schedules than from bar charts. Unfortunately, the LSM technique has the same disadvantage as a bar chart. It does not retain the logical relationships among activities so both planning and updating are very difficult in cases of large projects.

In the mid fifties a great deal of research was conducted to eliminate the disadvantages of the traditional scheduling techniques. The results of these efforts are the network-based scheduling techniques.

1.3 NETWORK-BASED SCHEDULING TECHNIQUES

The basic conception behind the development of network-based scheduling techniques is to eliminate the disadvantages of the traditional scheduling methods. The two most important problems were to create a technique where the logical relationships among activities were kept, and to divide the planning and scheduling part into two separate phases. Network-based techniques give solutions to these problems and a lot of additional benefits can be gained by using them.

The heart of network-based techniques is always a graph, a set of nodes and connected arrows. Originally activities were represented by arrows, and the nodes served as events, that is they described the logical relationship among activities. Observing Figure 1-3 it can be seen how simply and clearly the graph represents the network logic.

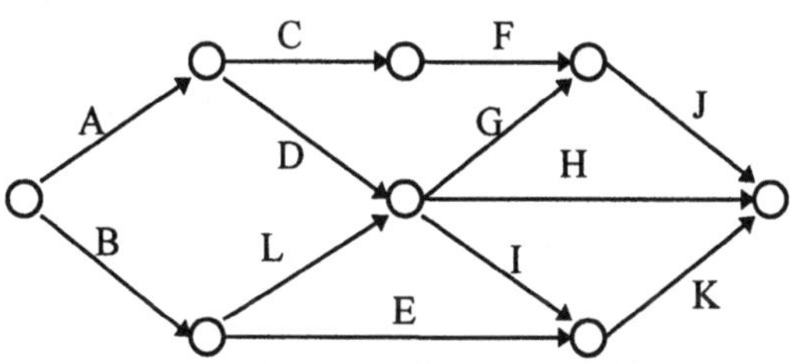

Figure 1-3 *A sample activity on arrow network to illustrate the network logic*

For example, in Figure 1-3, activity A has to be completed before activities C and D may start; activity B has to be completed before activity L can start, G,H and I can start after finishing after finishing activities D and L, etc. In short, the logic of the whole project can be easily followed. Generally we can say *that all activities preceding a given activity must be completed before the activity in question may start.* This network representation has limits, which will be discussed in Chapter 2, Pros and Cons of CPM Technique.

Another possible representation of project relationships is one where the nodes of the graph serve as activities, and the arrows describe the logical relationships between activities. The activity-on-node representation of Figure 1-3 can be seen in Figure 1-4. We had to draw two more nodes to correctly represent the network logic of the activity-on-arrow network. At this moment it may seem that activity-on-node representation is more complex and requires more drawing then activity-on-arrow representation. However, this is not the case. Activity-on-node representation has great advantages compared to the activity-on-arrow notation. We will discuss these problems thoroughly in the following chapters.

The third possible representation of network logic is the so called time-scaled diagram. This is in fact a Gantt chart that shows the logical relationships between activities.

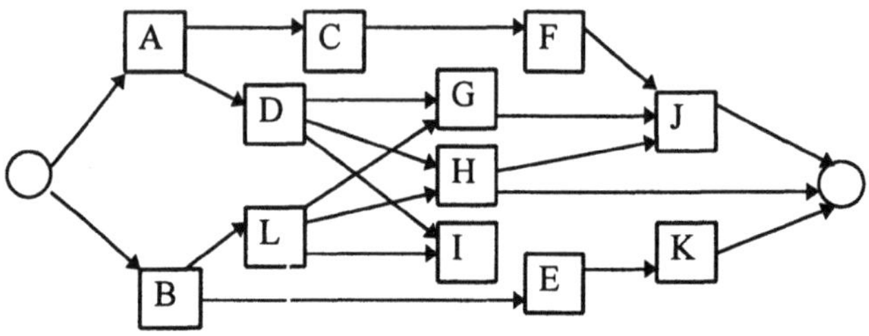

Figure 1-4 *Activity-on-node representation of* Figure 1-3

This type of a diagram assumes that we already finished the scheduling phase, so usually this technique is used to portray the results, and is not used in the planning phase. A time-scaled diagram can be seen in Figure 1-5.

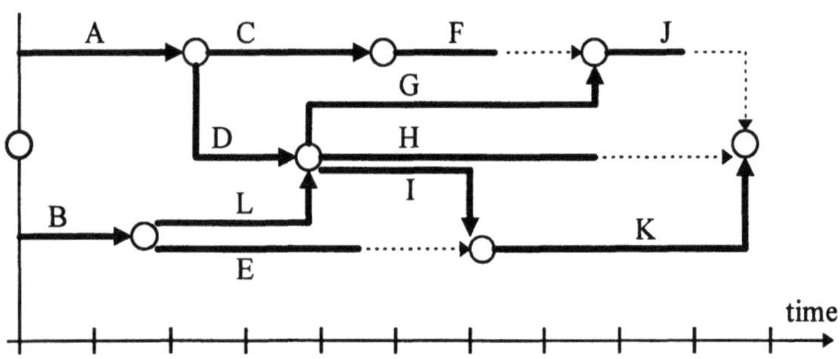

Figure 1-5 *Time-scaled network diagram of* Figure 1-3

It can be seen from Figure 1-3 and Figure 1-4 that by using network techniques we are able to represent the network logic without any information about accomplishing the entire project.

What do we have to do after this, what are the basic steps of network-based scheduling? We discuss these problems in the following chapter.

1.4 NETWORK-BASED PROJECT MANAGEMENT: BASIC STEPS

The network-based project management methodology embodies the following steps, in order of appearance:

Step 1. Defining Activities
Step 2. Defining Activity Interdependencies
Step 3. Drawing the Network
Step 4. Time and Resource Estimation
Step 5. Basic Calculations
Step 6. Advanced Calculations
Step 7. Project Control
Step 8. Project Review

Steps 1 to 4 are also known as the planning phase, Step 5 and Step 6 are known as the scheduling phase, Step 7 and Step 8 are known as the control phase. Each step contains numerous small steps, which can vary or are not necessary to accomplish for each project. For example, if we do not have resources for the network, we do not have to do resource allocation at the advanced calculation step.

We strongly have to emphasize that planning and scheduling is a dynamic process. If the results that were obtained in the scheduling phase are not satisfactory we have to go back to the first, second or fourth steps of the planning phase. This iteration process has to be done repeatedly until the results are satisfactory.

The steps presented above will be described in the following.

Step 1. Defining Activities

The first step of the network-based project management.

The steps of the planning phase are the most important steps in the whole project management process. If the benefit we have gained from using these tools is not satisfactory at the end of the project, it is usually due to the inadequate planning phase.

A project can consist of thousands of activities, so defining these activities is sometimes difficult. The first question, which has to be answered is, how detailed should the plan be? Different phases of the construction process and different levels of leadership require less ore more detailed plans. An efficient tool the so called WBS (Work Breakdown Structure) was developed to assist project managers in defining activities and producing networks in the necessary level of details. WBS is a tool, which helps the 'top-down' design, to go into details, to take into account as many activities as necessary. (We will discuss the WBS methodology in Chapter 9,

Art of Scheduling: Word Break Down Structure) The result of the Step 1 is a complete activity list and the structure of the activities.

Step 2. Defining Activity Interdependencies

The second step of the planning phase is to determine the interdependencies among activities. The result of this step is the so called preceding or succeeding activity list: the subject of Chapter 2, CPM Scheduling:Preceding (Succeeding) Activity Lists. The heart of this list is the activity list that has been prepared during step 1. It contains the activities together with all the predecessor or successor activities preceding or succeeding each activity.

This step demands the understanding of the whole project and the knowledge of the applied technologies. The basic rule in preparing the activity interdependencies is that the most knowledgeable person should take part at this stage. Practically, it means that a subcontractor's manager has to take part in this step when their work is under consideration. This is largely true in the whole planning phase.

Several kinds of interdependencies can exist among activities, and sometimes it is difficult to define them. Various kinds of network technique were developed, which help this step in different ways. For example, there are network techniques in which an interdependency means that after finishing the preceding activities the activity in question may start (see Figure 1-3). This feature may cause difficulties when modeling overlapping activities. The CPM technique -- the subject of Chapter 3 and Chapter 4 -- is able to handle only this type of relationship between activities. Another technique will be discussed thoroughly in this book from Chapter 4 to Chapter 7 the so called Precedence Diagramming Method, which recently gained wide acceptance. The main reason of its popularity is the possibility of describing many more type of interdependencies among activities.

Step 3. Drawing the Network

The third step of the planning phase is the drawing of the network. Recently this step has lost in its significance as the role of computer based project management software has been increasing. At first it does not seem to be difficult to draw a network if the relationships are given. In spite of this perception, the preparation of an activity-on-arrow network (see Figure 1-3) can be a challenge even for the most-experienced network experts. Drawing a network, which is correct and well-organized, is still a major problem for software developers and this is the subject of Chapter 2: Drawing of an Arrow Diagram.

Let us mention here that this step does not require knowledge about the project itself. The input of this step is the preceding activity list, and the result is a graphical portrayed set of activity relationships, i.e., the network itself.

Step 4. Time and Resource Estimation

The last step of the planning phase is the time estimation for each activity. The result of this estimation is the so called *activity duration.* As the estimation depends upon human judgment, some people can argue, that the result of a network-based calculation cannot be better than the result of traditional scheduling techniques. However this notation does not hold true. In network-based scheduling techniques the experts are not influenced by the desired completion date because they simply do not know what the result of their planning will be.

Estimating activity durations without bias is at least as important as establishing real network logic. Even the best experts cannot be objective and very often they try to provide an estimation, which might be acceptable for a couple of activities but certainly not for all of them. To avoid bias there are some rules, which preferably have to be kept during the estimation process.

1. Use a group of experts. This helps to eliminate the bias from the estimation process.
2. Time estimation has to be done activity by activity. Do not allow any preceding estimation affect the estimation of the activity in question. For example, assume for a given activity that all the materials, equipment, etc., will be available in time. Do not attempt to solve possible resource conflicts here. Parallel activities might be able to compete for the same resources but this is not the time and place for this kind of speculation.
3. Consider a normal level of labor and equipment for each activity in order to determine the 'normal' duration. If sophisticated software packages are available the so called 'crash duration' can also be determined with the use of the highest possible level of resources. The term 'crash duration' means the rationally fastest accomplishing time for an activity. The use of crash duration time is discussed in Chapter 3: CPM Least Cost Scheduling; in Chapter 6: Precedence Diagramming with Bounded Duration and in Chapter 7: PDM Least Cost Scheduling.
4. Use normal working days, specify where weekends and holidays should be considered as working days.

In network-based project management methodology different time estimation methods exist which are used in different types of network technique.

The simplest one is the time estimation method when one activity duration is given to each activity. Usually this is the normal duration of the activity. This is the simplest method, but it involves the highest level of human decision. Network techniques based upon this assumption are the most popular because of their simplicity.

The next method is the so called 'two time estimation method'. The two durations, which are given to each activity, are the normal activity duration and the shortest possible duration: the crash duration. Some network techniques, e.g., CPM Least Cost Scheduling, can handle these types of duration as a lower and upper bound, and all the durations can be chosen in this interval if necessary. So the estimation called two time estimation means many more activity durations in practice. Varying activity durations in a project can result in millions of logically identical but different in activity durations networks. In these kinds of networks some additional goals can be formulated and a computational procedure will decide on the length of activity durations according to the goal selected.

The third method used for time estimation is the 'tree time' or PERT estimation. The PERT (Program Evaluation and Review Technique) is a network technique based on the same network logic that has been shown in Figure 1-3, on page 5. Its time estimation method significantly differs from the previous ones. It is based on the assumption that activity durations follow a stochastic distribution. For example, a dewatering depends on the level of the groundwater. As the level of groundwater follows a stochastic distribution the activity duration of draining will also follow a stochastic distribution. Originally a β distribution was assumed in the PERT model. To simplify this probabilistic procedure three activity duration times were chosen from the distribution, the most likely, the most optimistic and the most pessimistic performance time. The definition of these activity durations are:

- most likely time: the duration that is likely to occur more often than any other duration
- most optimistic time: the duration that would be the fastest 19 times in every 20 occurrences
- most pessimistic time: the duration that would be the slowest 19 times in every 20 occurrences.

With this simplification the original PERT technique has been virtually reduced to a technique using the 'one time estimation method'. After finishing the calculations using, for example, the most likely durations, it has to be performed again with the most pessimistic and finally with the most optimistic duration. The other main problem with the original PERT assumption is that the original (historical) data usually not available, therefore the activity performance time distribution is usually hypothetical.

Step 5. Basic Calculations

This is the first step of the scheduling phase. We come to this step after finishing the planning phase. Doing the basic calculations we can get answers to the following questions:

What is the expected duration of the project?

What are the earliest start and finish dates for each activity?

What are the latest start and finish dates for each activity that does not interfere with the project completion time?

If the results are not satisfactory, for example the project completion date is beyond the deadline, then we have to go back to the planning phase to revise the network logic or the time estimation.

Step 6. Advanced Calculations

This step involves those calculation methods which can be used after the basic calculations. These are usually time cost trade-off procedures and/or resource connected calculations.

Time cost trade-off procedures are also known as least cost scheduling techniques. With the use of least cost scheduling techniques the following questions can be answered: How should the project manager choose the activity durations if he or she wants to accomplish the project by a given date at the lowest possible cost? Mathematical models that handle this problem can be found in Chapter 3, CPM Least Cost Scheduling, and in Chapter 7, PDM Least Cost Scheduling.

If we have the resources for each activity, then after the basic calculations we can produce the resource histograms. These histograms tell us that how much of a resource we need on a given work day. Usually these histograms do not reflect with our resources. Sometimes they show a larger demand for human resources than we can technically or economically provide, or show an unwanted fluctuation in demand from day to day. The big fluctuation of personnel on site can cause considerable problems in managing site facilities and organizing project activities. It also creates problems for the human resource manager of the firm who has to satisfy this request without causing losses for the firm. It is sometimes impossible to satisfy the sudden personnel demands on site. Therefore, fluctuations in resources are best to be avoided. Resource connected calculations can help to solve these problems. Two basic types of resource connected calculations are known, resource allocation and resource leveling.

Resource allocation is used when there are definite limitations on the amount of resources. During this process the project completion time derived from basic calculation process can be exceeded. This procedure is the subject of Chapter 8, Resource Allocation.

Resource leveling is used when sufficient amount of resources are available to carry out the project under consideration. The role of this computational method is to decrease the amount of changes in the resource histogram without increasing the project duration. We discuss this method thoroughly in Chapter 8, Resource Leveling.

If the results concerning the resources are still not satisfactory, then we have to go back to the planning phase and take into consideration the entire network logic and the time and resource estimation again.

Step 7. Project Control

After finishing the planning and scheduling phase the project can start. The role of network-based project management cannot finish at the start of the construction process. The project manager has to maintain the plan, modify it when new circumstances arise and update it on a regular basis. The control phase is also used to measure the project cost and for making comparisons with the original cash flow previsions. Project managers very often finish their work on the network after finishing the scheduling phase. This is a wrong practice. The real work starts in the control phase; planning and scheduling serve merely as a preparation phase for project control.

Step 8. Project Review

It is advisable to review the progress of the project on completion, and analyze our work from these points of view:

How accurate was the activity list, and the list of interdependencies?

How good was the time and resource estimation?

What were the reasons for changes during the project?

How effective was the project control?

1.5 HISTORICAL REVIEW

The history of network-based scheduling techniques began in 1931 when Karol Adamiecki developed his Harmonygraph [1],[50]. This tool can be considered as the forerunner of modern network techniques. It kept the activity relationships and gave answers to the first two questions listed in Step 5. Basic Calculations. This method had however no answers for the question about the latest possible start of activities. His work has been very much ignored and most of the authors count the history of network-based project methodology from the late fifties.

Network-based management systems have developed along at least three or more parallel streams. One of them was a military stream and had led to the development of PERT. The other was an industrial initiation the result of which was the development of CPM. The third stream was European, and the result of this research has served as a basis of today's Precedence Diagramming Method. There was definitely some other noteworthy and independent work at the same time mainly in Europe, but they never had such wide acceptance as those three systems mentioned above.

History of CPM has begun in 1956, when an American chemical firm, the E.I. du Pont de Nemours Company decided to develop a new scheduling tool to improve the company's engineering functions [35]. As a result of the work done by Morgan Walker from DuPont and James E. Kelley, Jr., from Remington Rand a new method had been developed [41],[42], later named the CPM. It was an activity on arrow network technique, in which money was tied to the activities and which produced time cost trade off calculations. The first phase of this calculation was a simple time analysis (see Chapter 2, CPM Calculation) and the second phase was the sophisticated time cost trade-off phase. However, construction industry experts could not cope with the whole method at the same time and they have assimilated it as two techniques, first as the time analysis phase as CPM, and later as the method of CPM Least Cost Scheduling. It is an acceptable technique, if we consider, how revolutionary this method was at the time of its conception. So, in today's construction practice the term CPM technique means the basic calculations, i.e., time analysis and the term CPM Time Cost Trade-off or CPM Least Cost Scheduling covers the original CPM technique.

The first test was made in 1958, when a scheduler group using the new technique and a conventional scheduler group were working independently on the construction management plans of the same project. The new group had finished its work faster and produced a better schedule than the group of traditional schedulers.

History of PERT (Program Evaluation and Review Technique) dates back to 1957, when the Polaris missile program was initiated. There were about 9000 contractors and subcontractors involved in the project. The Special Project Office (SPO) of the US Navy has managed the program who was cognizant of DuPont's work. The SPO was helped by a consultant firm of Booz-Allen and Hamilton and the Lockheed Missile System Division. In 1958, the SPO directed the research team, C. E. Clark, D. Malcolm, and J. H. Roseboom of Booz-Allen and Hamilton and R. Young and E. Lennen of Lockheed Missile System Division, to devise the Performance Evaluation and Review Technique. W. Fazar was in charge of the success of the project at SPO. The result of this effort was a new network technique which is known the PERT [45],[53] today. The structure of the graph and the logic was the same as in CPM but there were significant changes compared to the CPM technique. While CPM was activity oriented, PERT was event oriented. Events are distinguished phases of the project which have to be achieved by specific dates. The event approach was developed in order to comply with the Navy's reporting system. The PERT technique allowed the use of uncertainty in time estimates but took no account of cost. An additional great benefit of the work is the detailed and comprehensive methodology of the whole planning, scheduling and control phases. For example the Work Breakdown Structure (WBS) has been developed within the frame of this research.

Early history of Precedence Diagramming Method is not so radiant as the two above mentioned. It is dated back to 1958 when a French scientist, B. Roy, has developed his 'Method of Potentials'[58],[59]. His method used the activity on node representation and he introduced the Start-to-Start (SS) precedence relation with lead/lag factors. His method was so sophisticated that it allowed the use of maxSS precedence relationships! (Maximal type of relationships are the subject of Chapter 5, Advanced Precedence Diagramming) A response to his work was the HMN technique, which was developed in Germany and allowed the use of Finish-to-Finish (FF) and maxFF relations. These works were very much ignored especially in the United States but the non-English speaking Europe still uses the MPM (Metra Potential Method) name instead of Precedence Diagramming regarding to Roy's work. In the United States John Fondahl's work [18] is considered as a forerunner of Precedence Diagramming Method. In his classical work he introduced the activity-on-node representation of the CPM technique. The first complete description of Precedence Diagramming with several type of precedence relationships appeared in 1964 in an IBM user's manual [39]. One of the principal author of this work was J. Craig. Nevertheless, the birth of Precedence Diagramming is not definite, there were many parents and none could done as much for popularization as the US Navy or DuPont. Despite of this Precedence

Diagramming Method has so many advantages over the CPM and PERT that today this is the most frequently used network technique in construction practice.

There were hundreds of remarkable pieces of work that appeared in the sixties and seventies which were based on these three network techniques. Some of these works made significant improvements to the original techniques. The most interesting development was the addition of decision nodes to the network. The aim of these works were to handle several alternatives of the project in one network. One of the pioneers was Freeman [23] whose work was generalized by Eisner [15] and Elmgrabhy [16],[17]. The result of their work was the addition of three types of decision nodes: AND nodes; EXCLUSIVE-OR nodes and INCLUSIVE-OR nodes. Their works are considered as the basis of some research activities that were led to the development of GERT (Graphical Evaluation and Review Technique) The initial GERT developments were made by Alan Pritsker [55],[56],[57], who published his results in 1966. The growing of the GERT family has started with Pritsker's work and soon became huge. The Precedence-GERT that was able to handle several types of activity relationships was developed by Pritsker [54] in 1974. Arisawa and Elmagrabhy [4] devised a technique that was able to perform time cost trade-off in GERT network in 1972. In 1975 Herbert [36],[37] developed his R-GERT method, which solved resource constraint scheduling problems in GERT.

This list is no complete by any means but we are sure it covers almost all the industrial applications. In the next chapters we are going to discuss the deterministic network techniques. Practically all the applications in the construction industry based on CPM or Precedence Diagramming Method. Recently, the application of PDM has increased and CPM now bears the consequences of this. In this book the emphasis is placed on Precedence Diagramming to promote the wider spread and deeper understanding of use of PDM in construction practice.

We discuss the traditional CPM technique in the second chapter. In the third chapter we describe the CPM Least Cost Scheduling. In Chapter 4 to Chapter 7 Precedence Diagramming is our subject.

There is no special mathematical and engineering background necessary to understand this book and none is assumed by the author. The inevitable mathematical background is discussed in Appendix A.

Chapter 2

CPM SCHEDULING

In this chapter we are going to discuss the CPM network scheduling technique, also known as CPM/time technique, to make a clear distinction from Critical Path Method as it was originally developed. This chapter has been divided into ten subchapters.

In the first part, Introducing a Sample Project, we are going to establish our sample project. This sample will be used to illustrate the algorithms throughout the book.

In the second part (Basic Definitions. Drawing Rules), we give all the definitions necessary to understand the CPM structure.

In the next subchapter, Preceding (Succeeding) Activity Lists, we give a procedure how to filter unnecessary information from the preceding or succeeding activity lists. Practically, IPA and ISA lists are the only tools comparing networks to each other so it is necessary to learn how to produce IPA and ISA lists.

The problem of drawing a correct and proper activity-on-arrow network from preceding activity lists is the subject of the fourth subchapter. (Drawing of an Arrow Diagram)

In the fifth part, Levels of Network, we introduce the reader to the level of the network. This will form the foundation for basic calculations and some resource handling procedures.

The sixth part, CPM Calculations, contains the time analysis procedure of the CPM technique.

What kind of information can be gained from the calculation process? This is the subject of the seventh subchapter, Information Obtained From Calculations.

In the next part, Pros and Cons of the CPM Technique, we discuss some of the drawbacks of the CPM technique.

In the ninth part, Practical Problems and Solutions, we provide the reader with some practical problems and solutions of them, and in the last subchapter, Historical Review, we give some additional information about the history of CPM.

2.1 INTRODUCING A SAMPLE PROJECT

Hereinafter, we are going to establish a schedule for a simple, fictitious industrial construction project in which we have reduced the size of activities to a very small number.[i] The reason of using such an oversimplification is that we did not intend to produce a schedule of hundreds of activities. Our goal was to develop the smallest project network still acceptable as a construction project but, one with a size easily manageble by for the readers.

The project can be summarized as follows:

A covered depot has to be built from prefabricated standardized steel elements in order to protect a piece of large industrial equipment used for engine production in heavy industry. The machine is so big that a special mount has to be placed underneath. The assembly of the building structure can start only after the placement of the machine.

Suppose that we are the project managers directed to carry out this work. We have defined the following activities and time estimates for the project:

Preparations, 30 days (adaptation of the standardized plans, getting the necessary permits from authorities, choosing of machine)

A Order machine, 20 days
B Deliver machine, 30 days
C Order building materials, 10 days (prefabricated steel structure, mount for the machine, gravel and cement)
D Preparatory works on site, 14 days
E Delivery of bulldozers, 30 days

[i]The sample project has been developed by L. Soós, and P. Lánczos, associate professors of Department of Construction Management, Budapest University of Technology.

F	Delivery of gravel and cement, 20 days
G	Delivery of mount structure, 80 days
H	Delivery of prefabricated steel structure, 90 days
I	Groundwork, 10 days
J	Foundation for the depot, 30 days
K	Placing the mount for the machine, 20 days
L	Placing the machine on the mount, 5 days
M	Assembling the building structure, 20 days
N	Finishing works, project delivery, 15 days

Our task is to create a schedule. To achieve this goal, we have to follow the methodology that has been described in the introductory chapter, Basic Steps of Network-Based Project Management. For the sake of convenience, the steps of the planning and scheduling phases are repeated here.

Step 1. Defining Activities
Step 2. Defining Activity Interdependencies
Step 3. Drawing the Network
Step 4. Time and Resource Estimation
Step 5. Basic Calculations
Step 6. Advanced Calculations

Note that Step 1 and Step 4 have been combined: we have created the activity list and performed the time estimation in one step. In simple cases when time estimation is based on no resources, this is acceptable. Our next task is to define the activity relationships. We do this on such a way that we take each activity from A to O and try to answer the question: "*What activities have to be completedbefore the activity in question can start?*"

Answering this question, we can make the following interdependencies list:

- There are no activities that have to be done before preparations (A).

- Ordering of machine (B) is only possible after selecting it (A), that is, B is preceded by A.

- Before the delivery of machine (C) we have to order the machine (B) and finish the preparatory works on site (E).

- The ordering of building materials (D) can take place after preparations (A).

- Preparatory work on site (E) can start after preparations (A).
- Delivery of bulldozers (F) can start after preparations (A), and does not depend on the preparatory work on site (E).
- Delivery of gravel and cement (G) can start after completing the of preparatory work on site (E) and after ordering the building materials (D).
- Delivery of mount structure (H) can start after the finish of preparatory work (E) and after ordering the materials (D).
- Delivery of prefabricated steel structure (I) can start after the finish of preparatory work (E) and after ordering the materials (D).
- Groundwork (J) can start after delivery of bulldozers (F) and after preparatory work (E).
- Foundation (K) can be done after finishing groundwork (J) and after completing the delivery of gravel and cement (G).
- Placing of mount structure (L) can start after delivery of mount structure (H) and after foundation work (K).
- The machine can be placed on mount (M) after the delivery of the machine (C) and after the placement of the mount structure (L)
- Assembling of prefabricated steel structure (N) can start when machine's placement (M) and the delivery of the steel structure (I) has ended.
- Project delivery (O) can begin after the assembling of the building (N) has finished.

The results of these dependencies are summarized in the second column of Table 2-1. This list is called the immediate preceding activity list (IPA).

Yet another method exists to develop the possible interdependencies of the project. In that case, the following question has to be answered: "*What activities have to be done after the activity in question has finished?*"

By answering the question, we can establish the activity interdependencies summarized in the second column of Table 2-2, Immediate Succeeding Activity List.

Activities	Immediate Preceding Activity List (IPA)	Complete Preceding Activity List (CPA)
A	-	-
B	A	A
C	B,E	B,E,A
D	A	A
E	A	A
F	A	A
G	D,E	D,E,A
H	D,E	D,E,A
I	D,E	D,E,A
J	F,E	F,E,A
K	G,J	J,G,F,E,A,D
L	K,H	K,J,G,H,D,E,F,A,
M	C,L	A,B,C,D,E,F,G,H,I,J,K,L
N	M,I	A,B,C,D,E,F,G,H,I,J,K,L,M
O	N	A,B,C,D,E,F,G,H,I,J,K,L,M,N

Table 2-1 *Preceding activity lists for the sample project*

- After finishing preparations (A), the ordering of the machine, the delivery of bulldozers (F), the preparatory work on site (E) and the ordering of materials (D) can start, that is, after A can start B,F,E, and D.
- After ordering the machine (B) can start the delivery of the machine (C).
- Placement of machine (M) can start after it is delivered (C).
- After ordering of building materials (D) can start the delivery of gravel and cement (G), the delivery of mount (H) and the delivery of steel structure (I).
- Groundwork (J), delivery of machine (C), gravel and cement (G), mount (H) and steel structure (I) can start after the preparatory work on site (E).
- Groundwork (J) can start after the delivery of bulldozers (F)
- Foundation (K) can start after the delivery of gravel and cement (G).
- Mount can be placed (L) after its delivery (H).
- Assembling of steel structure (N) can start after its delivery (I).

- Foundation (K) can start after the groundwork (J).
- Placement of mount structure (L) can start after the foundation (K)
- Machine can be placed (M) after the placement of mount structure (L)
- Assembling of steel structure (N) can start after the machine has been placed on mount (M)
- Project delivery (O) can start after the assembling the steel structure (N)
- There is no activity after the project delivery (O)

Activities	Immediate Succeeding Activity List (ISA)	Complete Succeeding Activity List (CSA)
A	B,D,F,E	B,C,D,E,F,G,H,I,J,K,L,M,N,O
B	C	C,M,N,O
C	M	M,N,O
D	G,H,I	G,H,I,K,L,M,N,O
E	C,J,G,H,I	C,J,G,H,I,K,L,M,N,O
F	J	J,K,L,M,N,O
G	K	K,L,M,N,O
H	L	L,M,N,O
I	N	N,O
J	K	K,L,M,N,O
K	L	L,M,N,O
L	M	M,N,O
M	N	N,O
N	O	O
O	-	-

Table 2-2 *Succeeding activity lists for the sample project*

Comparing the IPA (Immediate Preceding Activity) list to the ISA (Immediate Succeeding Activity) list it is hard to tell at the first sight if these lists describe the same activity interdependencies or not. If we want to compare these lists we have, to have a method that will enable us to make conversions between these lists. What makes the situation more difficult is that here we produced immediate preceding (succeeding) activity lists seemingly effortlessly, but in reality it is a result of a long iteration process. Indeed, it is very hard to establish a preceding (succeeding) activity list that contains solely preceding (succeeding) activities. The produced lists usually contain more information than necessary. This is not a fault but the efforts result in a set of needless information. For illustration, here is an example.

Suppose that the project manager of our sample project tries to answer the question: "What activities do I have to finish before the placement of mount (L) can start?" in the following way:

- Placement of mount structure (L) can start after delivery of mount structure (H), after foundation (K) and after groundwork (J)

This answer is correct, but the inclusion of groundwork (J) in this list is completely unnecessary. Certainly, groundwork precedes the placement of mount if it precedes foundation, as it was stated when we have made the IPA list.

Creating a PA (Preceding Activity) or SA (Succeeding Activity) list has two extreme cases: when only immediate predecessors and successors are included or when all the preceding and succeeding activities are included. These are called the CPA (Complete Preceding Activity) and the CSA (Complete Succeeding Activity) lists. The CPA list of the sample project can be found in the third column of Table 2-1, while the CSA list is in the third column of Table 2-2.

All the four lists describe the same activity interdependencies. The reader is encouraged to verify if these lists really contain the same information. It surely is not an easy task even though we only have fifteen activities. Of course, a project manager always makes every effort to exclude the redundant information from the activity list and to produce immediate activity lists in the first run, but this is almost impossible. So when we produced a list, which we feel to be correctly describe all the activity interdependencies, then we had to exclude all the unnecessary information and produce an immediate interdependencies list.

We can now start with the drawing of the network. Unfortunately, different networks can be drawn based on the same activity list, which shows the same relationship between activities. These networks can differ either in layout or in structure, so it is hard to tell whether different networks describe the same project or not. The best thing we can do if we want to compare two different plans, is to use the immediate predecessor or successor activity list. Figures 2-1;2-2;2-3;2-4 depict networks based on the same IPA list (Table 2-1). However, they do not look to be the same at first. When the CPM diagram is drawn, we can continue with the calculations. During the calculation process we define the total project duration, the possible earliest start of activities, and the latest possible start that does not interfere with the project duration.

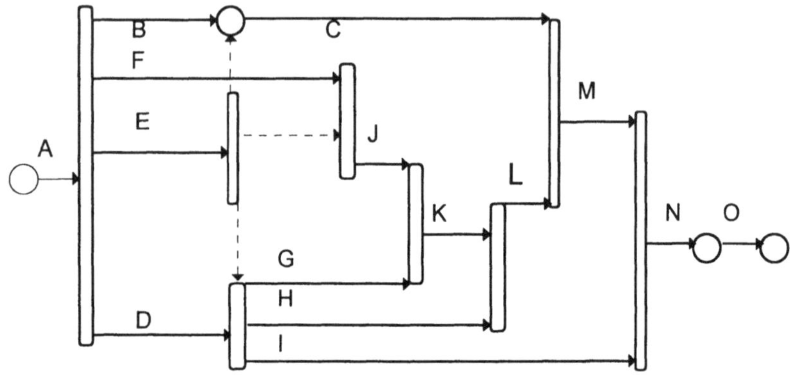

Figure 2-1 *CPM diagram for the sample project. #1*

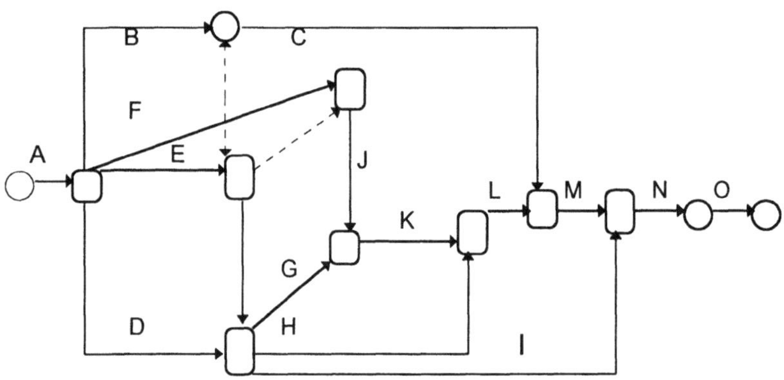

Figure 2-2 *CPM diagram for the sample project. #2*

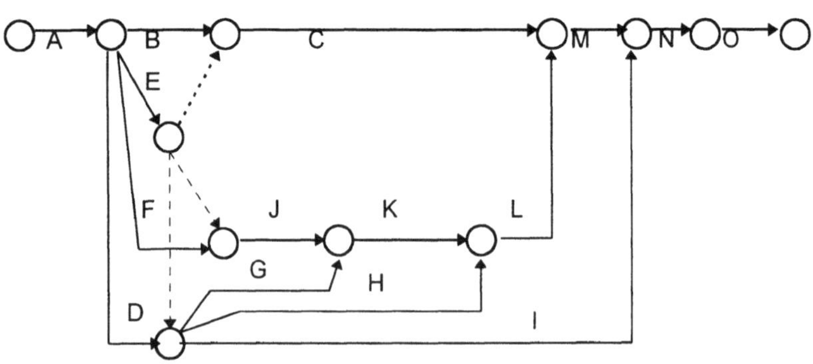

Figure 2-3 *CPM diagram for the sample project. #3*

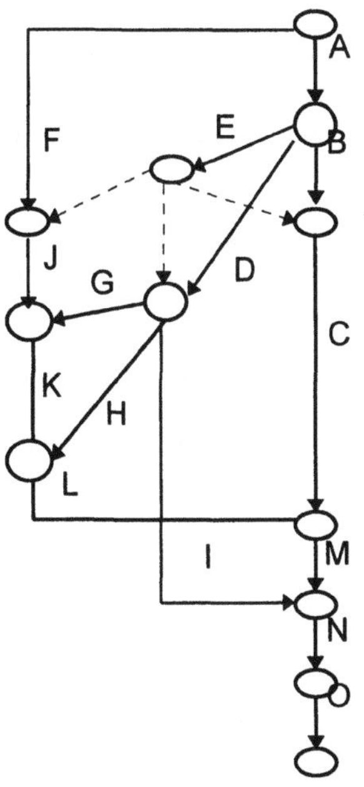

Figure 2-4 *CPM diagram for the sample project. #4*

Of course, it is too early for the reader to fully understand the CPM diagrams, how to convert activity lists, how to filter surplus information from the lists, etc. before being able to draw, calculate, evaulate and analyze networks, some basic rules and definitions have to be discussed. In the following subchapter we provide the reader all the necessary rules and definitions of basic terms that are necessary to understand the CPM technique.

2.2 BASIC DEFINITIONS. DRAWING RULES

In mathematical sense, a network schedule is a directed graph. A directed graph is a structure formulated by nodes and tied together by directed arrows. We denote the set of nodes with **N** (**N**=$\{1,2,\ldots\ldots,i,\ldots\ldots,j,\ldots\ldots,n\}$, and the set of directed arrows with ***A*** (*A*=$\{(1,2),\ldots\ldots\ldots\ldots,(i,j),\ldots\ldots\ldots,(n\text{-}1,n)\}$. We identify an arrow using the two nodes connected by the arrow, the tail and the head node. The graph is denoted by the symbol [**N**,***A***]. This mathematical structure will be used as a basis of the CPM technique to depict activity interdependencies.

The arrows of the graph represent the activities the nodes represent the events. An event is a certain moment in time which occurs when all the preceding activities have been accomplished. According to this representation a CPM diagram is often called the activity-on-arrow diagram[i]. The definitions of these two basic components and the definition of a network schedule follow.

Definition 2-1 *Activities of a network schedule represent individual processes that have to carry on continuously in time without interruption and with the same intensity*[ii].

The time of accomplishing activity (*ij*) is denoted by $\tau_{ij.}$

Definition 2-2 *An event, which is a certain moment in time, can occur when all the predecessor activities have been accomplished. Successor activities can start after the occurrence of the event.*

Events have no duration. The moment of occurrence is called event time and denoted with π_i (for event *i*). We denote the graph in which activity durations are given with $[\mathbf{N}, A, \tau]$.

Definition 2-3 *We call the* structure $[\mathbf{N}, A, \tau]$ *as a network schedule if it is loopless.*

When drawing a CPM network, one has to keep in mind the following rules:

Rule 1. There can only be one initial event (node) and one terminal event (node) They are denoted by s (source) and t (terminal).

Rule 2. Loops are not allowed in the network. Existance of a loop means that there is a node from which there is a path along the directed arrows back to the node in question. It entails that an occurrence of an event would depend on the occurrence of a logically successor event.

Rule 3. The network cannot contain double or multiple activities.

These rules are the original rules of the CPM technique and there are lots of extensions that relax or omit one of these rules. However, no extension of the CPM technique relaxas Rule 2. In fact, this is the most important rule.

Let us examine Rule 1. This rule means that we have to organize our project so that there is a certain event, which indicates the start of the project, and another event, which indicates the end of the project. This rule helps us to establish a correct network. There are some extensions, which omit this rule.

---

[i] Another way to depict an IPA list is the activity-on-node representation. We going to discuss this later.

[ii] There are extensions when activity interruption is allowed. This kind of techniques are also discussed Chapter 4.

What is the reason for establishing the third rule? What do double and multiple activities mean? First we answer the second question. A double or multiple activity means that two or more activities (arrows) connect two nodes (events). The reason that it is not allowed in CPM technique can be found in the identification of activities, which is given by the tail and head events. We can identify only one activity with the help of the events, and if we have more activities between the same nodes, the *(i,j)* identification is not unambiguous.

What is the solution if double activities are not allowed?

There is an another kind of activity used in CPM diagramming: the 'dummy' activity. 'Dummy' activities have neither real meaning nor duration. They are used to solve some structural and logical problems in the network like the problem of double activities. In the left side of Figure 2-5, activity A and B are double activities. To resolve this problem a new event, *i1*, and a dummy activity are necessary between *i1* and *j*. We do have to emphasize that C,D, and E remain immediate successors of B because *(i1,j)* is not present on the activity list.

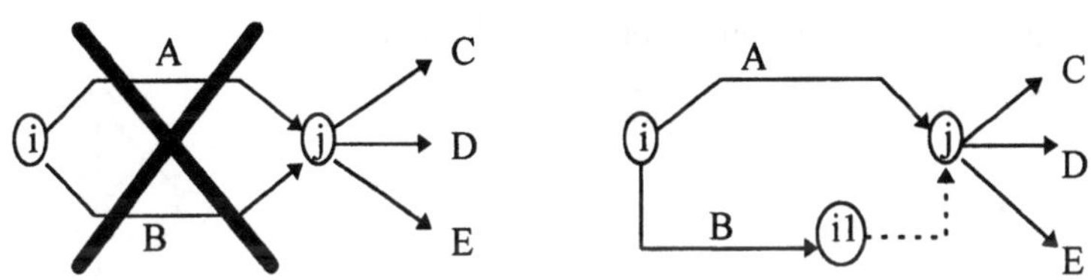

Figure 2-5 *Avoiding double activities with the use of 'dummy' activities*

Dummies are usually depicted with dotted or dashed lines to clearly distinguish them from real activities.

Another important use of dummies is that we can solve some of the logical problems can arise from IPA lists. Consider the following example, which contains four activities and is a part of an IPA list (See Table 2-3).

activities	IPA
D	B
C	A,B

Table 2-3 *A small part of an IPA list*

A possible solution can be seen in Figure 2-6a, which is not correct. The reason is that the IPA list, according to this figure, contains more information: not only B, but A is also an immediate predecessor of D. A correct drawing of this part of the IPA list is shown in part b) of Figure 2-6.

We can summarize the role of dummy activities as that necessary for avoiding double or multiple activities, and for correctly depicting logical relations.

Another possible way of handling double or multiple activities is identifying them by their names.

As it can be seen in Figure 2-5, if activities are identified by their names, then using of double or multiple activities is allowed as it does not cause problems in activity identification. In this chapter, we use names for identifying activities which is much more convenient, but for the sake of historical fidelity, we keep Rule 3.

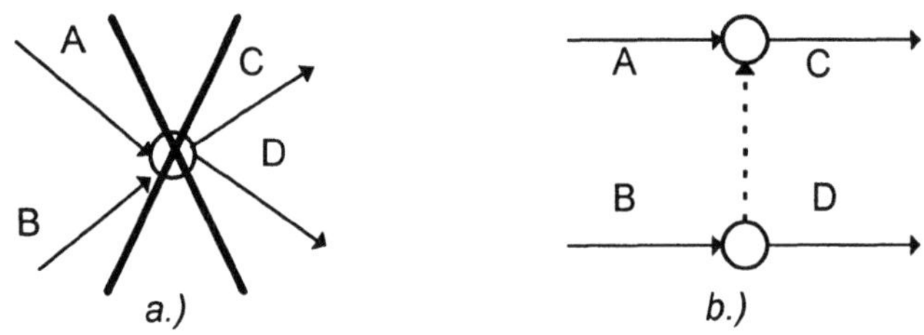

Figure 2-6 *Depicting activity interdependencies in CPM diagram from* Table 2-3

In Figure 2-7 we show right and wrong CPM diagrams.

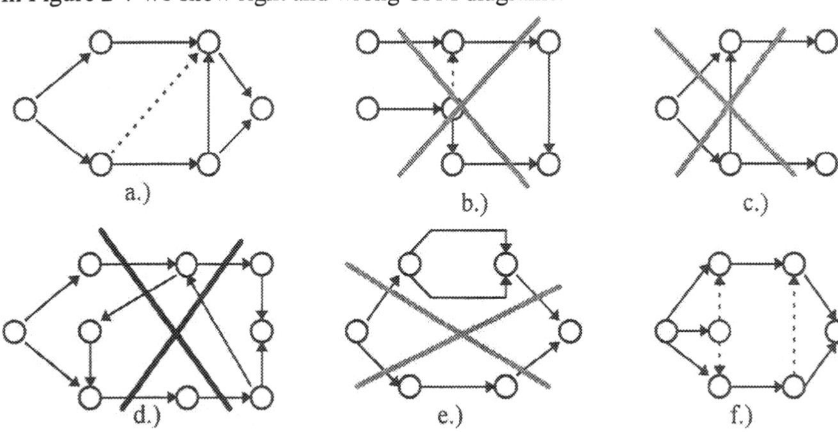

Figure 2-7 *Right and wrong CPM diagrams*

Only a.) and e.) are correct, b.) has two initial events, c.) has two terminal events, d.) contains a loop, and e.) has double activities. As we mentioned before, there are some extensions according to which b.), c.) and e.) can be a correct CPM diagrams.

In the following subchapter we give a procedure how to filter unnecessary information from preceding and succeeding activity lists. Practically, IPA and ISA lists are the only tools of comparing networks to each other.

2.3 PRECEDING (SUCCEEDING) ACTIVITY LISTS

As we mentioned earlier, an immediate list, IPA or ISA, is completely enough for drawing a network. If the graphical portrayal of the network is made from a non-immediate list, it will contain numerous unneeded dummies, which are totally unnecessary. To avoid including redundant information in the graphical representation, it is advisable to check the PA or SA list whether they are immediate lists or not. If they are not immediate lists, filter out unnecessary data from the PA or SA lists.

In the following, we are going to show how to make an IPA list from a PA list. The procedure has to be slightly modified when making an ISA list from an SA list.

Let the activities be indexed during the process with *i*. (*i*=1...n) Two kinds of labels will be used for activities, a '-' means that this activity is labeled but not examined yet, a '+' means that this activity is labeled and examined so we do not have to deal with it anymore. There are activities that have no labels at the beginning of the procedure. At the end, all activities have to be labeled with a '+'.

The steps of the procedure can be summarized in the following manner:

Step 1. Let i=1, and let the label of the activity i be a '-'. There are no labels added to the other activities.

Step 2. Let a '-' label be added to all the activities that are included as predecessors of i. Let '+' the label of activity i.

Step 3. Choose a '-' labeled activity j. If there is no such an activity, then go to Step 5.

Step 4. Delete all the activities from among the predecessors of i that are also among the predecessors of j. Label '-' those activities that are designated as predecessors of activity j. Change the label of activity j to '+'. Go to step 3.

Step 5. Delete all the labels. Let $i=i+1$. If $i \leq n$, then let i be labeled negative and go to Step 2. If $i>n$, then stop.

Illustrating this algorithm in practice, we show the transformation of the CPA list of Table 2-1 to an IPA list. The solution can be followed in Table 2-4. The iteration process is shown in the columns of the table. For example, in iteration 3 -- when the third activity, that is, activity C is in question -- activities C, B, and A are being labeled.

In iteration 1, when activity A is in question, there are no activities being deleted among the predecessors of A.

In iteration 2, when activity B is in question, there are no activities being deleted among the predecessors of B.

In iteration 3, when activity C is in question, activity A will be deleted among the predecessors of C.

In iteration 4, when activity B is in question, there are no activities being deleted among the predecessors of D.

In iteration 5, when activity E is in question, there are no activities being deleted among the predecessors of E.

In iteration 6, when activity F is in question there are no activities being deleted among the predecessors of activity F.

In iteration 7, activity A will be deleted among the predecessors of G.

In iteration 8, activity A will be deleted among the predecessors of H.

In iteration 9, activity A will be deleted among the predecessors of I.

In iteration 10, activity A will be deleted among the predecessors of J.

In iteration 11, activities D, E, F and A will be deleted among the predecessors of activity K.

In iteration 12, activities J, G, F, E, D and A will be deleted among the predecessors of activity L.

In iteration 13, activities A, B, D, E, F, G, H, J and K will be deleted among the predecessors of M.

In iteration 14, activities A, B, C, D, E, F, G, H, J, K, L will be deleted among the predecessors of N.

In iteration 15, activities A, B, C, D, E, F, G, H, I, J, K, L, M will be deleted among the predecessors of O.

	CPA list	Iteration process. Act. *i* in question															IPA
		1	2	3	4	5	6	7	8	9	10	11	12	13	14	15	List
A	-	+	+	+	+	+	+	+	+	+	+	+	+	+	+	+	-
B	A		+	+										+	+	+	A
C	B E A			+										+	+	+	B E
D	A				+			+	+	+		+	+	+	+	+	A
E	A			+		+		+	+	+	+	+	+	+	+	+	A
F	A						+				+	+	+	+	+	+	A
G	D E A							+				+	+	+	+	+	D E
H	D E A								+					+	+	+	D E
I	D E A									+				+	+	+	D E
J	F E A										+	+		+	+	+	F E
K	J G F E A D											+	+	+	+	+	G J
L	K J G H D E F A												+		+	+	K H
M	A B C D E F G H J K L													+		+	C L
N	A B C D E F G H I J K L M														+	+	M I
O	A B C D E F G H I J K L M N															+	N

Table 2-4 *Creating an IPA list from a CPA list.*

The following comments are necessary to add to the conversion procedure of the PA list:

- The order of taking activities in question is completely indifferent from the point of view of the algorithm.
- If the activity list is undoubtedly complete, then the 'Label '-' those activities that are designated...' sentence can be omitted from step 4, because all the predecessors are listed among predecessors of *i*.

Sometimes there can be a need for all the predecessors of the activities. In that case, we have to transform the PA list to a CPA list. The algorithm of making CPA lists from PA lists is summarized below.

The same kind of labeling method is used as in making IPA lists from PA lists. Let the activities be indexed during the process with i (i=1...n). Two kinds of labels will be used for activities, a '-' means that this activity is labeled but not examined yet, and a '+' means that this activity is labeled and examined so we do not have to deal with it anymore. There are activities that have no labels at the beginning of the procedure. However, at the end of the procedure, all the activities will be labeled by a '+'.

The steps of the procedure are summarized below.

Step 1. Let i=1, and let the label of activity i be a '-'. There are no labels added to the other activities.

Step 2. Let the '-' label be added to all the activities that are included as predecessors of *i*. Let '+' be the label of activity *i*.

Step 3. Choose a '-' labeled activity *j*. If there is no such an activity, then go to Step 5.

Step 4. Put all the activities in among to the predecessors of *i* that are also among the predecessors of *j*, and they are not the part of the predecessors of *i*.. Label '-' those activities that are among the predecessors of activity *j*. Change the label of activity *j* to '+'. Go to Step 3.

Step 5. Delete all the labels. Let i=i+1. If $i \leq$n, then let i be labeled negative and go to Step 2. If i>n, then stop.

For the illustration of this procedure, we show the transformation of the IPA list of Table 2-1 to a CPA list. The solution of this problem can be followed in Table 2-5. The iteration process is shown in the columns in the same way as it was represented in Table 2-4. For example, in iteration 3 -- when the third activity, that is, activity C is in question -- activities C, B, and A are being labeled.

	IPA List	Iteration process. Act. *i* in question															CPA list
		1	2	3	4	5	6	7	8	9	10	11	12	13	14	15	
A	-	+	+	+	+	+	+	+	+	+	+	+	+	+	+	+	-
B	A		+	+										+	+	+	A
C	B E			+										+	+	+	B E A
D	A				+			+	+	+		+	+	+	+	+	A
E	A			+		+		+	+	+	+	+	+	+	+	+	A
F	A						+				+	+	+	+	+	+	A
G	D E							+				+	+	+	+	+	D E A
H	D E								+					+	+	+	D E A
I	D E									+				+	+	+	D E A
J	F E										+	+		+	+	+	F E A
K	G J											+	+	+	+	+	J G F E A D
L	K H												+		+	+	K J G H D E F A
M	C L													+		+	A B C D E F G H J K L
N	M I														+	+	A B C D E F G H I J K L M
O	N															+	A B C D E F G H I J K L M N

Table 2-5 *Creating a CPA list from an IPA list.*

In iteration 1, when activity A is in question, there are no activities being added to the predecessors of A.

In iteration 2, when activity B is in question, there are no activities being added to the predecessors of B.

In iteration 3, when activity C is in question, activity A will be added to the predecessors of C.

In iteration 4, when activity D is in question, there are no activities being added to the predecessors of D.

In iteration 5, when activity D is in question, there are no activities being added to the predecessors of E.

In iteration 6, when activity F is in question, there are no activities being added to the predecessors of F.

In iteration 7, when activity G is in question, activity A will be added to the predecessors of G.

In iteration 8, activity A will be added to the predecessors of H.

In iteration 9, activity A will be added to the predecessors of I.

In iteration 10, activity A will be added to the predecessors of J.

In iteration 11, activities D, E, F and A will be added to the predecessors of K.

In iteration 12, activities J, G, F, E, D and A will be added to the predecessors of activity L.

In iteration 13, activities A, B, D, E, F, G, H, J and K will be added to the predecessors of M.

In iteration 14, activities A, B, C, D, E, F, G, H, J, K, and L will be added to the predecessors of N.

In iteration 15, activities A, B, C, D, E, F, G, H, I, J, K, L, and M will be added to the predecessors of O.

The following remarks are necessary to understand this conversion procedure:

- The order of taking activities in question is completely indifferent from the point of view of the algorithm.
- Both of the above mentioned procedures result in the same pattern of the iteration matrix.

In this chapter, algorithms were given about how to produce an IPA list from any kind of a PA list, and how to produce a CPA list from any kind of a PA list. Sometimes other kinds of conversations have to be performed, e.g., producing an ISA or a CSA list from any SA list, or converting an IPA or a CPA list to an ISA or a CSA list, or vice versa. By understanding the above discussed procedures, solutions for these problems can be devised by the reader.

Now we are able to transform these lists to each other and produce an IPA list from any PA list. The IPA list does not contain unneeded information about the project. The next step of network-based project management is the graphical portrayal of the IPA list, that is, drawing the network. We present two possible ways of network drawing in the following.

2.4 Drawing of an Arrow Diagram

As it has been discussed earlier, all kinds of preceding or succeeding activity lists are suitable for drawing a CPM diagram, but avoiding superfluous work IPA or ISA lists are recommended. It is not trivial task to draw a network of only twenty or thirty activities much less when there are hundreds of activities in a project. This is the most time-consuming step in the network-based project management methodology, one which requires lot of patience and attention.

After the logically correct portrayal of the IPA list, the task has not yet been accomplished as the proper layout of activities and events are also important from the interpretability point of view. In the following we give two different methods for drawing an activity-on-arrow CPM diagram. Both of them are time-consuming. The first requires more manual work but less thinking. The second method is much faster, but more difficult. We use our sample project to illustrate these methods.

Method 1

The first method can be summarized in the following manner:

Step 1. Put down on a piece of paper all the individual activities. It is worth to adopt the methodology that activities following each other in time -- as it can be estimated at this stage -- should follow each other on paper too. The result of this step can be seen in Figure 2-8.

Step 2. Following the IPA list, link all the activities that are related to each other with a dummy (see Figure 2-9). The result of this step is a correct graphical representation of the IPA list, which, of course, has too many dummy activities and may have more than one terminal or initial events.

Step 3. Reduce the number of terminal and initial events to one each. Check if double or multiple activities have emerged. In this case add dummies to cancel them.

Step 4. Eliminate as many dummies as possible by answering the following question for a chosen dummy: "Will the IPA or ISA list change in case of deleting this dummy?" If the answer is positive, then do not delete. If the answer is negative, delete the dummy in question if multiple activities would not arise after deleting. Repeat this step for all the dummies. The result of this step can be seen in Figures 2-10 through 2-18.

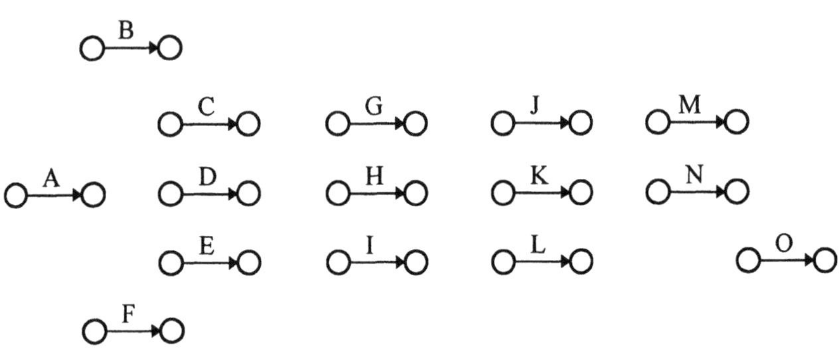

Figure 2-8 *Depicting activities individually.*

Activities are depicted individually in Figure 2-8. The dummies are shown in Figure 2-9. We numbered the dummies from '1' to '23'. In the fourth step, we take dummies into account in ascending order. The result of this step can be followed in Figures 2-10 through 2-18.

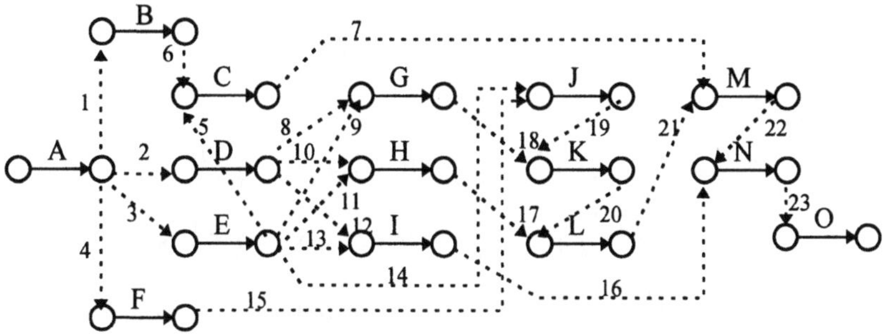

Figure 2-9 *Depicting dummies between activities.*

Dummy '1' between activity A and activity B can be deleted as the original IPA list does not change after the removal (see Figure 2-10).

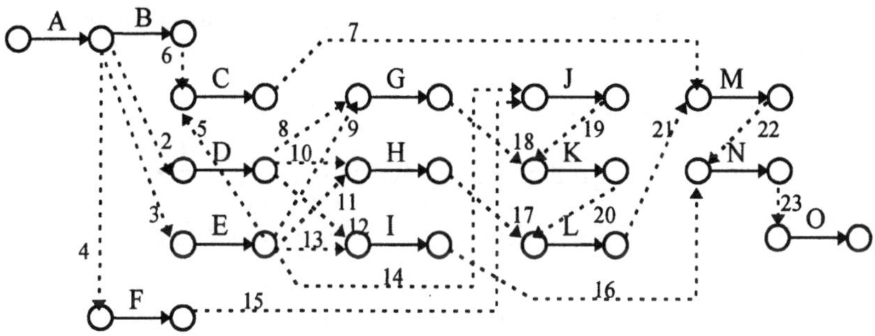

Figure 2-10 *Deleting dummy '1'*

Dummy '2' between activity activity A and activity D can be deleted as the original IPA list does not change after the deletion (see Figure 2-11).

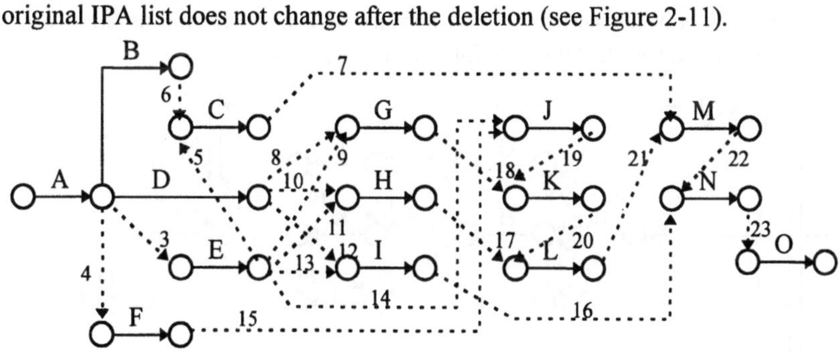

Figure 2-11 *Deleting dummy '2'*

Dummy '3' between activity A and activity E, as well as dummy '4' between activity A and activity F can be deleted. (see Figure 2-12.)

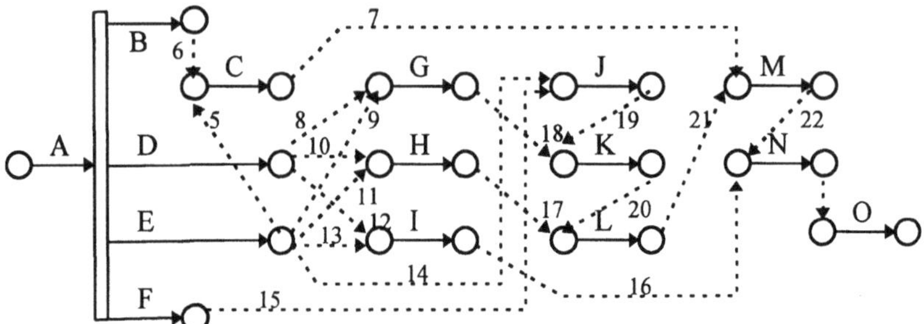

Figure 2-12 *Deleting dummies '3', '4'*

Dummy '5' cannot be deleted because the original IPA list would change. (Activity B would be an immediate predecessor of G, which has not been stated in the original IPA list.) Dummy '6' between activity C and activity B can be deleted as the original IPA list does not change after the deletion (see Figure 2-13).

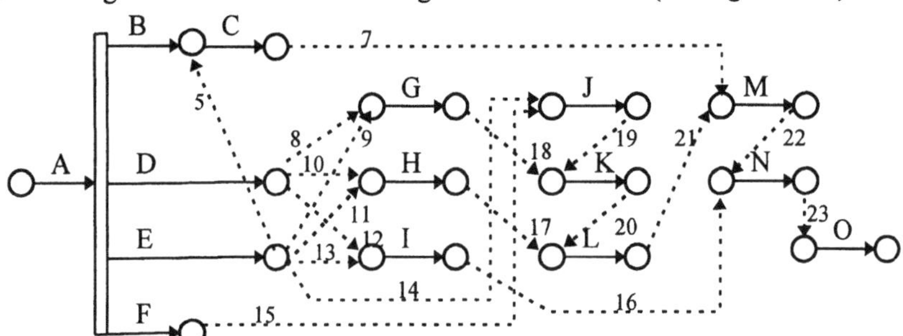

Figure 2-13 *Deleting dummy '6'*

Dummy '7' between activity C and activity M can be deleted as the original IPA list does not change after the deletion (see Figure 2-14).

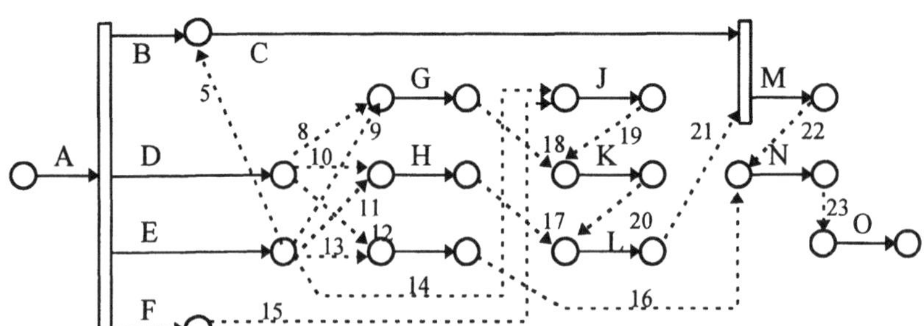

Figure 2-14 *Deleting dummy '7'*

Dummy '8' between activity D and activity G can be deleted as the original IPA list does not change after the deletion (see Figure 2-15).

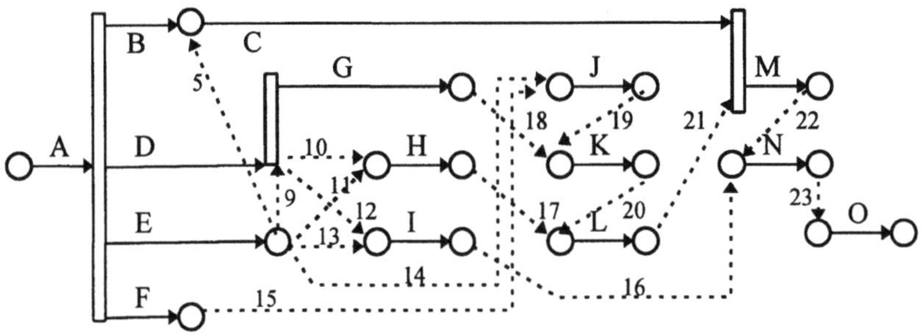

Figure 2-15 *Deleting dummy '8'*

Dummy '9' cannot be deleted as C would form a double activity with D and the original IPA list would change. (Activity D would be a predecessor of C)

Dummy '10' between activity D and H, as well as dummy '11' between activity D and I can be deleted, but dummy '12' and dummy '13' also have to be deleted to avoid forming multiple dummy activities with dummy '9'. The result of this step is shown in Figure 2-16.

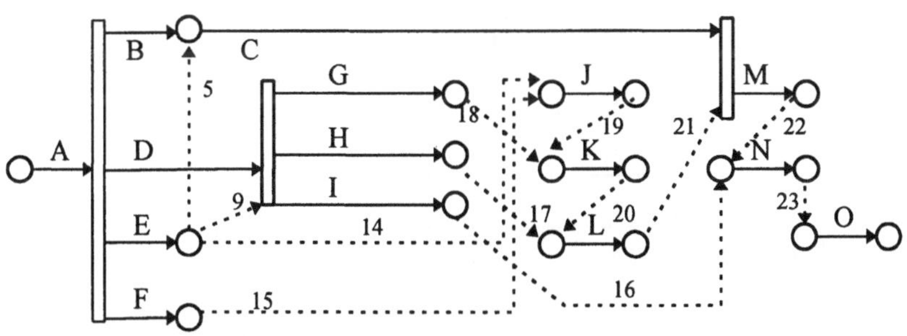

Figure 2-16 *Deleting dummies '10', '11', '12', '13'*

Dummy '14' cannot be deleted as the original IPA list would change. (Activity F would be an immediate predecessor of I, H, and G, which has not been stated in the original IPA list.

Dummy activity '15' between activity J and F, '16' between activity I and N, '17' between activity L and H, '18' between activity K and G, and dummy '19' between activity K and J can be deleted without causing any changes in the original IPA list. The result of these deletions can be seen in Figure 2-17.

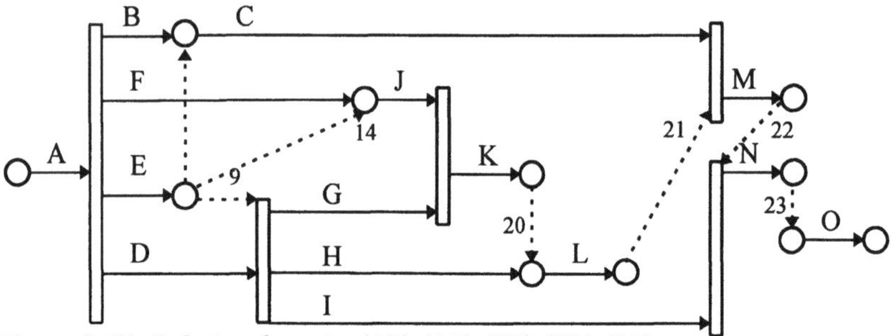

Figure 2-17 *Deleting dummies '15', '16', '17', '18', '19'*

Dummy activity '20' between activity L and K, '21' between activity M and L, '22' between activity M and N, and dummy '23' between activity O and N can be deleted without causing any changes in the original IPA list. The result of these deletions is shown in Figure 2-18.

The iteration process has finished. The final result can be seen in Figure 2-18.

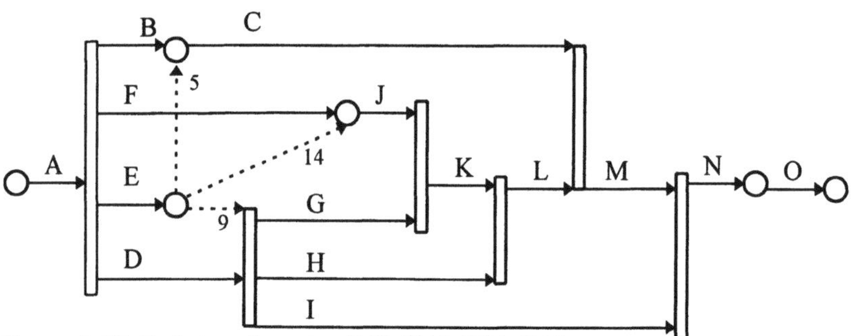

Figure 2-18 *Deleting dummies '20', '21', '22', '23'*

This method is indeed time consuming, but only routine work. Sometimes dozens of dummies can be eliminated between two drawings.

Note that taking dummies into account in different orders will result in logically correct but different networks. All these networks will represent the same IPA list, but the number and the place of dummies can differ from each other. We will illustrate this feature in a small sample.

On the left side of Figures 2-19 through 2-22 we show the same sample network from which we intend to eliminate the unnecessary dummies. The order of taking dummies into consideration follows their identifying number. As a result of four different orders we get four different, but logically identical networks.

That is why we strongly emphasize that IPA or ISA lists are the best tools for comparing different looking, but perhaps logically identical same networks.

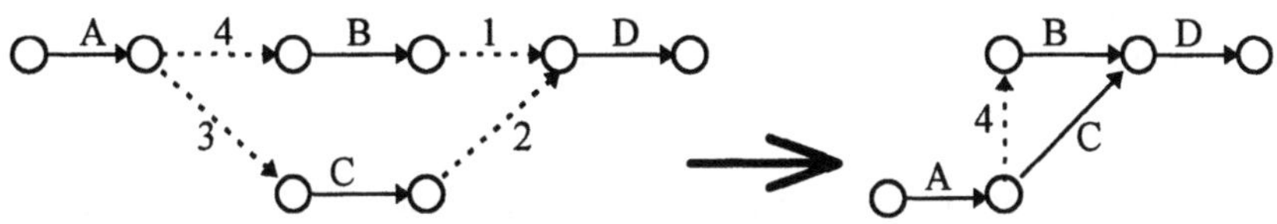

Figure 2-19 *Elimination of dummies in different orders. #1*

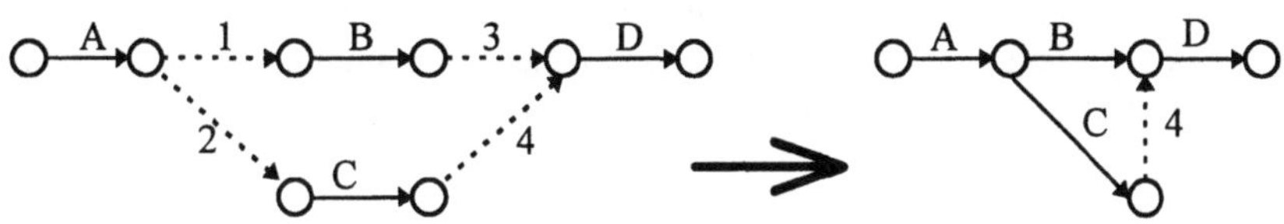

Figure 2-20 *Elimination of dummies in different orders. #2*

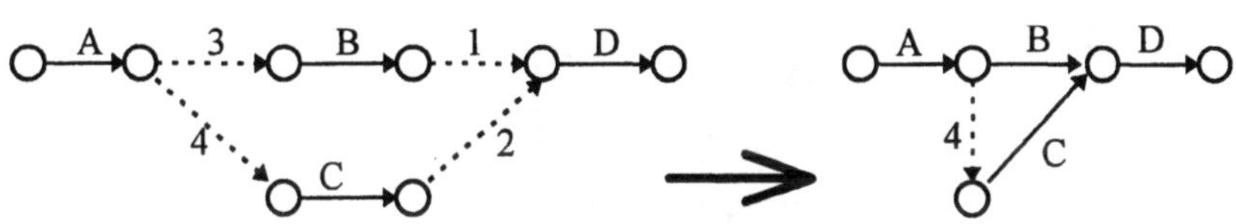

Figure 2-21 *Elimination of dummies in different orders. #3*

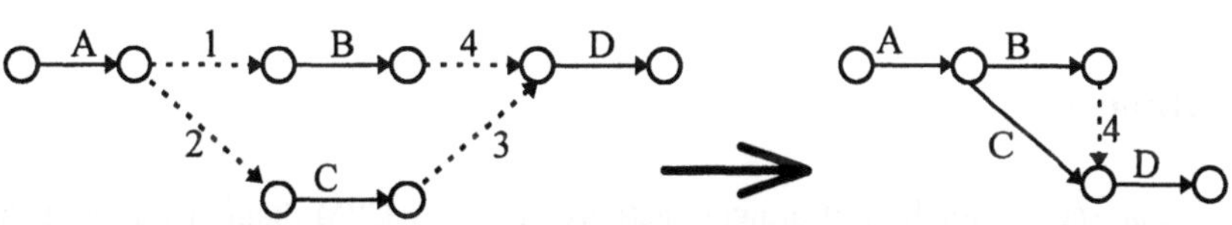

Figure 2-22 *Elimination of dummies in different orders. #4*

This method gave the basic idea for an another type of graphical portrayal of IPA lists, which is called the activity-on-node representation. An activity-on-node representation can be performed in an easy, two step way. As it was mentioned when we introduced the four steps drawing procedure, after the second step, we have a logically correct network, which contains redundant dummies. In the activity-on-node representation we use the first two, but slightly modified steps.

This easy procedure can be summarized in two steps:

Step 1. Put down all the activities as individual nodes.
Step 2. Following the IPA list, link all the activities which are related to each other.

The result of this procedure can be seen in Figure 2-23. Another significant difference between this method and the above mentioned activity-on-arrow drawing procedure is that the structures of different networks that represent the same IPA list are always the same as they do not depend on the order of taking activities into account.

A new and nowadays widely used network technique has emerged from such a network representation: the so called Precedence Diagramming Method. PDM is the subject of Chapter 4, Chapter 5, Chapter 6, and Chapter 7.

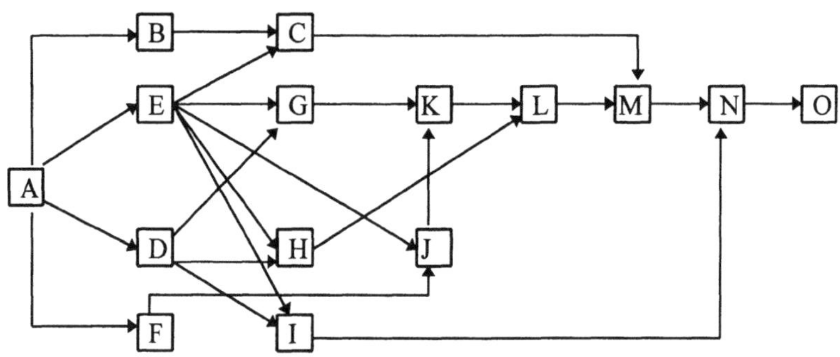

Figure 2-23 *Activity-on-node diagram of the sample project.*

Method 2

The second method of drawing activity-on-arrow CPM diagrams is faster but requires more attention. The basis of this method is the IPA list, as it is of the first method as well.

The steps of this method are listed in the following. Two kinds of labels will be given to the activities of the IPA list during the method. A negative, '-' label for an activity means that this activity can be depicted, a '+' label means that this activity in question has been drawn. There are activities which have no labels at the beginning. However, at the end of the process all the activities have to be labeled positive.

Step 1 Choose all the activities which have no predecessors, and label them with a '-'.

Step 2 Draw these activities from a common initial event, and label them with a '+'.

Step 3 Choose activities from non-labeled activities which have only positively labeled predecessors. These activities have to be labeled with a '-'.

Step 4 Order these negatively labeled activities by following these rules:

Rule 1 Mark in the predecessor list of these activities all the activities, that can be found among the predecessors more than once.

Rule 2 Order these negatively labeled activities in ascending order based on the number of marks.

Rule 3 Make Rule 3.a. and b. for all the negatively labeled activities. Take them into account based on the above defined order.

a.) *If* the entire predecessor list of an activity *i*, and none of its parts, can be found among the predecessors of negatively labeled activities, then unmark all the predecessors of *i*. *This means that if i has only one marked predecessor it will be unmarked automatically.*

b.) *If* the entire predecessor list of activity *i* can be found among the predecessors of negatively labeled activities, but a small part of this list recurs somewhere among the predecessors of negatively labeled activities, then the other part of the predecessors of *i* has to be unmarked.

At the end of this step order them again by the number of repetitive predecessors.

Step 5 Draw the negatively labeled activities in the ascending order of repetitive predecessors. All the marked predecessors have to be connected to the activities with dummies. Change the label of the depicted activities to '+'.

Step 6 If there are activities without labels, then go to Step 3.

Step 7 Make one terminal event. Use dummies to avoid multiple activities. If there are multiple dummies, eliminate the unnecessary ones.

We illustrate the algorithm with our sample project. For the sake of convenience, we have copied the IPA list from Table 2-1 (see Table 2-6). In the first step, we can choose activity A as the only activity without predecessors. In the second step, we draw this activity and label it with a '+'. The result of this step can be seen in Figure 2-24.

In the third step, we can choose B, D, E, F as these activities have only positively labeled predecessors. We mark all the activities based on Rule 1. Rule 2 has no result in this case. Based on Rule 3.a., we unmarked all the predecessors of

B, then D, then E, then F. In step 5 we depict these activities without dummies. The result cis shown in Figure 2-25.

Act.	A	B	C	D	E	F	G	H	I	J	K	L	M	N	O
IPA List	-	A	B E	A	A	A	DE	DE	DE	FE	GJ	KH	CL	MI	N

Table 2-6 *IPA list of the sample project*

As we have non-labeled activities, then according to Step 6, we have to go back to Step 3. We label activities C, G, H, I and J with a '-' according to Step 3. Step 4 can be followed in Table 2-7.

Activities. with '-'	Rule 1. (Marked predecessors)	Rule 2. # of marked act. and order		Rule 3. Deleting marks in ascending order	Drawing order
C	B ***E***	1	1	B ***E***	1
G	***D E***	2	3	D ***E***	2
H	***D E***	2	4	D ***E***	3
I	***D E***	2	5	D ***E***	4
J	F ***E***	1	2	F ***E***	5

Table 2-7 *Step 4 of the second drawing method*

As activity E can be found more than once in the precedence list, all of them have to be marked. Activity D also has to be marked. Following Rule 2, activities have to be ordered by the number of marked activities. So the right order is C, J, G, H, I as it is shown in the second column of Rule 2. (If the number of marked activities is the same, we can choose any order.) In Rule 3, we try to unmar as many activities as possible. We take them into account in ascending order. According to Rule 3 a) and b), no activity can be unmarked among the predecessors of C. Also there are no changes among the predecessors of J.

Observing the predecessors of activity G, we can see that all of them are repeated among the predecessors of H and I. However, a part of this list, which is activity E, is repeated among the predecessors of C and J, so according to Rule 3 b), only D can be unmarked. The result is the same when we examine the predecessors of H and I. After this only E remains marked in the preceding activity list.

We define the drawing order again by the number of marked activities. As this number is the same for all the activities we can choose an arbitrary order.

After this we draw the activities following their order. Marked activities will be connected by dummies. The result of this step can be seen in Figure 2-26.

Figure 2-24 *Drawing of activity 'A'*

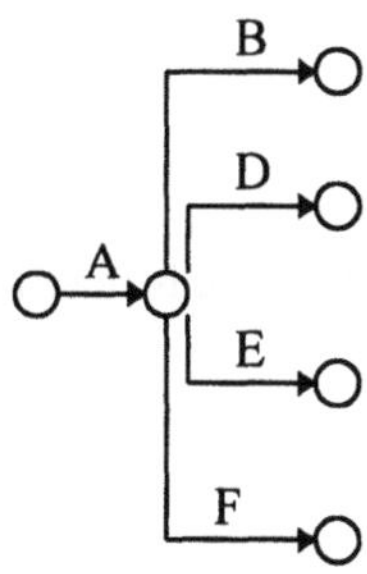

Figure 2-25 *Drawing of activities 'B', 'D', 'E', 'F'*

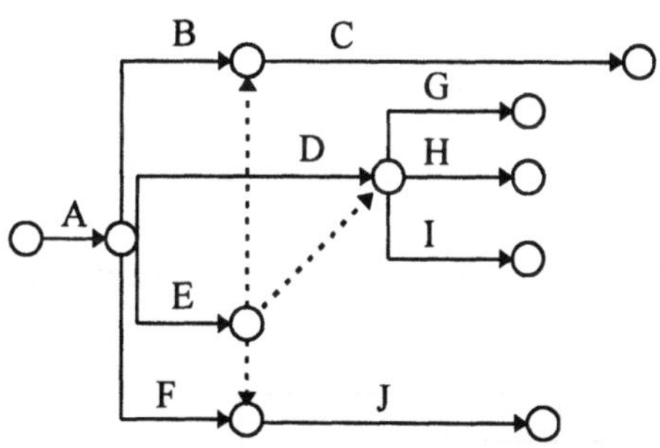

Figure 2-26 *Drawing of activities 'C', 'G', 'H', 'I', 'J'*

After the drawing of activities C, G, H, I, and J we have to change their labels to '+' and go back to Step 3. In this step we have to choose activities which have only positively numbered predecessors. In this case we find activity K. As this is the only activity, there is no meaning of doing Step 4. We can go directly to step 5 and draw the activity. The result of this iteration is shown in Figure 2-27.

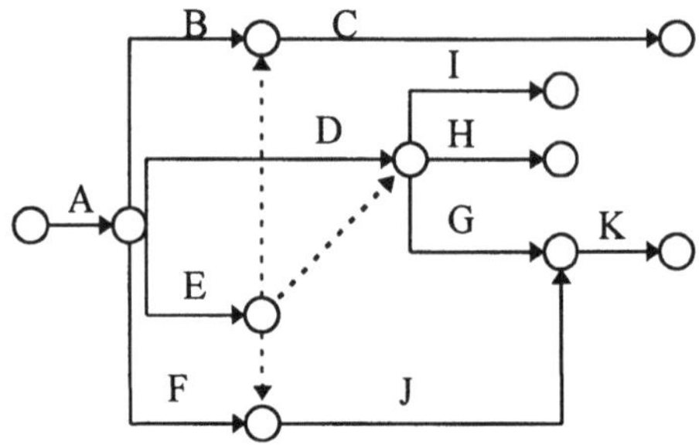

Figure 2-27 *Drawing of activity 'K'*

In the following iteration we can choose activity L. The result of this iteration is presented in Figure 2-28.

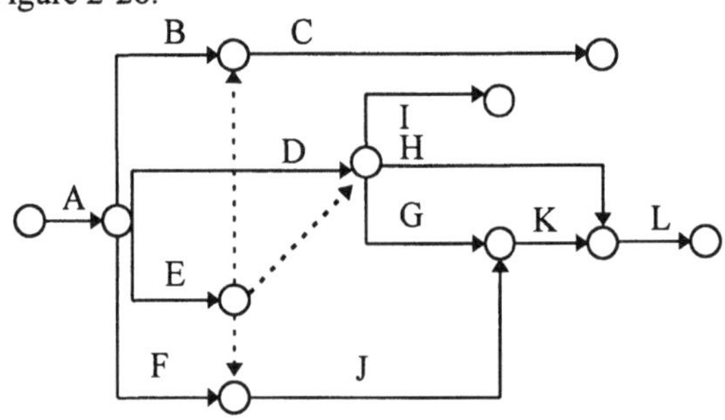

Figure 2-28 *Drawing of activity' L'*

Starting the new iteration with Step 3, we can mark only M, and we can draw it immediately (see Figure 2-29).

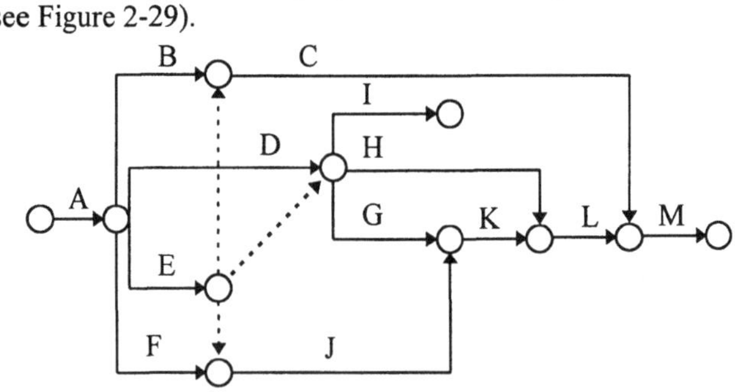

Figure 2-29 *Drawing of activity 'M'*

In the following iteration we can mark activity N, and in the last iteration, activity O. The result of the last two iterations can be seen in Figure 2-30.

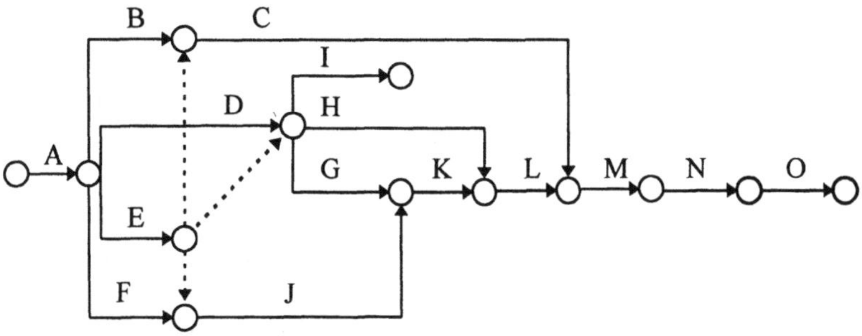

Figure 2-30 *Drawing of activities 'N' and 'O'*

In this method we had certain activities that could be handled together. These are: activity A in the first iteration, activities B, D, E, F in the second iteration, activities C, G, H, I, J in the third iteration, K in the fourth, L, M, N, and O in the fifth, sixth, seventh and eighth iteration. These groups of activities form the so called *levels of the network.* Levels of the network is the subject of the next subchapter.

Both of these methods can be used for drawing networks, but both of them take a long time. The activity-on-node representation is much more convenient.

As we mentioned before, recently activities have been identified by their names, not by the their tail and head events. Despite this, it is recommended to give identifiers to the events, because some of them have a special importance in the project schedule: they are referred to as 'milestones'. There are computer programs that still cannot build the network from the IPA list and require the events (nodes) and their codes to be input data.

The worst case when they require that head event numbers have to be grater than tail event numbers for all the activities. This can cause a lot of confusion when we want to modify the network, e.g., to add one or more activities. Sometimes, in these cases, event numbering has to be done repeatedly. The reason for this rule was to avoid loops in the network. Modern computer software can do loop detecting automatically; this is not the users' job any more.

For numbering the events, we have to know the level of the networks. Levels can be used for numbering the activities and this is the basis of the time analysis and resource leveling as well.

In the next subchapter we introduce the reader to the levels of the network and show how to number events using levels.

2.5 LEVELS OF NETWORK

The levels of the network can be used for numbering the events in ascending order, and it is also used in time analysis and in resource leveling. In the following we give a definition of the kth level.

Definition 2-4 *The kth level of the network contains those events for which the largest distance from the initial event is k events, including the initial event itself.*

Definition 2-5 *The level of an activity is equal to its tail event level.*

It derives from the above definitions that no activity can be among events on the same level, and no activity can lead from level k to l if $l<k$. Let the initial event's level be level 0.

Based on the above definitions, the numbering of the events can be done as follows.

Step 1. The first level containing the initial event, is level 0. Let k indicate the actual level. Number this event with an arbitrary number.

Step 2. Mark all the events for which all the predecessor events are on the kth level. These events are the events of the k+1 level. Number the events so that the smallest number on this level has to be greater than the largest number on level k. Do not use the same number for different events.

Step 3. Let k=k+1. If k contains the terminal event then the algorithm has finished else go back to Step 2.

The numbering of events using this algorithm is shown in Figure 2-31.

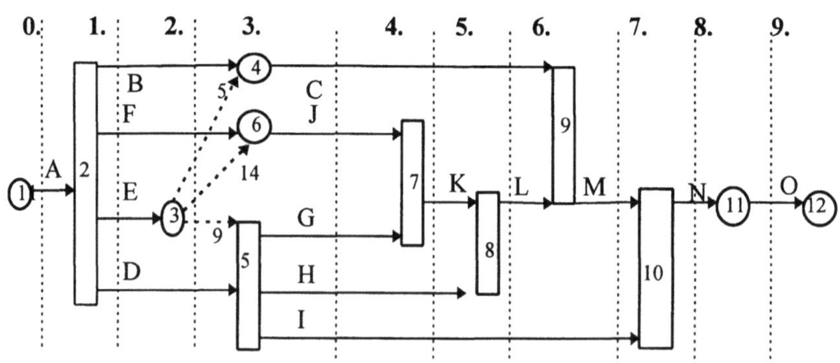

Figure 2-31 *Levels of the sample network.*

This is indeed an easy algorithm, and practically the same as Step 3 of the second drawing method. Now the reader can see that in the second drawing method we have drawn the activities in order of their levels. In each iteration we have depicted all the activities of the same level.

With this step we have completed all the necessary preparations for basic calculations. First we made the activity list, then we prepared a preceding or succeeding activity list. We showed how to filter unneeded information from the PA list in order to produce an IPA list. We also showed how to draw the network, and how to number the events in ascending order.

Among these steps only the preparations of the activity and the preceding activity lists require the understanding of the project. The creation of the IPA list, the drawing the network, and the numbering the events only require network technique skills. No wonder therefore that engineers have extended a lot of effort to reduce the time required by these steps. The use of activity-on-node representation dramatically reduces the duration of drawing. Identification of activities by their name instead of their tail and head events makes the events numbering process practically unnecessary in activity-on-arrow notation. Furthermore, we can make changes in the network much faster if we do not have to renumber the events each time.

2.6 CPM CALCULATIONS

The Critical Path Method (CPM) originally was a two-phased method. The first phase was the so called time analysis phase. The second phase was the sophisticated time cost trade-off phase. The construction industry assimilated the original CPM technique as two different network techniques. The first phase of the original technique became today's CPM technique, and the second phase became today's CPM Least Cost Scheduling Technique. In this chapter, we introduce the reader to the CPM time analysis, which is the first phase of the original method.

During the time analysis, we want answers to these basic questions:

- What is the duration time of the whole project?
- What is the earliest and the latest possible occurrence of the events which does not interfere with the project duration?
- What is the earliest possible start and the latest possible finish of the activities which does not interfere with the project duration?

If the reader remembers the definitions of activities (Definition 2-1) and events (Definition 2-2) then probably notices some kind of a connection between the second and the third question. The former is called the *event oriented calculation,* while the latter the *activity oriented calculation.*

In this chapter we present two algorithms to answer these questions. Both of them lead to the same algorithmical steps. The first will probably be preferred by engineers, while the second will be loved by mathematicians. The second method is also recommended to those interested in time cost trade-off procedures. In the second algorithm we introduce an interesting approach, the so called primal-dual methodology which is a great tool for solving optimization problems. We show that even a simple time analysis is an optimization problem, however, very few of us are aware of this fact.

In the algorithms we use the following notations:

N	- set of events
A	- set of activities (including dummies)
s, t	- initial and terminal events
i	- optional event
(i,j)	- activity from event i to j
τ_{ij}	- duration of activity (ij)
E_i	- early event time of the ith event.
L_i	- late event time of event i
π_i	- arbitrary event time of i between E_i and L_i
ES_{ij}	- earliest start of activity (ij)
EF_{ij}	- earliest finish of activity (ij)
LS_{ij}	- latest start of activity (ij)
LF_{ij}	- latest finish of activity (ij)
p	- project duration

Now we give the exact definition of the new terms:

Definition 2-6 (E_i - *Early event times*) *The earliest possible occurrence of an event during the project.*

Definition 2-7 (p - *project duration*) *The accomplishing time of the project, which is equal to the earliest possible occurrence of the terminal event.* $p=E_t$

Definition 2-8 (L_i- *Late event time*) *The latest possible occurrence of event i which does not interfere with the project duration.*

Definition 2-9 (ES_{ij} - *Early start*) *The earliest possible start of activity* (ij).

Definition 2-10 (EF_{ij} - *Early finish*) *The earliest possible finish of activity* (*ij*).

Definition 2-11 (LS_{ij} - *Late start*) *The latest possible start of activity (ij) which does not interfere with the project duration*

Definition 2-12 (LF_{ij} - *Late finish*) *The latest possible finish of activity (ij) which does not interfere with the project duration.*

It comes from the definition of an event, early event time (E_i) and early start of an activity (ES_{ij}) (Definition 2-2, Definition 2-6, Definition 2-9) that the earliest possible start of an activity is equal to the earliest occurrence of its tail event, that is, $ES_{ij}=E_i$.

It also comes from the definition of an event, late event time (L_i) and late finish of an activity (LF_{ij}) (Definition 2-2, Definition 2-8, Definition 2-12) that the latest possible finish of an activity is equal to the latest occurrence of its head event, that is $LF_{ij}=L_j$.

Finally, it derives from the definition of activities (Definition 2-1) and the activity times that early activity start defines early activity finish and vice versa, that is $ES_{ij}+\tau_{ij}=EF_{ij}$. We can state the same for the late activity times too: $LS_{ij}+\tau_{ij}=LF_{ij}$.

Now it can be seen, that it does not makes any difference whether we speak about event oriented calculations or activity oriented calculations, since they are practically the same. The only difference is that at the end only event or only activity times will be shown.

In the following we present two procedures to determine the activity and the event times. As practically they are the same, we will call them the usual and the primal-dual procedure.

Usual Procedure

This procedure is a two phased procedure. In the first phase we can determine the earliest occurrence of events (E_i) and the early start and finish of activities. Do not forget that from early event times we can determine early start times as $E_{ij}=E_i$, and from early start times we are able to determine early finish times, $EF_{ij}=ES_{ij}+\tau_{ij}$

The basis of the first phase can be summarized as follows. We can determine the earliest start, which is equal to the early event time, of an (*ij*) activity if we know the earliest finish and start of all the predecessor activities. We can determine the earliest activity times of the predecessor activities if we have these data for the predecessors of predecessor activities.

Therefore, for the calculation we need an event for which the early event time is known. This will be the initial event. Let the early event time for *s* be 0. As we start the calculation from the initial node, this phase is also known as forward calculation.

Forward calculation:

Step 1. Let $E_s=0$

Step 2. Look for an event *j* for which the occurrence of all the immediate predecessor *i* events are known.

$E_j=\max(E_i+\tau_{ij})$ for all (ij) immediate predecessor of i [i]

(The following equivalent formulas can also be used:

$E_j=\max(ES_{ij}+\tau_{ij})$ $\forall\ (ij)\in\mathbf{A}$ or $E_j=\max(EF_{ij})$ $\forall\ (ij)\in\mathbf{A}$)

Step 3. If $j\equiv t$ then stop else continue with Step 2.

The occurrence of event *t* gives the *p* project duration by definition. The first phase gave answers to the question, which are the earliest occurrence of events and the earliest start time of activities. Obviously, there are activities that may start later without interfering with the project duration. The latest event times and latest activity times are the result of the second phase.

The basis of the second phase can be summarized as follows. We can determine the latest finish of (ij) , which is equal to the late event time of *i*, if we know the latest start and finish of all the successor activities. We can determine the latest activity times of the successors if we have these data for the successors of successor activities.

Therefore, for the calculation we need an event for which the late event time is known. This will be the terminal event. Let the late event time for *t* be the project duration.

Backward calculation:

Step 1. Let $L_t=E_t=p$

Step 2. Look for an event *i* for which the occurrence of all the immediate successor *j* events are known.

$L_i=\min(L_j-\tau_{ij})$ for all (ij) immediately succeeding i

(The following equivalent formulas also can be used:

$L_i=\min(LF_{ij}-\tau_{ij})$ $\forall\ (ij)\in\mathbf{A}$ or $L_i=\min(LS_{ij})$ $\forall\ (ij)\in\mathbf{A}$)

Step 3. If $j\equiv$s then stop else continue with Step 2.

---

[i] We can use the '$\forall\ (ij)\in\mathbf{A}$' expression instead of the term 'for all immediate (ij) predecessors of *i*'. The correct interpretation of this expresson would be 'for all (ij), which are elements of the activity set, but as event *i* is under examination , in this case it can be interpreted as 'immediate predecessors of *i*'.

In mathematical sense, the first phase is a longest path problem from *s* to *t*. The second phase is a longest path problem from *t* to *s*.

The longest path from the initial event to the terminal event is called the *critical path*. If we have more than one paths with the same length, then we have multiple critical paths. The critical path has a great importance in the project management process. This is the subject of the next subchapter, Information Obtained From Calculations.

To illustrate this algorithm, we show the time analysis for the sample project. The results of the calculation are shown in Figure 2-32.

Event numbers can be seen inside nodes, early event times and late event times can be seen above and under events.

Activity durations are written under or near activities using *italics*. The activities on the critical path are depicted with bold arrows.

Do not forget to take into account the dummies with zero activity duration!

Forward calculation

Events can be taken into account in this order: '1';'2';'3'; ('4';'6';'5';) '7'; '8'; '9'; '11'; '10'; '12'. The order of parenthetic events is indifferent because they are on the same level (see Levels of Network).

$E_{s=1}=0$
$E_2=E_1+\tau_{12}=0+30=30$
$E_3=E_2+\tau_{23}=30+14=44$
$E_4=\max\{E_2+\tau_{24};\ E_3+\tau_{34}\}=\max\{30+20;44+0\}=50$
$E_5=\max\{E_2+\tau_{25};\ E_3+\tau_{35}\}=\max\{30+10;\ 44+0\}=44$
$E_6=\max\{E_2+\tau_{26};\ E_3+\tau_{36}\}=\max\{30+30;\ 44+0\}=60$
$E_7=\max\{E_6+\tau_{67};\ E_5+\tau_{57}\}=\max\{60+10;44+20\}=70;$
$E_8=\max\{E_7+\tau_{78};E_5+\tau_{58}\}=\max\{70+30;44+80\}=124$
$E_9=\max\{E_4+\tau_{49};\ E_8+\tau_{89}\}=\max\{50+30;124+20\}=144$
$E_{10}=\max\{E_9+\tau_{910};\ E_5+\tau_{510}\}=\max\{144+5;44+90\}=149$
$E_{11}=\max\{E_{10}+\tau_{1011}\}=149+20=169$
$E_{12}=\max\{E_{11}+\tau_{1112}\}=169+15=184$

Backward calculation

Events can be taken into account in this order. '12'; '11'; '10'; '9'; ('4'; '8';) '7'; ('6'; '5';) '3'; '2'; '1'. The order of events in parentheses is indifferent because they are in the same backward level. (Backward level means that levels of the network have been defined from the terminal events.)

$L_{t=12}=E_{t=12}=184$
$L_{11}=\{L_{12}+\tau_{1112}\}=184-15=169$
$L_{10}=\min\{L_{11}-\tau_{1011}\}=169-20=149$
$L_9=\min\{L_{10}-\tau_{910}\}=149-5=144$
$L_4=\min\{L_9-\tau_{49}\}=144-30=114$
$L_8=\min\{L_9-\tau_{89}\}=144-20=124$
$L_{7=}\min\{L_8-\tau_{78}\}=124-30=94$
$L_6=\min\{L_7-\tau_{67}\}=94-10=84$
$L_5=\min\{L_7-\tau_{57};\ L_8-\tau_{58};\ L_{10}-\tau_{510}\}=\min\{94-20;\ 124-80;\ 149-90\}=44;$
$L_3=\min\{L_4-\tau_{34};\ L_6-\tau_{36};\ L_5-\tau_{35}\}=\min\{114-0;\ 84-0;\ 44-0\}=44$
$L_2=\min\{L_4-\tau_{24};\ L_6-\tau_{26};\ L_3-\tau_{23};\ L_5-\tau_{25}\}=\min\{114-20;\ 84-30;\ 44-14;\ 44-10\}=30$
$L_1=\min\{L_2-\tau_{12}\}=\min\{30-30\}=0$

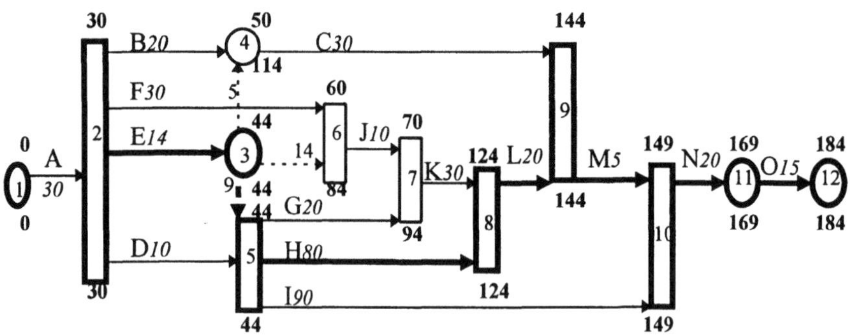

Figure 2-32 *Result of CPM calculations in the sample project*

Using the $E_i=ES_{ij}$, $L_j=LF_{ij}$, and $ES_{ij}+\tau_{ij}=EF_{ij}$ and $LS_{ij}+\tau_{ij}=LF_{ij}$ formulas, we can define the activity times as well. Recall, that we had to define the activity times during the calculation. The elements of the $E_j=\max(E_i+\tau_{ij})\ \forall\ (ij)\in\mathbf{A}$ formula in Step 2 of the forward calculation give the early finish times of the activities as it can also be seen from the equivalent expressions, $E_j=\max(ES_{ij}+\tau_{ij})\ \forall\ (ij)\in\mathbf{A}$ or $E_j=\max(EF_{ij})\ \forall\ (ij)\in\mathbf{A}$. The elements of the $L_i=\min(L_j-\tau_{ij})\ \forall\ (ij)\in\mathbf{A}$ formula in Step 2 of the backward calculation give the late start times as it can also be seen from its equivalents, $L_i=\min(LF_{ij}-\tau_{ij})\ \forall(ij)\in\mathbf{A}$ or $L_i=\min(LS_{ij})\ \forall\ (ij)\in\mathbf{A}$.

Remarks: This algorithm is basically the same as the one in the defining of the levels of the network. If we consider all the activity durations to be one unit, we will get the number of the levels as the result of the forward analysis.

If we keep the order of taking activities into account, we can use this order backward at the backward calculation phase. This makes the algorithm faster.

If the network contains a loop or loops, the algorithm will calculate the early event time of those events, which are on a smaller or equal level than the events from the loop. The events of the loop and the events on a higher level cannot be calculated.

Primal-Dual Procedure

In the following, we show that the CPM time analysis can be formulated as an optimization problem. This approach will lead to the same calculation procedure as the above given algorithm, but the interpretation is more mathematical. We recommend reading this part to those who want to learn about time cost trade-off procedures too. We use the same approach there, but here it is much more understandable as the reader is already familiar with the applied algorithm.

The following approach -- in which we examine the CPM problem from the investor and from the contractor points of view -- will help to understand the mathematical logic of the primal-dual approach.

Let the CPM network of a construction be given. The contractor provided the activity durations and would like to know how long it takes to complete the project. For this, he looks for the longest path from the initial to the terminal event as this path determines the necessary project duration. Therefore, the contractor's problem is a longest path problem.

The investor can follow the project with the help of events. His task is to define a possible event (milestone) system which is acceptable for the contractor. This means that each of the activities could be accomplished between their tail and head event times. There are millions of possible event systems but he is interested in only one solution, which gives the smallest difference between the start and the terminal event, but still can be accomplished with the given activity durations.

Before establishing the mathematical model, we provide the necessary definitions and notations.

The set of the nodes are denoted by $N=\{1,2,...i,..j....n\}$. The set of the activities are denoted by $A=\{(1,2),...(i,j),...(n-1,n)\}$. The graph is denoted by the symbol $[N,A]$. A possible event time is denoted by π_i, activity durations are denoted by τ_{ij}.

Definition 2-13 *Let s and t be two distinguished points of the graph. Let* $(s\equiv 0, 1,...i....m\equiv t)\in \mathbf{N}$, *a special set of points to which every* $(i-1,i)\in \mathbf{A}$, $i=1...m$. *In this case the* $(0, 1,...i, ...m)$ *set is called a path from s to t. We denote the path with the* $P(s,t)$, *or shortly, with the symbol* P.

Definition 2-14 *(Length of a path): If there is a value τ_{ij} given to all arrows, then the length of the path, which is denoted by the symbol $\tau(P)$, is defined as* $\tau(P)=\sum_{(ij)\in P}\tau_{ij}$

The mathematical model of the problem from the investor's point of view is:

$$\pi_j-\pi_i\geq \tau_{ij} \qquad \forall\ (i,j)\in \mathbf{A} \qquad \mathbf{2.1}$$
$$\pi_t-\pi_s\rightarrow \text{minimum} \qquad \mathbf{2.2}$$

This model is called the primal task of the CPM problem. (**2.1** is called the set of constraints, while **2.2** is called the objective function.)

The contractor would like to know how much time is necessary to accomplish the project. So he looks for a path

$$P=\{s\equiv 0,\ 1,....i,......m\equiv t\} \qquad \mathbf{2.3}$$

for which the following objective function results in a maximum value

$$\tau(P)=\sum_{(ij)\in P}\tau_{ij}\rightarrow \text{maximum} \qquad \mathbf{2.4}$$

This problem is called the dual problem of CPM.

Using a lemma, first we show that there is a strong connection between the two models, that is, the value of **2.2** is always greater or equal than the value of **2.4**, and when the two values are the same, then both solutions are optimal. After this we show in a theorem that there exist solutions to the two problems when the values of the objective functions are equal. The proof of this theorem will give the algorithm too. In the last theorem, we prove that if there are optimal solutions for each model, then the values of their objective functions are the same.

Lemma 2-1 For of any schedule π satisfying **2.1** and any path $P(s\rightarrow t)$,

$$\tau(P)\leq \pi_t-\pi_s \qquad \mathbf{2.5}$$

holds, and equilibrium exists if and only if the following is satisfied:

$$\pi_j-\pi_i=\tau_{ij} \qquad \forall\ (i,j)\in P \qquad \mathbf{2.6}$$

Proof:

$$\tau(P)=\sum_{i=1}^{m}\tau_{i-1,i}\leq\sum_{i=1}^{m}\pi_i-\pi_{i-1}=(\pi_1-\pi_{0=s})+(\pi_2-\pi_1)+.....+(\pi_i-\pi_{i-1})+(\pi_{m=t}-\pi_{m-1})=\pi_t-\pi_s,$$

Equality holds if and only if **2.6** is satisfied along the path.

An important consequence of the lemma is presented hereby.

Theorem 2-1 (weak form of equilibrium) If there exist a policy π that satisfies **2.1** and a path $P(s \to \tau)$ from *s* to *t* and $\tau(P)=\pi_t-\pi_s$, then both solutions are optimal.

Proof: (Indirect way) Suppose that $\tau(P)=\pi_t-\pi_s$ is satisfieed but there exists a path P* for which $\tau(P^*)>\tau(P)$. In this case $\tau(P^*)>\tau(P)=\pi_t-\pi_s$, but this is a contradiction according to the lemma. Suppose that $\tau(P)=\pi_t-\pi_s$ satisfies but there exists a system π^* for which $\pi_t^*-\pi_s^*<\pi_t-\pi_s$. In this case $\pi_t^*-\pi_s^*< \pi_t-\pi_s=\tau(P)$, but this is a contradiction according to the lemma.

Theorem 2-2 If there exists policy π that satisfies 2.1, and a path $P(s\to t)$, then exists path P and system π for which 2.6 satisfies, that is, $\pi_t-\pi_s$ minimal and $\tau(P)$ is maximal. (The proof is constructive, it gives the algorithm as well.)

Proof 1: Let us constitute policy π that satisfies **2.1** in the following way. Divide **N** into two disjunct sets of **S** and **T** with these characteristics of **S**:

- **2.1** has to be satisfied for all the arrows in **S**
- let $s \in S$,
- if an $i \in S$, then there should be path $P(s,i)$, along which $\pi_j-\pi_i=\tau_{ij}\ \forall\ (i,j)\in P\ i,j\in S$
- there is no $(i,j)\in A$ where $j\in$ **S** and $i\in$ **T**

The $S=\{s\}$ set satisfies this construction in the beginning. If $S\equiv N$, then, according to the construction of **S**, the solution is optimal. If $S\neq N$, then enlarge **S** so that the above defined attributes remain valid.

For this, look for a $j\in$ **T** point to which paths lead only from **S**. There should be such a point because in the opposite case a loop would be in the network. The event time of this point will de defined in the following way:

$$\pi_j := \max_{\substack{i\in S \\ (i,j)\in A}}[\pi_i + \tau_{ij}]$$

By enlarging **S** with *j*, the attributes of **S** remain. After *m-1* enlargements we reach the $S\equiv N$ state.

The theorem is thus proved.

Proof 2. Let us constitute policy π that satisfies **2.1** in the following way. Divide **N** into two disjunct sets of **S** and **T** with the following characteristics of **T**:

- **2.1** has to be satisfied for all the arrows in **T**
- let $t \in$ **T**,
- if a $j \in$**T** then there should be a path P(j,t), along which
  $\pi_j - \pi_i = \tau_{ij}$ $\forall$ $(i,j) \in$P $i, j \in$T
- there is no $(i,j) \in A$ where $i \in$**S** and $j \in$**T**

The **T**=$\{t\}$ set satisfies this construction in the beginning. If **T**$\equiv$**N**, then, according to the construction of **T**, the solution is optimal. If **T**$\neq$**N**, then enlarge **T** so that the above defined attributes remain valid.

For this, look for a point $i \in$**S** from which arrows lead only to **T**. The event time of this point will de defined in the following way:

$$\pi_i := \min_{\substack{j \in T \\ (i,j) \in A}} [\pi_j - \tau_{ij}]$$

By enlarging **T** with i, the attributes of **T** remain. After m-1 enlargements we reach the **T**$\equiv$**N** state.

The theorem is thus proved.

System π we have after the first construction is nothing else but the system of *early event times*. System π we have after the second construction is, in fact, the system of *late event times*. In the first construction, any π_s value can be used. It is expedient to use the zero value. (π_s=0)

In the second construction, any π_t value can be used. It is expedient to choose the π_t value of the first construction. (L_t=E_t)

As the reader can see, the construction of S leads to the procedure of forward calculation, and the construction of T leads to the procedure of backward calculation.

Theorem 2-3 If there exists a minimal policy π satisfying **2.1** and a maximal length path P, then their values are the same.

Proof: According to Theorem 2-2, there is an optimal system π* and an optimal path P*, that is, $\pi_t^* - \pi_s^* = \tau(P^*)$. As π_i and P are also optimal, then $\pi_t^* - \pi_s^* = \pi_t - \pi_s$ and $\tau(P^*) = \tau(P)$, but then $\pi_t - \pi_s = \tau(P)$.

The constructive proof of Theorem 2-2 gives the same algorithm shown in Chapter 2 (CPM Calculations: Usual Procedure). Nevertheless we are going to show it again. This time we will show the algorithm in a matrix format. The rows and columns of the matrix show the events by their numbers, usually in ascending order.

The cells of the matrix represent the activities of the network. The cell of the ith row and j^{th} column represents the (ij) activity of the network. Cells contain the activity durations.

The left-sided column of Table 2-8 contains the results of the forward calculation.

The backward calculation results are put in a separate table (Table 2-9). The row under the matrix shows the late event times.

We start with the forward calculation. In the beginning, we know the early event time for the initial event only. In the next step, we enlarge the **S** set with the events which have predecessors only in **S**. For this, we cover the rows and columns of events which are elements of **S**, that is, their early event times are known. We can take into further account those events for which the uncovered cells of their column are empty. For example, at the start of the calculation, we cover the row and column of event '1'. We can take event '2' into further account because its column's cells are empty at the uncovered part. (This procedure is nothing else but the determination of the next level of the network) The event time for an event in question can be determined from the $\pi_j = \max(\pi_i + \tau_{ij})$ formula. Now this event will be the element of S and the procedure can be continued till the **S≡N** situation appears. We recommend that the reader follows this procedure in the table (Table 2-8). The rows and columns are not covered for the sake of lucidity.

		1	2	3	4	5	6	7	8	9	10	11	12
0	1		30										
30	2			14	20	10	30						
44	3				0	0	0						
50	4									30			
44	5							20	80	90			
60	6							10					
70	7								30				
124	8									20			
144	9										5		
149	10											20	
169	11												15
184	12												

Table 2-8 *CPM calculation: Early event times*

At the beginning of the backward calculation, the late event time of the terminal event is known only. In the next step we enlarge the T set with event *i*, which has successors only in **T**. For this we cover the rows and columns of events which are elements of **T**, that is, their late event times are known. We can take into further account those events for which the uncovered cells of their rows are empty.

For instance, at the start of the calculation, we cover the row and column of event '12'. We can take event '11' into further account, because its row's cells are empty at the uncovered part.

The event time for event i in question can be determined from the $\pi_i = \min(\pi_j - \tau_{ij})$ formula. Now this event will be the element of **T** and the procedure can be continued till the **T≡N** situation appears. The whole procedure can be followed on Table 2-9. The rows and columns are not covered for the sake of lucidity.

	1	2	3	4	5	6	7	8	9	10	11	12
1		30										
2			14	20	10	30						
3				0	0	0						
4									30			
5							20	80	90			
6							10					
7								30				
8									20			
9										5		
10											20	
11												15
12												

0	30	44	114	44	84	94	124	144	149	169	184

Table 2-9 *CPM calculations: Late event times*

The next topic discussed in this book will be about what kind of information can be obtained from calculations and what they are good for in the project management practice.

2.7 Information Obtained From Calculations

As it has become clear from the preceding pages the time analysis of a project is separated into two parts. Forward calculations give the earliest occurrence of events and the earliest start and finish of activities. Backward calculations give the latest occurrence of events and the latest start and finish of activities. No events can occur earlier or later -- without affecting the project itself -- as it determined by the forward and backward calculations. Events can occur any time between their early and late event times and activities can also take place any time between ES_{ij} and LF_{ij}. The so called floats or slacks are used for measuring these possible delays during the project. There are several types of floats/slacks. Originally, the term slack was used for events and the term float was used for activities. Unfortunately, today the terms slack and float are loosely applied both for jobs and events. If there is any risk of ambiguity in practice, the terms 'event-slack/float' and 'activity-float/slack' should be used.

There are certain events for which the early event time and late event time are the same after the calculations. This means, if the occurrence of this kind of an event is delayed, then the project accomplishing time will be longer too.

These events are the *critical events*. Critical events are all on the longest path or paths from the initial to the terminal event. This path determines the length of the project and is called the *critical path*. Activities on this path cannot be delayed without causing delays in the project deadline, and they are called the *critical activities*. In the following, we will give the definition of the critical events, critical path, critical activities and slack. Slack shows how much an event can slip without interfering with the project deadline.

Definition 2-15 *(Critical event) An event is called critical if its early and late event times are the same.*

Definition 2-16 *(Slack or event float) A slack of an event i is the time difference between its late and early event time. We denote it with* F_i *and define it as* $F_i=L_i-E_i$

Definition 2-17 *(Critical path) The longest path or paths that leads from the initial to the terminal event is called the critical path or paths. (All the critical events have to be on a critical path.)*

Definition 2-18 *(Critical activity) An activity is called critical if it is on the critical path.*

If we have multiple longest paths, then we have multiple critical paths. These critical paths can be independent from each other or connected to each other. Some of the possibilities are depicted in Figure 2-33.

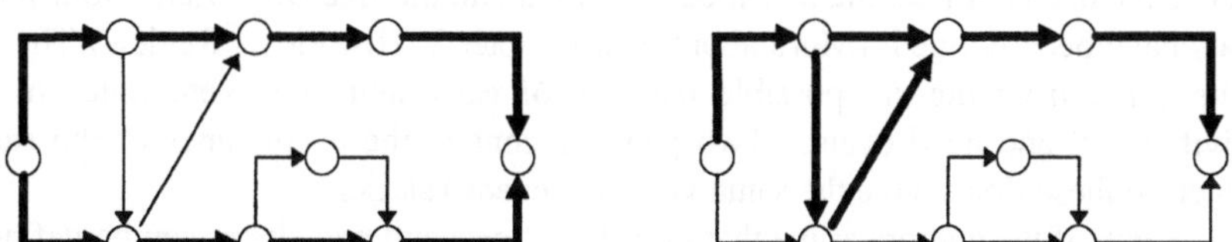

Figure 2-33 *Separate and connected critical paths*

The equality of early and late event times of critical events comes from the fact that the longest path from *s* to *t* is the same as the longest path from *t* to *s*, and we start the backward calculation from $L_t=E_t$ event time. There are some extensions when backward calculation starts from the desired deadline. If it is smaller than E_t, then we will have *negative slacks*. These negative slacks show how much these events delay the desired deadline. If project completion date is beyond E_t, then we will have only greater than zero slacks. In case of applying these extensions, the definition of critical events has to be modified as follows.

Definition 2-19 *($L_t \neq E_t$, Critical event) An event is called critical if its slack is the smallest among all.*

The definition of critical events is not enough to determine the critical path. The critical path passes through all the critical event of the network. However, it is a necessary but not a sufficient condition. Connection of critical events can result more path and sometimes they are not all critical. In Figure 2-34 a) we show the result of the time analysis of a small network. As it can be seen, all the events are critical. We have shown four possible connections of critical events, but only version a) shows a real critical path. (The supposed critical paths are drawn in bold.) Therefore, to determine the critical path, we also have to determine the critical activities.

How can we determine critical activities in an easy and fast way?

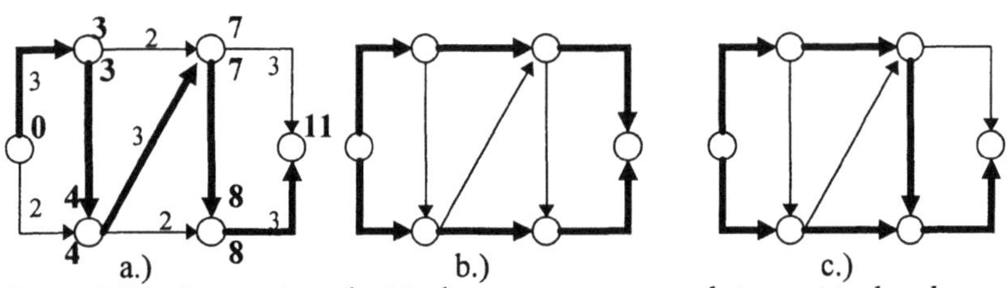

Figure 2-34 *Connection of critical events may not result in a critical path*

To answer this question, first we have to learn what activity floats are.

There is quite a big confusion in this field. Originally, two types of activity floats have been defined, the so called total float and the free float. Later, four more floats have been defined. (More about it in Chapter 2: Historical Review) The six floats come from the six possible pairings of early and late event times of an activity's tail and head event. Two pairings contain the event times of the same events, so these floats have the same value as (event) slacks.

Some of the authors argue that only four floats and two slacks can be defined. This is acceptable because two from the six defined floats come about the same way as slacks. The problem is that some authors confused about the floats and they define four floats out of which one or two are the same as slacks. Fortunately, there is no arguing about the two most important floats, the total float and the free float.

In the following we give the definition of all six activity floats originally defined, but we use the names most common in practice today.

Definition 2-20 *(Total float)Total float of an activity denotes the amount of time by which the start of the activity in question can exceed its earliest start without affecting its latest finish value, that is without affecting the project duration. This value is given by* $TF_{ij}=L_j-E_i-\tau_{ij}$ or $TF_{ij}=LF_{ij}-ES_{ij}-\tau_{ij}$ or $TF_{ij}=LF_{ij}-EF_{ij}$

Definition 2-21 *(Free Float) Free float of an activity denotes the amount of time by which the start of the activity in question can exceed its earliest start without affecting the earliest start of succeeding activities. This value is given by*
$FF_{ij}=E_j-E_i-\tau_{ij}$ or $FF_{ij}=ES_{jh}-EF_{ij}$ or $FF_{ij}=ES_{jf}-ES_{ij}-\tau_{ij}$ or $FF_{ij}=ES_{jh}-LF_{gi}-\tau_{ij}$

This float is often called activity float. Battersby defined this type of float as *early free float.* (References will be given in the Historical Review)

Definition 2-22 *(Independent Float) Independent float of an activity denotes the time by which the start of activity can change if one of or all the predecessors finish at their late finish time and one of or all the successors start at their early start time. This value is given by* $IF_{ij}=E_j-L_i-\tau_{ij}$ or $IF_{ij}=ES_{jh}-LF_{gi}-\tau_{ij}$

Definition 2-23 *(Conditional Float) Conditional float of an activity denotes the time by which the start of the activity in question can change if predecessors finish at their late finish time and the activity does not affect its latest finish, that is, the project duration. This value is given by* $CF_{ij}=L_j-L_i-\tau_{ij}$ or $CF_{ij}=LF_{ij}-LF_{gi}-\tau_{ij}$
(Battersby has defined this float as *late free float.*)

Definition 2-24 *(Early Interfering Float) Early interfering float is the difference of total and free float, that is* $EIF_{ij}= L_j-E_i-\tau_{ij}-(E_j-E_i-\tau_{ij})=L_j-E_j$. (***Slack of j!***)

Early interfering float is the original definition by Battersby, but this is the same as the slack of event *j*. Recently the term of event slack *j* is used instead of EIF.

Definition 2-25 *(Late Interfering Float) Late interfering float is the difference of total and conditional flow, that is* $LIF_{ij}=L_i-E_i$. ***(Slack of i!)***

Late interfering float is the same as the slack of event *i.* We can use the terms of event slack instead of the last two definitions. For better understanding, we have shown the graphical interpretation of floats and slacks in Figure 2-35.

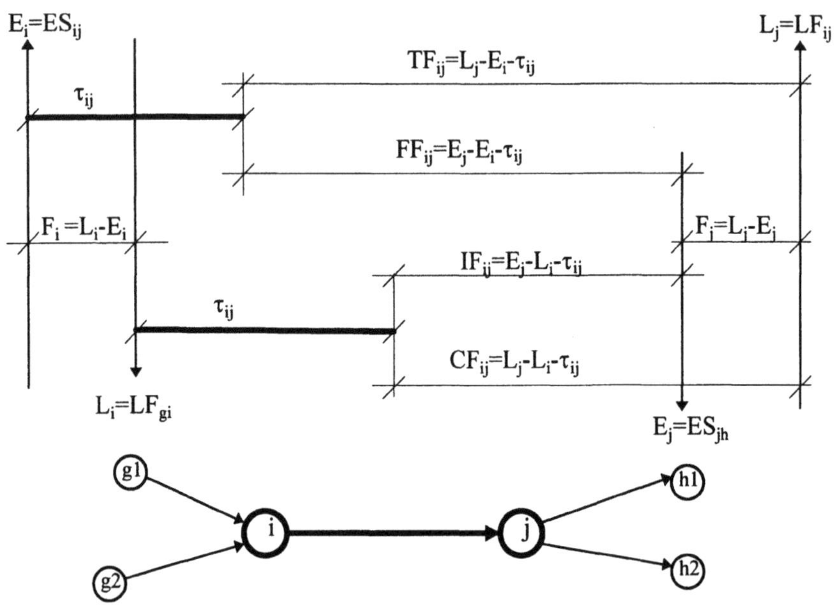

Figure 2-35 *Interpretation of floats and slacks.*

If the backward calculation starts from the $L_t=E_t$ value, then slacks can never be negative. Among the four activity floats, only independent float can be negative the rest have to be greater or equal to zero. If the backward calculation starts from the desired deadline, which is less than E_t, then all the floats and slacks can be negative. In this case a negative slack shows how much earlier the late event time has to be accomplished without delaying the defined deadline. Knowing the definitions of floats, we can give equivalent definitions to Definition 2-17 and Definition 2-18 (critical path and critical activities):

Definition 2-26 *(Critical activity) An activity is critical if its total float is the least among all. (Using Lt=Et computation rule, this means a zero float.)*

Definition 2-27 *(Critical path) A path from s to t, along which all the activities have the least total float, that is, all the activities are critical.*

The critical path has a great importance in the project management process. If an activity on the critical path is delayed, this will affect the project duration by the same amount of delaying. Therefore, special attention has to be paid to these activities to accomplish them without delays.

It was mentioned at the beginning of this chapter that millions of possible ways exist to accomplish a project between its earliest and latest time policy. Special tools were developed in construction practice to control the activities between their time span.

These tools are called as constraints. The following types of constraints, which may be assigned to activities:

As Soon As Possible (ASAP)	Activity has to start as soon as possible.
As Late As Possible (ALAP)	Activity has to finish as late as possible
Must Start On	Activity has to start exactly on a given date.
Must Start On or Later	Activity has to start on a given date or later.
Must Start On or Earlier	Activity has to start on a given date or earlier.
Must Finish On	Activity has to finish on a given date
Must Finish On or Later	Activity has to finish on a given date or later.
Must Finish On or Earlier	Activity has to finish on a given date or earlier.
Must Start Between...	Combination of fourth and fifth constraints.
Must Finish Between...	Combination of seventh and eighth constraints

We give a comprehensive discussion of use of constraints in Chapter 5: Constraints in PDM. Now we show a possible way of accomplishing the sample project. First, in Figure 2-36, we show the Gantt chart of our sample project where all the activities begin at earlier start time. This is nothing else but a use of ASAP constraint for all the activities. In Figure 2-37, we show the Gantt chart of our sample project, when all the activities finish at late finish time. This is nothing else but a use of ALAP constraint for each activity.

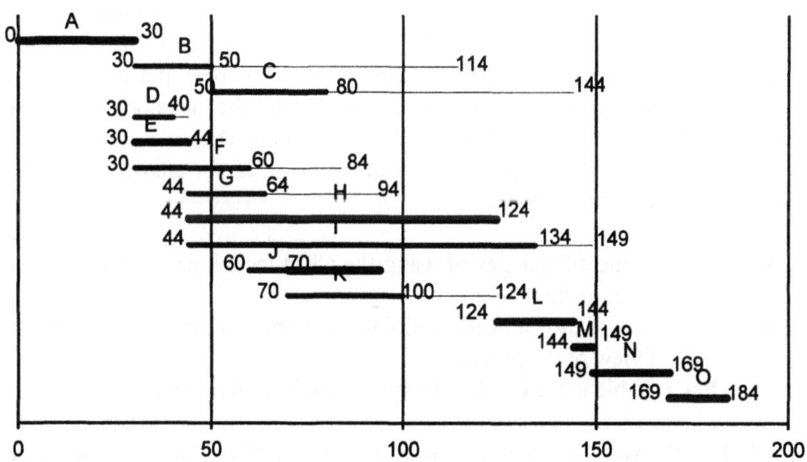

Figure 2-36 *Gantt chart for the sample project: Activities start at* ES_{ij}

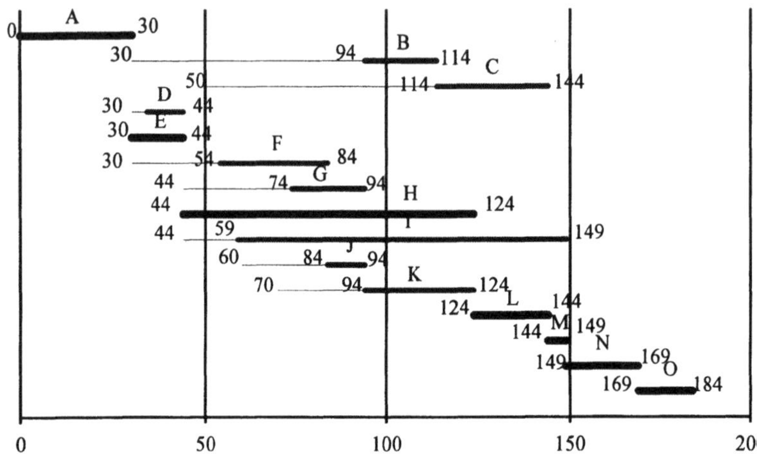

Figure 2-37 *Gantt chart for the sample project: Activities start at* LS_{ij}

2.8 PROS AND CONS OF THE CPM TECHNIQUE

Although this is the only network technique presented in this book so far, we can already compare it to the non-network techniques. We can also determine what forces the users to do unnecessary work during the preparation of a network or during the calculation process. A great deal of effort has been put in the past into eliminating the shortcomings. This effort has led to the development of some extensions. These extensions will also be discussed in this chapter.

In this chapter, we are going to examine what is the answer of the CPM technique to these three questions.

- What are the advantages of using the CPM technique over the traditional non-network techniques?
- What parts of the CPM methodology demand unnecessary work from the point of view of the project?
- How suitable is the CPM technique to model real projects?

CPM, as the very first complete network-based methodology, has brought a lot of new features that could not be found in traditional non-network techniques before. The first and most important advantage over the non-network techniques is that CPM is able to keep the project logic due to its graph interpretation. This makes the schedule easy to survey and modify. The whole graph structure makes the methodology easy to be computerized.

The second advantage is that satisfying the network rules during the planning phase forces the users to think through the project in a more thoroughly than ever before. The schedulers are compelled to make more advanced planning just because of using this new technique. This involves the split of scheduling and planning phases, which have been considered jointly in the traditional techniques.

The third advantage is that a lot of useful information can be obtained as a result of this technique. The critical path, which determines the project length, has never been defined before just as the possible delays of activities and events, floats and slacks.

The CPM technique, due to its revolutionary features and relative simplicity, has gained a wide acceptance in every segment of the industry and a great percentage of project managers has become familiar with it in a short time. Unfortunately few of them was forced to use it in his own practice. In the late 1960's, for managing large scale projects, the CPM technique has become the network technique of choice as an alternative to non-network techniques.

As the CPM was the first technique among the network-based techniques it had a lot of shortcomings which have been revealed during applications. The first group of problems that CPM sometimes requires much more work to satisfy all its rules than it is necessary. There are steps in the planning phase which are completely unnecessary from technical point of view, but which are essential because of CPM rules.

As we have mentioned before, the network based project methodology can be divided into eight steps, from which the first four are related to the planning phase.

Step 1. Defining Activities
Step 2. Defining Activity Interdependencies
Step 3. Drawing the Network
Step 4. Time and Resource Estimation
Step 5. Basic Calculations
Step 6. Advanced Calculations
Step 7. Project Control
Step 8. Project Review

Among these steps, 3, 5, and 6 do not require special knowledge of the project. It is an understandable demand on the user's part to simplify these time consuming steps and leave to the computers as much as possible.

Drawing of a network, based on preceding activity lists is one of the most demanding tasks in the CPM methodology. Yet all it provides is a graphical representation of a printed list. Furthermore, the numbering of events in ascending order seems to be a useless effort from the project managers's point of view. Recall the drawing algorithms of a CPM network presented earlier and imagine how much work is involved when a project has thousands of activities. Inserting a new activity can trigger a new renumbering process and it may very well have to be repeated hundreds of times. The first extensions of the CPM technique have been developed to eliminate these shortcomings.

The elimination of ascending numbering of events was the first extension of the original technique. First, it was left to the computer to perform this renumbering in the background. The next step was an even better solution: identification of activities has been done by their names, not by their tail and head events. This made possible the use of multiple events. In today's practice, we identify the activities by their names, but we keep the drawing rule, which forbids the users to make multiplied activities.

The third and most revolutionary step in reducing the time of drawing process was the introduction of the activity-on-node notation. Using this notation, the time to produce a network has been dramatically decreased.

Using these extensions, the applicability of the CPM technique has become easier. In contrast, its widespread use has declined. The reason: with its basic definitions and rules, it is hard to model real projects. It does not give enough flexibility to describe all the possible activity interdependencies that can arise in practice. We are going to examine this question closer.

As it is evident from the definitions of activities and events of the CPM technique (Definition 2-1, Definition 2-2,) activities follow each other in a successive order. This means that an activity can start only if all its predecessors have finished.

These basic CPM rules can cause lots of significant problems in modeling. In practice, activities often have to be accomplished concurrently or partly concurrently. CPM has a serious difficulty to describe overlapping with its structural definitions. To prove this point let us examine a very much simplified project.

A 1 kilometer-long sewer with a pumping plant has to be built. Activities have to be carried on continuously, without interruption, due to the resource limits. The project can be divided into four activities:

A- groundwork (10 days)
B- pipe laying (8 days)
C- backfilling (10 days)
D - construction of pumping plant (20 days)

Activity C has to be preceded by activity B, and activity B can start after activity A has finished. Activity D does not depend on any other activities.

The network, and the result of the time analysis can be seen in Figure 2-38. The Gantt chart of the project is presented in Figure 2-39.

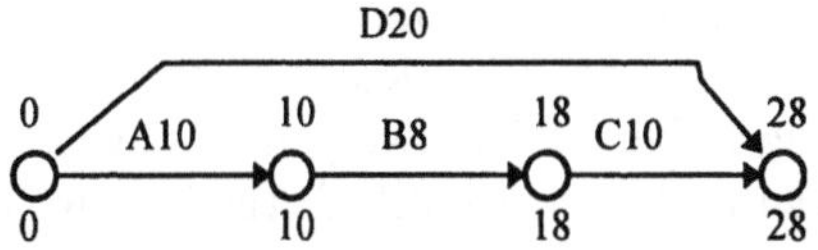

Figure 2-38 *CPM diagram of a sewer construction*

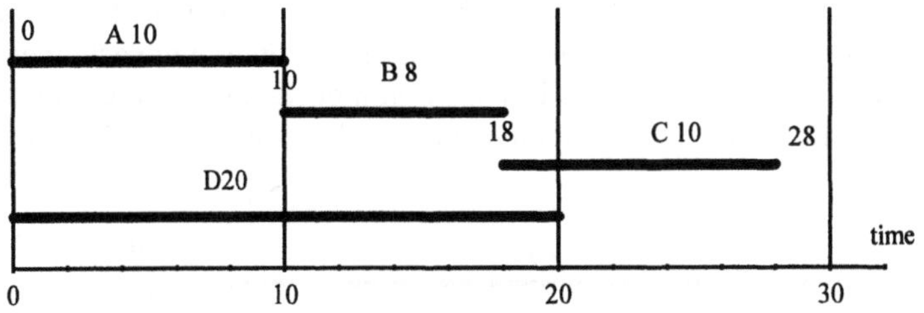

Figure 2-39 *Gantt chart of a sewer construction*

This schedule is bad. In practice, A, B and C are carried out in an overlapping fashion, partially simultaneously. This means that we should not have to wait until the end of groundwork to start with pipe laying. When groundwork has sufficiently progressed, say, two hundred meters ahead, we can start with pipe laying. When a part of activity B has been completed, say, the first two hundred meters, then activity C can start. A Gantt chart showing the overlapped schedule can be seen in Figure 2-40. At the start of activity B, activity A is in 400m distance but to insure two hundred meters between the two activities in every stage of the project, we have to start activity B on day 4. It can also be seen from the figure that the total project duration time has been reduced to 20 days, from ead of 28 days, using activity overlapping.

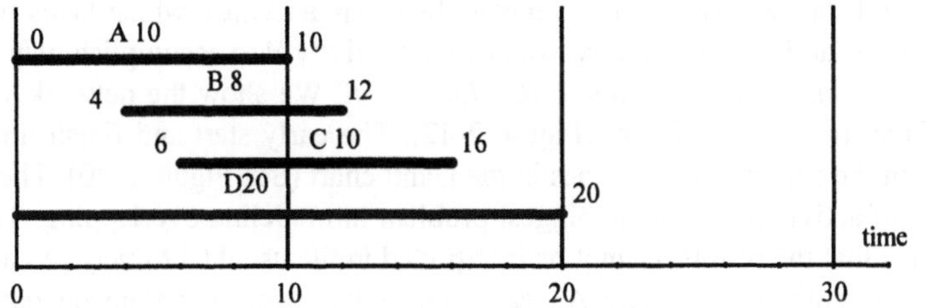

Figure 2-40 *Gantt chart of sewer construction: Activities are overlapped*

Overlapping usually results in faster project finish. This type of activity interdependency is widely used in construction practice.

Now, let us try to describe this activity overlapping using the CPM diagram. A possible solution can be seen in Figure 2-41. We tried to force the overlapping between activity A and B by dividing activity A into two parts. Activity B can start when the first part of A (which takes two days) has finished. The last, two-days long part of activity B can start after finishing A. This logic has to be used to ensure the necessary two days overlapping between activity B and C. As a result of this argument, a network consisting of eight activities and six events has emerged. This is twice the original number of activities. We could have had more activities if overlappings had not been the same. The CPM representation of these kinds of activity interdependencies sometimes can result in ten times more activities than on the activity list. The size of a CPM diagram in modeling of a large project can be enormous. Unfortunately, this is not the only problem with this kind of overlapping representation.

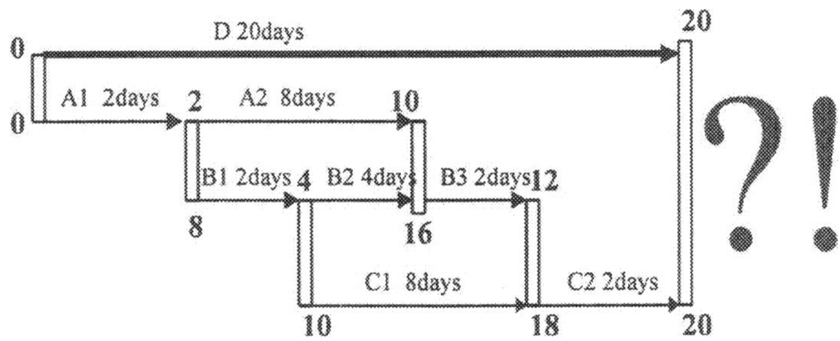

Figure 2-41 *A possible way of modeling overlapped activities with CPM*

Observing the CPM diagram (Figure 2-41), it can be seen that the early finish and start of activity B and C are not the same as in Figure 2-40. The reader can argue that we wrongly divided the activities: activity A should be divided into a 4- and 6-days part, and activity C should be divided into a 6- and 4-days part. This division may give better results, but surely does not come natural way. If the prescribed distance is two hundred meters between activities, which takes two days for activity A and C and less than two days for activity B to accomplish, *why should we divide A and C into a 4- and 6-day long part?* We show the network with this kind of partition in next figure (Figure 2-42). The early start and finish times now agree with the start and finish times in the Gantt chart (See Figure 2-40). The correct partition of activities is not the biggest problem in modeling overlapping. The main problem is that the results cannot be interpreted in the usual CPM way. As activities has become splittable, activity A2 can start at its latest start time on day 8, but according to CPM rules, B1 can start at its earliest start on day 4, and B2 can start on day 6. This means that pipe laying is at five hundred meters on day 8, when the groundwork is at four hundred meters, which is impossible. In CPM we cannot avoid these kinds of situations due to the CPM definitions. Activities originating from the same events have no relationships to each other, but only to their predecessors.

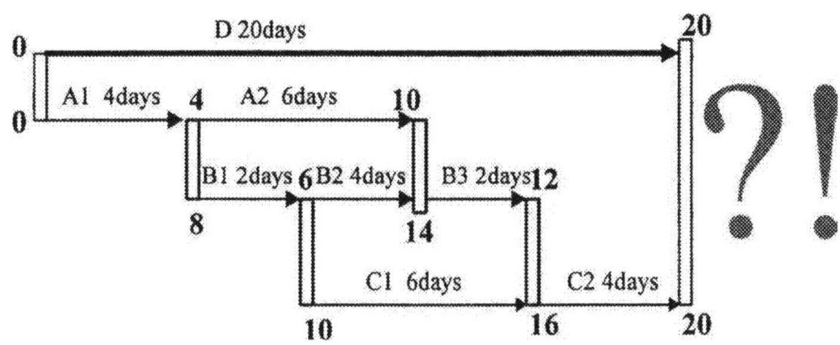

Figure 2-42 *Another possible way of representing overlapped activities in CPM*

Summarizing the shortcomings of the above used overlapping representations, we can say that it is hard to define the right division of activities without knowing the final results, i.e., the Gantt chart. The biggest problem is that the result cannot be interpreted in the usual CPM way. The reason for this is, that activities are divided into smaller parts and CPM handles these parts as individual activities .

Another possible way of depicting overlapped activities is the use of '*ladder*' activities. These activities are served to describe similar jobs like A1 and A2 or B1, B2 and B3. In fact, these are nothing else than the original activities, which have to be overlapped. Overlapping is realized by lead and lag activities, which prescribe the necessary time between the head and tail events of activities. A diagram made this way is called a '*ladder diagram*'. A ladder diagram can be seen in Figure 2-43. Dotted line arrows represent lead and lag times between activities.

A ladder diagram seems to be an easy tool to describe overlapping without activity division. Unfortunately, it has the same drawbacks as the ones mentioned before. According to CPM interpretation, activity A can start at its latest start, on day 6, and activity B can start at its earliest start, on day 0. This means that pipe laying down can start and finish earlier than groundwork! The situation is the same when we examine activity B and activity C. Activity C can start earlier than activity B which is also impossible.

The reason for this also can be found in the CPM logic. A dotted activity between the tail events of activity B and C does not imply a relation between activity B and C. It simply describes that activity C can start after two days of the occurrence of the tail event of B, that is, after two days of finishing the preceding activities of the tail event of B. In this case, the preceding activity is also a dotted line lead activity. *However, there is no any relationship between B and C (see Definition 2-2)!*

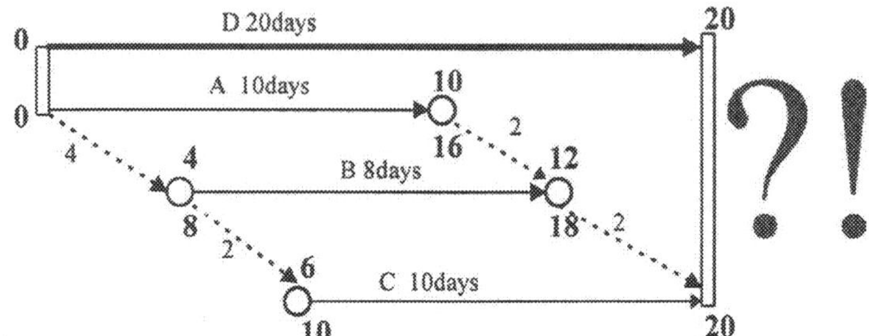

Figure 2-43 *'Ladder' diagram to describe overlapped activities*

In summary, overlapping cannot be modeled in adequate way in the CPM technique. Activities can be divided or ladder diagrams can be used to get the same early start and finish times as they can be gained from a Gantt chart. These kinds of diagrams never can be interpreted the same way as a CPM diagram, as they are not entirely consistent with the traditional arrow diagrams. These facts should be kept in mind by the planner.

Despite of these serious drawbacks, CPM was very popular and the most widely used network technique in the 1960's and 1970's. If there are no overlapping activities in the planned projects, it is a perfect and easy-to-understand tool for project managers. In the construction industry, as it derives from the nature of this field, sometimes hundreds of parallel and overlapped activities have to be carried out in a project. Proper schedule of these projects with CPM is almost impossible, that is why today CPM has less and less role in construction. Nevertheless, there are some fields where CPM can be used without displaying any of these shortcomings.

2.9 PRACTICAL PROBLEMS AND SOLUTIONS

Problems

1. Develop an IPA list based on below given PA list!

Activities	A	B	C	D	E	F	G	H
Predecessor activities	-	-	B	B,A	A,B	D,C,B	E,D,C	F,C

2. Develop an IPA list based on below given PA list!

Activities	A	B	C	D	E	F	G	H	I	J
Predecessor activities	-	A	B	E,F	F	-	-	A	F	H,G,I,A

3. Develop an IPA list based on below given PA list!

Activities	A	B	C	D	E	F	G	H	I	J	K	L	M	N	O	P
Predecessor activities	-	F	J,E D	O	-	-	A	B,J D,F	G,J,B F,D	E	I,B	C,D	H	P	-	D

4. Develop a CPA list based on the PA list of 1.!

5. Develop a CPA list based on the PA list of 2. !

6. Develop a CPA list based on the PA list of 3.!

7. Develop an ISA list based on below given SA list!

Activities	A	B	C	D	E	F	G	H
Succeeding activities	E,D,F	E,D,C	F,G,H	F,G,H	G	H	-	-

8. Develop an ISA list based on below given SA list!

Activities	A	B	C	D	E	F	G	H	I	J
Succeeding activities	H,B,C	C	-	-	D	I,E,J	J	J	J	-

9. Develop an ISA list based on below given SA list.

Activities	A	B	C	D	E	F	G	H	I	J	K	L	M	N	O	P
Succeeding activities	G	H,I	L	P,C H,I L	J,H	B,I	I,K	M	K	C,I H	-	-	-	-	D,I	N

10. Develop an ISA list based on PA list of 1.!

11. Develop an ISA list based on PA list of 2.!

12. Develop a CPA list based on SA list of 8.!

13. Draw the network from the below given IPA list!

Activities	A	B	C	D	E	F	G	H	I
IPA list	-	A	B	E	F	-	-	G	H

14. Draw the network from the below given IPA list!

Activities	A	B	C	D	E	F	G	H	I
IPA list	B,E,D	-	E,D	-	-	D	F	C	A

15. Draw the network from the below given IPA list!

Activities	A	B	C	D	E	F	G	H	I
IPA list	-	-	-	G,H,I	H,I	I	A	A,B,C	C

16. Draw the network from the below given IPA list!

Activities	A	B	C	D	E	F	G	H	I	J
IPA list	-	A,D	D,I	-	-	J,B	C,B	F	-	E,D

17. Draw the network from the below given ISA list!

Activities	A	B	C	D	E	F	G	H	I
ISA list	B	C	-	-	D	E	H	I	-

18. Draw the network from the below given ISA list!

Activities	A	B	C	D	E	F	G	H	I
ISA list	I	A	H	F,C,A	A,C	G	-	-	-

19. Draw the netwok from the below given ISA list!

Activities	A	B	C	D	E	F	G	H	I
ISA list	G,H	H	I,H	-	-	-	D	D,E	F,ED

20. Draw the network from the below given ISA list!

Activities	A	B	C	D	E	F	G	H	I	J
ISA list	B	G,F	G	B,C,J	J	H	-	-	C	F

21. Calculate the early and late activity and event times on the below given network! Define all four floats for activity A and H! (Activity durations can be found near the name of activities.)

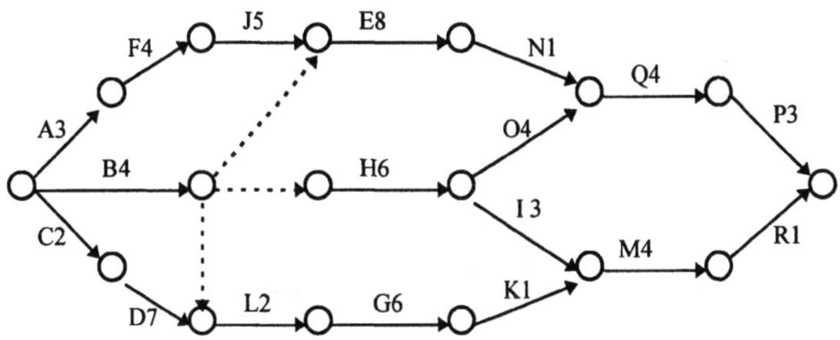

22. Perform a CPM calculation on the below given network! Define the critical path and all the four floats for activity E and H!

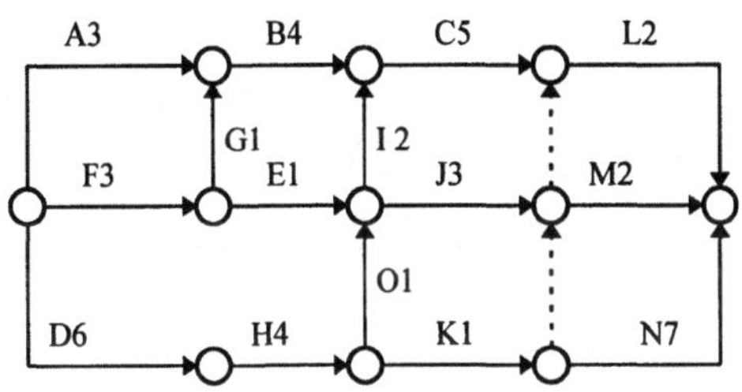

23. Perform a CPM calculation using the matrix representation!

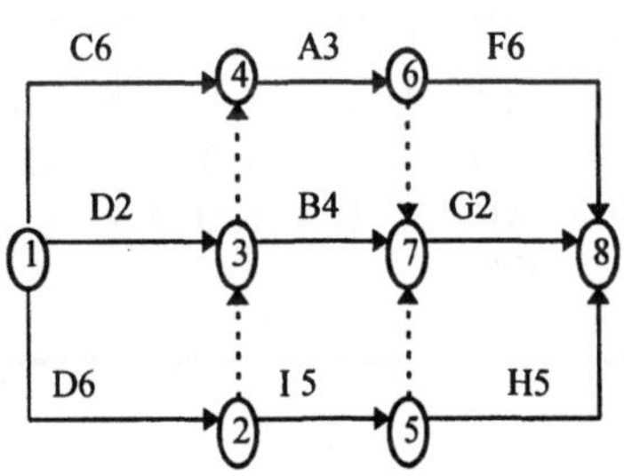

Solutions

1.

Activities	A	B	C	D	E	F	G	H
IPA list	-	-	B	B,A	A,B	D,C	E,D,C	F

2.

Activities	A	B	C	D	E	F	G	H	I	J
IPA list	-	A	B	E	F	-	-	A	F	H,G,I

3.

Activities	A	B	C	D	E	F	G	H	I	J	K	L	M	N	O	P
IPA list	-	F	J,D	O	-	-	A	B,D J	G,J B,D	E	I	C	H	P	-	D

4.

Activities	A	B	C	D	E	F	G	H
CPA list	-	-	B	B,A	A,B	D,C A,B	E,D C,A,B	F,D,C A,B

5.

Activities	A	B	C	D	E	F	G	H	I	J
CPA list	-	A	B,A	E,F	F	-	-	A	F	H,G,I A,F

6.

Activities	A	B	C	D	E	F	G	H	I	J	K	L	M	N	O	P
CPA list	-	F	J,D E,O	O	-	-	A	B,D J,F E,O	G,J B,D AF E,O	E	I,G B,J D,A F,E O	C,P J,D E,O	H,B F,J E,D O	P,D O	-	D,O

7.

Activities	A	B	C	D	E	F	G	H
ISA list	E,D	E,D,C	F,G	F,G	G	H	-	-

8.

Activities	A	B	C	D	E	F	G	H	I	J
ISA list	H,B	C	-	-	D	I,E	J	J	J	-

9.

Activities	A	B	C	D	E	F	G	H	I	J	K	L	M	N	O	P
ISA list	G	H,I	L	P,C H,I	J	B	I	M	K	C,I H	-	-	-	-	D	N

10.

Activities	A	B	C	D	E	F	G	H
ISA list	E,D	E,D,C	F,G	F,G	G	H	-	-

11.

Activities	A	B	C	D	E	F	G	H	I	J
ISA list	H,B	C	-	-	D	I,E	J	J	J	-

12.

Activities	A	B	C	D	E	F	G	H	I	J	K	L	M	N	O	P
CSA list	-	F	J,D E,O	O	-	-	A	B,D, J,F E,O	G,J B,D A,F E,O	E	I,G B,J D,A F,E,O	C,P J,D E,O	H,B F,J E,D O	P,D	-	D,

13.

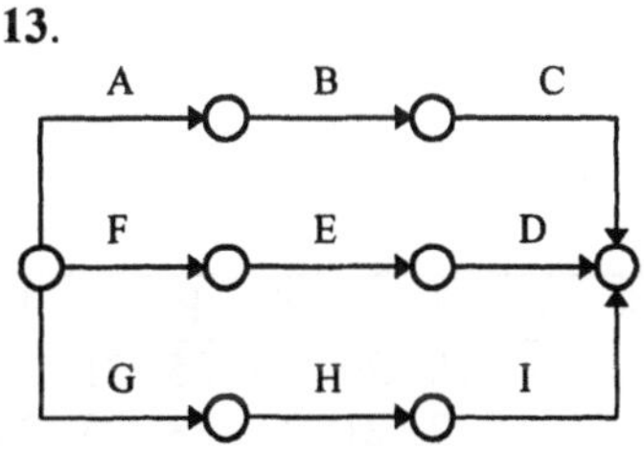

14.

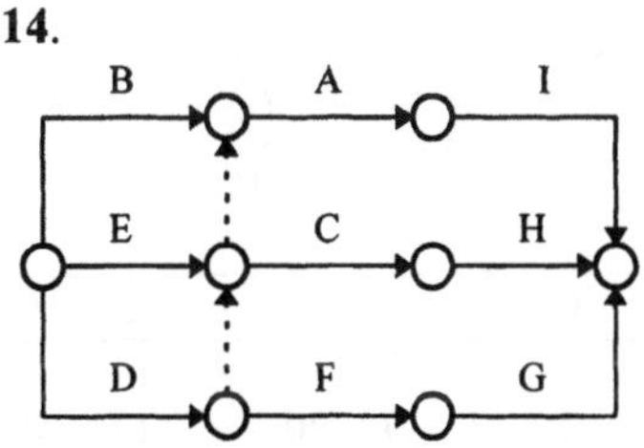

15.

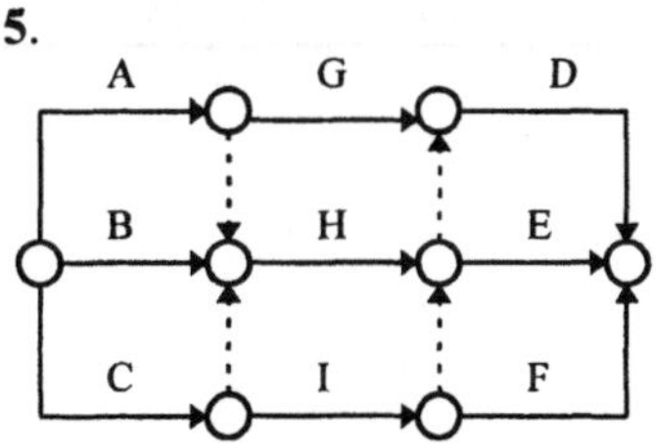

16.

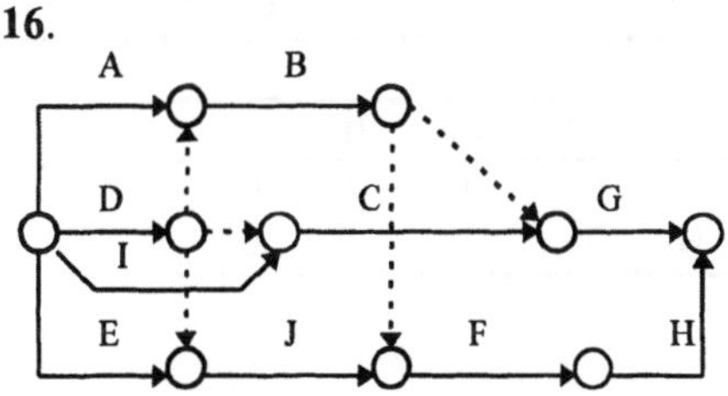

17.

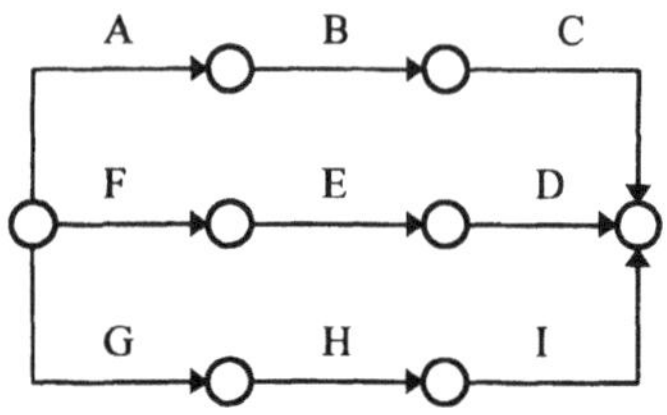

18.

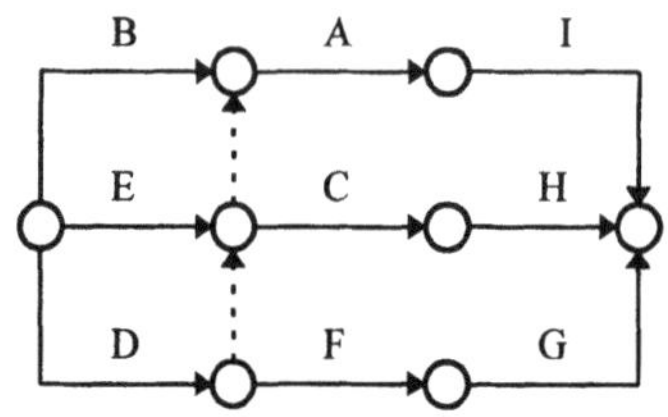

19.

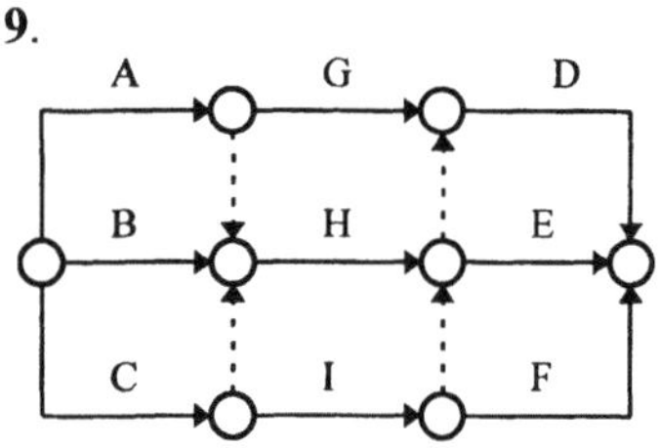

20.

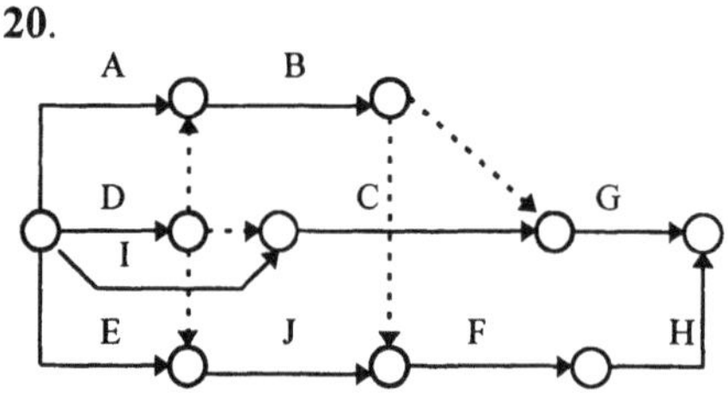

21.

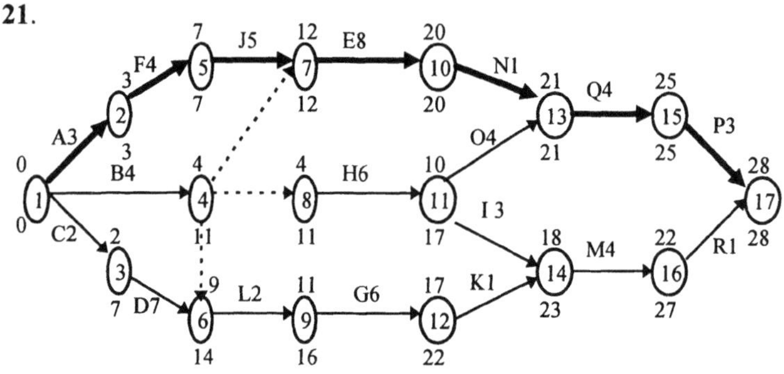

	A	B	C	D	E	F	G	H	I	J	K	L	M	N	O	P	Q	R
ES	0	0	0	2	12	3	11	4	10	7	17	9	18	20	10	25	21	22
EF	3	4	2	9	20	7	17	10	13	12	18	11	22	21	14	28	25	23
LS	0	7	5	7	12	3	16	11	19	7	22	14	23	20	17	25	21	27
LF	3	11	7	14	20	7	22	17	22	12	23	16	27	21	21	28	25	28

	TF	FF	IF	CF
A	0	0	0	0
H	7	0	-7	0

22.

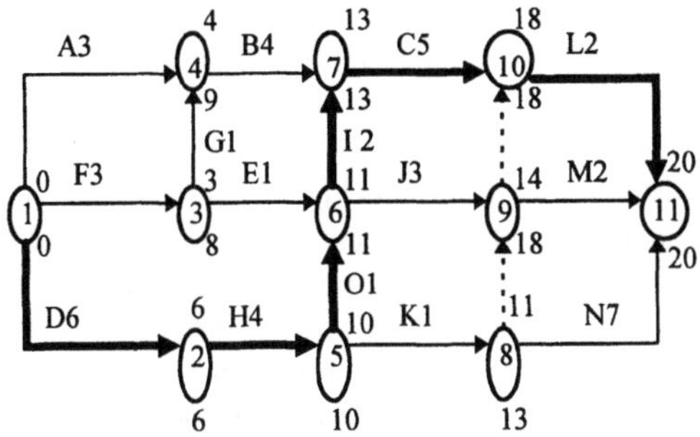

	A	B	C	D	E	F	G	H	I	J	K	L	M	N
ES	0	4	13	0	3	0	3	6	11	11	10	18	14	11
EF	3	8	18	6	4	3	4	10	13	14	11	20	16	18
LS	6	9	13	0	10	8	9	6	11	15	12	18	18	13
LF	9	13	18	6	11	5	8	10	13	18	13	20	20	20

	TF	FF	IF	CF
E	7	7	2	2
H	0	0	0	0

23.

		1	2	3	4	5	6	7	8
0	1		1	2	6				
1	2			*		5			
2	3				*			4	
6	4						3		
6	5							*	5
9	6								6
9	7								2
15	8								
		0	**5**	**6**	**6**	**10**	**9**	**13**	**15**

2.10 HISTORICAL REVIEW

As it has been discussed in Chapter 1 (Historical Review), the history of CPM has begun in 1956 with the work has been made at the E.I. du Pont de Nemours Company [35]. The new method was developed by Morgan Walker from DuPont and James E. Kelley, Jr., of Remington Rand [41],[42]. The first phase of this calculation was a simple time analysis and the second phase was the sophisticated time cost trade-off phase. First there were different names used for the new technique (CPM [52], CPS [60], CPA [9], LESS [5]). However, construction industry experts could not cope with the whole method at the same time and they have assimilated it as two techniques. The first, that is, the time analysis phase of the original technique has been assimilated as CPM technique and the whole method has been assimilated later as CPM Least Cost Scheduling technique. So, the term of CPM Time Cost Trade-off or CPM Least Cost Scheduling covers the original CPM technique.

There are hundreds of papers and books about CPM and CPM methodology, so our list cannot be complete by all means. The first comprehensive evaluation of the CPM method is a work of Anon [3], who elaborated and evaluated different cases when CPM has been used. A detailed description of the network drawing process has been given by McLaren and Buesnel [46]. This is a more detailed description of Method 1. The activity-on-node notation has been introduced by John Fondahl [18] in 1961. Our node numbering procedure is the same as Fulkerson's procedure [24] from 1962. The CPM calculation with primal-dual approach can be found in Klafszky's [43] and Hajdu's [30] work. A linear programming primal-dual approach can be found in the work of Wiest and Levy [66]. Their dual problem is a flow problem as they have derived it from the primal linear programming model of CPM. Definitions of total and free floats are parts of the original CPM papers. Definitions of all the possible floats and slacks can be found in Battersby's [6] excellent work. From the six floats, defined by Battersby, two is equal with tail and head event slack, therefore in today's construction practice we use four types of floats and the event slacks.

A comprehensive discussions of CPM can be found in a large number of books. A detailed description of history of CPM and other network techniques can be found in O'Brien's [51] work. In his book he emphasizes the methodology. A classical work in the field of network techniques is the book of Battersby, which has been cited before. His book is about 30-years old, but at parts more up-to-date than books published in the nineties. Moder et al.'s [50] book contains one of the most comprehensive description of network based techniques.

Chapter 3

CPM LEAST COST SCHEDULING

In this chapter we introduce the reader to the original CPM problem and its several solutions. The original CPM problem is a cost optimization problem. Nowadays, the simple CPM time analysis problem, which is the first phase of the original CPM problem, is called the CPM problem. For the cost optimization model we use the CPM/cost, CPM time-cost trade-off or CPM least-cost scheduling terms.

Time and cost are the two most important components in a project. The CPM time/cost problem makes a connection between these two important factors in a very elegant way.

In the first subchapter, Introducing a Sample Project, we draft a sample problem through which the algorithms are going to be illustrated.

After this, in Heuristic Solutions, we give some simple heuristic algorithms, which lead to an optimal or close-to-optimal solution.

In the third part, in Exact Solution, we give the correct mathematical model of the problem, and the solution of it. In our opinion, the given algorithm is the simplest among the existing algorithms.

In the fourth part, Maximal Cost Solution, we show how much the project accomplishing cost can differ if CPM time-cost trade-off is not used. Here we give an algorithm to determine the maximal cost schedule for a given project duration. This algorithm has not much practical use, but the difference between the minimal and maximum cost solutions can convince the reader about the usefulness of CPM/cost method.

In the fifth part, Which Is Better?, we compare the heuristics and the exact algorithms.

In subchapter 'Special Cases,' we show how the algorithm changes if the activity time versus activity cost curve is not linear as it is assumed in the original CPM/cost problem.

In the seventh part, Practical Problems and Solutions, we give some problems to be solved along with their solutions.

In the last part, Historical Review, we give a survey of the most significant works of this field.

To understand the mathematics of this chapter, we recommend to read these parts of APPENDIX A:

- The maximal flow-minimal cut problem for understanding the exact algorithm (Maximal Flow-Minimal Cut).
- The first "K" longest path through a network for understanding the algorithm for the maximal cost solution, and the Siemens' method, which is a heuristic method for CPM/cost problem (The First "K" Longest Path).

3.1 INTRODUCING A SAMPLE PROJECT

Let us consider the sample project that was described in the first part of Chapter 2, Introducing a Sample Project, on page 18. A covered depot has to be built from prefabricated standardized steel elements, in order to protect a large industrial equipment is used for engine production. The project has been divided into fifteen processes from A to O. The activity interdependencies are the same as in Chapter 2, but another method than before is used for estimating activity durations. The basic assumption of the CPM/cost model is that a possible time interval can be given for each activity, within the accomplishing time for an activity has to be.

Most of the accomplishing times of processes can vary depending on the number of the crew, the productivity of the applied machines and the length of the workday. Project managers usually assume a normal size of crew, normal workdays, and 'owned' equipment in deciding the activity durations. Activity duration times that result from these arguments are called *normal durations*. If time is not really important, management usually utilizes its resources on this, normal level.

There is an another way of estimating time in the CPM/cost model, one in which the management counts with the fastest possible accomplishing time of the activities. If the deadline is very strict, management can apply the best of its resources and schedule work around the clock. The fastest possible accomplishing time for an activity is called the *crash duration.* In the CPM/cost model it is assumed that no activity can be carried out faster than its crash duration. When management utilizes its resources on the highest possible level to accomplish each of the tasks at their crash duration, then this schedule will result in a shorter project duration.

Changing the activity durations will affect the project duration and the project cost. From the project's point of view, we can distinguish between two types of cost.

The first type of costs is called the *direct cost,* and indicates the cost that is directly attributable to a task. For example, labor, machine and material costs of a certain activity belong directly to the activity. If the entire activity is being performed by a subcontractor, then the activity direct cost is equal to the price of the subcontract. The sum of the direct costs of a project's activities is called the *project direct cost.*

The other type of costs is called the *indirect* or *overhead* cost, also known as *fixed* cost. This type includes the cost factors that are not easily attributable to activities, but rather belong to the whole project. The elements of indirect cost are, for example, the amounts of rent of social facilities, management salaries, interest charges, etc. These costs are usually linear, that is, one day saved on a project saves one day's fixed costs, ten days saved on a project saves ten times more. The sum of project direct cost and indirect cost is called the *project total cost.*

In the CPM/cost model there is a special emphasis on the change of the direct cost. The direct cost that is related to normal activity duration is called the *normal cost.* Shortening the duration of an activity to its crash duration usually result in higher accomplishing cost. Premium for overtime will increase the accomplishing cost of an activity. In order to shorten the activity, material costs can also increase. For example, in concrete work the use of expensive rapid hardening cement increases the costs. The cost related to crash duration is called the *crash cost.* The nature of change of the cost can be seen in Figure 3-1. There is a lower time limit -- the crash duration -- which no amount of expenditure can shorten. There is an upper limit, which is the normal duration, and one can make no more savings by letting the task go slowly. The change of the cost between the normal and the crash time points is usually convex, i.e., the costs shortening the activity duration by a new additional day are higher than the extra cost increment of the previous shortening.

In the CPM/cost model a linear approximation[i] is used instead of the convex activity time-cost curve. Using the linear approximation the so called *cost slope* can

---

[i]There are other types of cost curves when the assumption of linearity cannot be used. We discuss those cases in the chapter "Special Cases".

also be defined. The cost slope is the amount of money required for shortening an activity duration by one day.

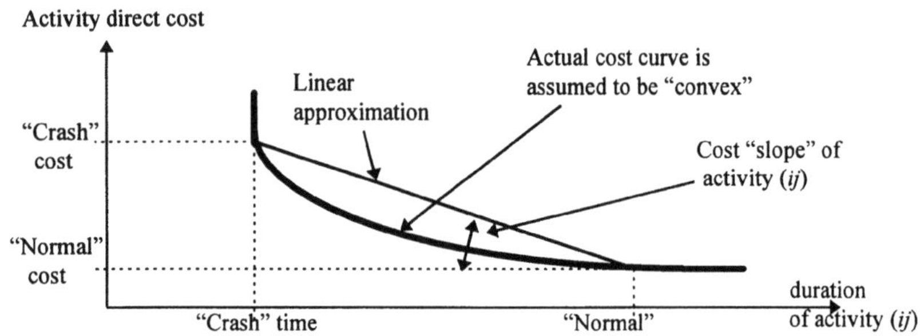

Figure 3-1 *Activity cost-time relationship in CPM*

In a network where an activity duration can change between its lower and upper bound, different project durations are possible. What is more, the same project duration can be achieved in hundreds of different ways. If all the activities are considered to be accomplished in their normal time, then we get the maximal project duration. If all the activities are considered to be accomplished in their crash time, we get the minimal project duration. Many project durations between these two extremes can be achieved and all of them can be achieved in hundreds of different ways which result in hundreds of different project direct costs. The set of the possible solutions is shown in Figure 3-2. Two points of the set of feasible solutions come automatically. If time analysis is done using normal durations, we will get point B which is the optimal solution for p_b. If time analysis is done using crash durations, we will get point D which is the most expensive solution for p_a. Point B is also called the trivial optimal solution.

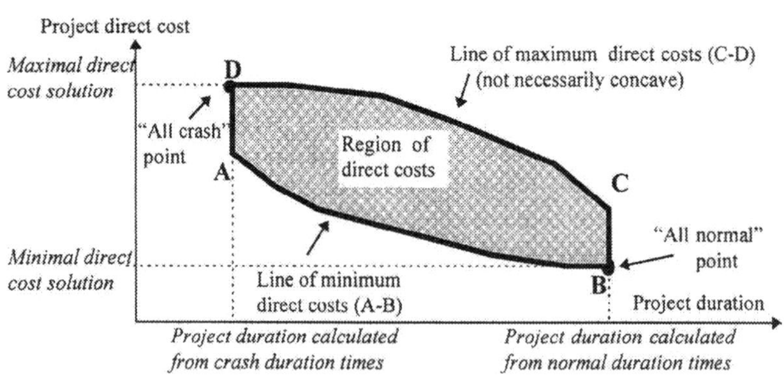

Figure 3-2 *Project duration versus project direct cost in CPM/cost model*

The goal in the CPM/cost model is to define the curve of the minimum direct costs, which represents the minimum direct cost solutions. If the line of the minimum direct cost is defined, then with the help of the line of indirect cost, the least cost schedule of the project can be easily determined. As it can be seen in Figure 3-3, the least cost schedule is the one with the minimal project total cost, which means that both lengthening and shortening the project duration would increase the total cost.

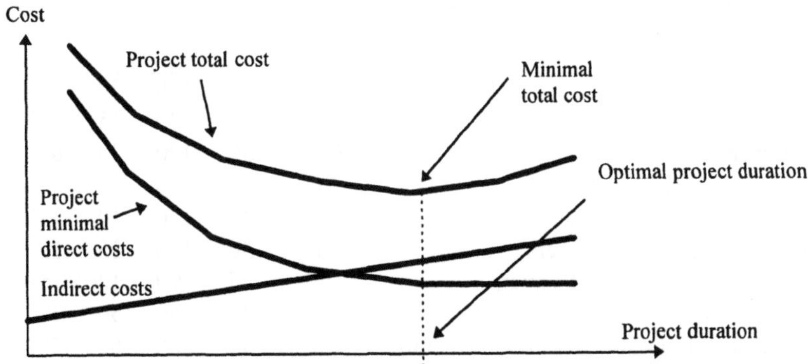

Figure 3-3 *Determining the project schedule for minimum total cost.*

The critical issue is how to determine the minimal direct cost curve for a project. Some of the available computational tools are the topics of this chapter.

Summarizing the above mentioned, in the CPM cost model a lower and upper limit is used for time estimation. The upper limit is called the normal duration, the lower limit is called the crash duration. The direct cost of the fastest performance is called the crash cost and it is assumed to be higher than or the same as the cost of the slowest performance which is called the normal cost. The change of direct the cost between its bounds is assumed to be linear. The goal of the CPM model is to determine the minimal direct cost for a given project duration.

We are going to use the following new notations during the chapter:

b_{ij} - normal duration for activity (ij)
a_{ij} - crash duration time for activity (ij)
Cn_{ij} - normal direct cost for activity (ij)
Cc_{ij} - crash direct cost for activity (ij)
Ca_{ij} - actual cost which corresponds to the actual duration, τ_{ij}
c_{ij} - cost slope for activity (ij)
p - duration of the project
p_b - project duration calculated from normal duration times
p_a - project duration calculated from crash duration times

In the sample project we have kept the activity durations of Chapter 2 (Introducing a Sample Project) as normal duration times. We have also kept the network logic of the original sample project. The normal costs for normal durations and the crash costs for crash durations can be seen in Table 3-1. The cost slopes, which come from the evaluation of the $c_{ij}=Cc_{ij}-Cn_{ij}/b_{ij}-a_{ij}$ formula are indicated as well, in the last column of Table 3-1.

Activities	IPA list	b_{ij}	Cn_{ij}	a_{ij}	Cc_{ij}	c_{ij}
A Preparations	-	30	100	20	200	10
B Order machine	A	20	30	15	50	4
C Deliver machine	B,E	30	400	20	600	20
D Order building materials	A	10	30	6	50	5
E Preparatory works on site	A	14	150	10	230	20
F Conveyance of bulldozers	A	30	120	25	195	15
G Delivery of gravel and cement	D,E	20	90	12	154	8
H Delivery of mount structure	D,E	80	160	50	370	7
I Delivery of prefabricated steel structure	D,E	90	280	50	520	6
J Groundwork	F,E	10	600	5	720	24
K Foundation for the depot	G,J	30	1200	25	1500	60
L Placing the mount for the machine	K,H	20	150	15	350	40
M Placing the machine on the mount	C,L	5	200	5	200	0
N Assembling the building structure	M,I	20	800	12	1200	50
O Finishing works, project delivery	N	15	150	10	210	12

Table 3-1 *Input data for the sample project in CPM/cost model*

In the next subchapter we show some of the solutions of the problem. These solutions are heuristic, that is, they sometimes result in non-optimal solutions.

3.2 HEURISTIC SOLUTIONS

Lots of heuristic methods have been developed since the 1960's. In this chapter two different methods will be introduced to the reader. In both methods we start the calculations from the trivial optimal solution, which comes from the time analysis calculated from the normal durations. In the first method we reach the desired project duration step by step, so that the minimal direct cost curve will be determined in the p_b and desired project duration interval. In the second method we reach the desired project duration in one step. We name the methods after the developers, the former as Fondahl's[i] method, the latter as Siemens' method.

Fondahl's Method

The procedure starts from the trivial solution which comes from the time analysis with normal durations. The idea behind the procedure can be summarized as follows. To reduce the project duration, the length of critical path or paths have to be reduced. If we only have one critical path, then the activity with the smallest cost slope has to be shortened. The size of the shortening is the smaller of the smallest total float among the activities that are not on the critical path and the possible size of the reduction of the activity in question. The latter value is derived as the difference of the actual duration and the crash duration. If we have more critical paths, then we have to reduce the length of all of them. For this, we have to choose the activities which have the smallest sum of the cost slope and will affect the length of all the critical paths. If the critical paths are linked, that is, they have common parts, then it is worth to check the activities which are the elements of more critical paths, even if their cost slopes are relatively high. Reducing the duration of such an activity will reduce the lengths of multiple critical paths. This is sometimes more economical than to reduce the duration of activities that are, despite their small cost sensitivity, parts of different critical paths.

The above step will result in a reduction of the project duration by the same value that all the critical paths have been shortened by. After this step there will be either more critical paths or at least one more activity which cannot be further shortened. The increment of the project direct cost is derived by multiplying the decrement of the project duration and the sum of the cost slopes of the shortened activities. The change of the project direct cost will be linear in this interval.

In the next step one has to look for the critical paths again, and this procedure has to be repeated until the desired project duration or the minimal project duration have been reached.

The algorithm consists of three steps and can be described as follows:

---

[i]Fondahl originally has developed three heuristic methods for the solution of the CPM time/cost problem. One where the procedure starts from the "all normal" point, one where the procedure starts from the "all crash" point, and one where the procedure starts from an arbitrary non-optimal but feasible solution. More details can be found in the chapter "Historical Review".

Step 1. Choosing activities for shortening.

- The activity has to be compressible.
- Activities chosen for shortening have to be on the critical path.
- Activities have to be chosen from all the critical paths. (One activity can be a part of more critical paths.)
- The sum of cost slopes of chosen activities has to be the smallest among all the possible combinations.

This is simple to do if there is only one critical path. If there are more, checking all the possible combinations is not easy. Originally a procedure was developed to help taking into account all the possible combinations. Above a certain but relatively small number of activities, this procedure requires more computational steps than providing the exact solutions in the maximal and minimal project duration intervals.

Step 2. Defining the size of compression.
Here, we have to keep in mind two rules.

- The possible speeding up is equal or less than the difference of the actual duration and crash duration. (If more activities have to be shortened, the minimal value is the determinant.) The value that comes from this rule is denoted by $\Delta 1$.
- All the critical paths have to remain critical, but there can also be new paths after the compression. To ensure, this we have to examine some of the non-critical activities which are not parts of any paths leading from s to t through an activity which has been chosen for shortening. The possible compression is equal to or less than the smallest total float of these non-critical activities. (Failing to comply with this rule and taking all the non-critical activities into consideration will not result in bad data, but can cause a longer computational process.) The value that comes from this rule is denoted by $\Delta 2$. The size of shortening is the result of the next formula $\Delta = \min\{\Delta 1, \Delta 2\}$.

Step 3. Time analysis, determination of increment of project direct cost.

- Decrease the durations of the chosen activities by a Δ value. This will decrease the project duration by the same amount. The increment of direct cost is the result of the multiplication of Δ and the sum of cost slopes of chosen activities.
- New time analysis has to be performed to determine the new event times. Experienced professionals can use the results of the previous calculations.

After this, the procedure has to be continued with Step 1 and repeated until the desired or the minimal project duration has been reached.

To illustrate the algorithm, we show the solution of the sample project. The application of the algorithm can be followed in the following figures. The early event times are shown above events. The three consecutive numbers below or beside an activity are the crash duration, the actual duration and the cost slope. The critical activities and events are depicted in bold face. The durations and the slopes of dummies are equal to zero.

The starting trivial solution can be seen in Figure 3-4. The project direct cost is the sum of activity direct costs, which is 4460 units. The project duration is 184 days.

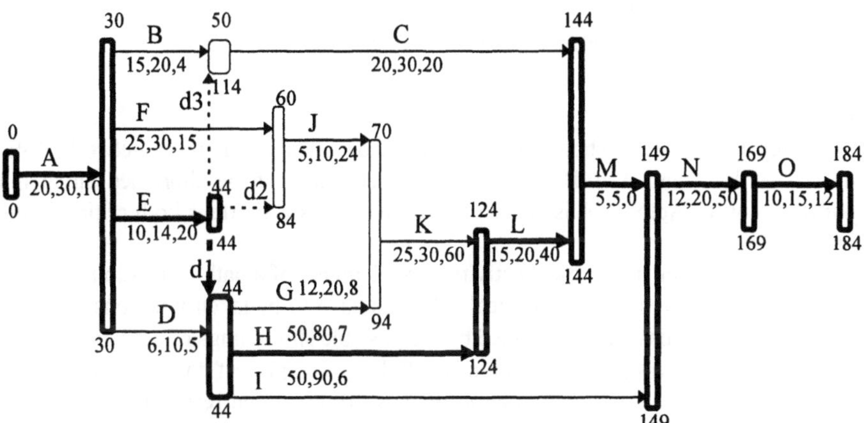

Figure 3-4 *Decreasing the project duration from 184 to 169 days.*

Iteration I.

Step 1. Figure 3-4 shows that there is only one critical path, A-E-d1-H-L-M-N-O. All the activities can be shortened along this path except M.

Step 2. All the non-critical activities, except activity D, have to be taken into consideration. (Activity D is the element of a path from s to t through the chosen activity, so its total float will not change after the compression.) Activity I has the smallest total float (15 days) among the appropriate non-critical activities. Activity H can be shortened by 30 days. The determinant value is the smallest, that is, 15 days.

Step 3. The new project duration is 169 days. The project direct cost increment is 15*7=105 units. The result of this iteration can be seen in Figure 3-5.

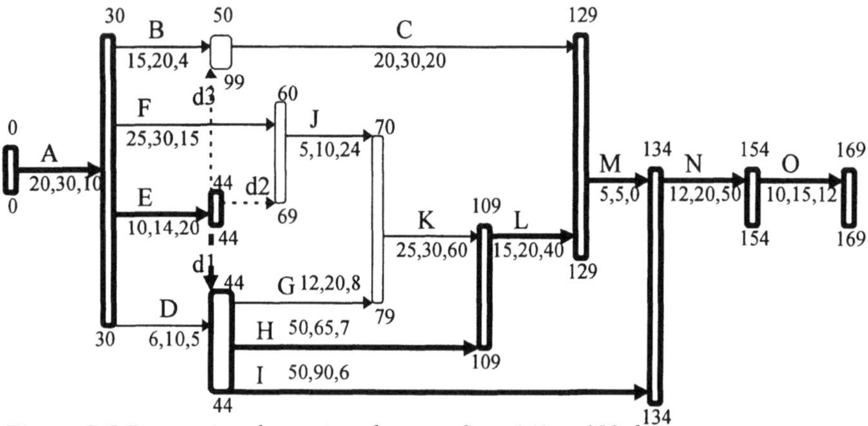

Figure 3-5 *Decreasing the project duration from 169 to 159 days.*

Iteration II.

Step 1. Figure 3-5 shows that there are two critical paths in the network, A-E-d1-H-L-M-N-O and A-E-d1-I-N-O. All the activities can be shortened along the paths except M. Activity A is chosen for shortening as it shortens both critical paths by 10 unit/day.

Step 2. All the non-critical activities are elements of a path from *s* to *t*, through the chosen activity, so no non-critical activities have to be taken into consideration. The Δ value is 10 days as determined by the possible shortenings of activity A.

Step 3. The new project duration is 159 days. The project direct cost increment is 10*10=100 units. The new duration of A is 20 days. The result of this iteration can be seen in Figure 3-6.

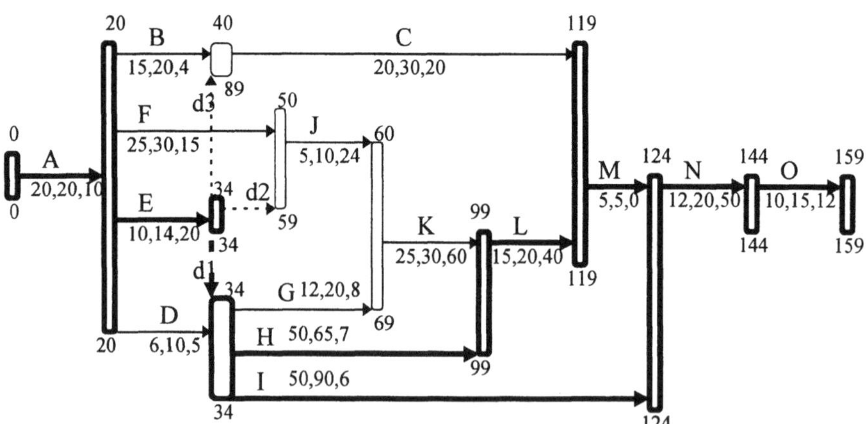

Figure 3-6 *Decreasing the project duration from 159 to 154 days.*

Iteration III.

Step 1. Figure 3-6 shows that we have the same two critical paths to shorten. Activity M and A cannot be compressed further. Shortening of activity O results in the most economical project compression.

Step 2. Activity O can be compressed by 5 days. Non-critical activities do not have to be examined as all of them are part of a path from *s* to *t* which includes activity O.

Step 3. The new duration for activity O is 10 days. There is no further possibility to shorten activity O. The increment in project direct cost is 5*12=60 units. The project duration after this compression is 154 days. The new event times can be seen in Figure 3-7.

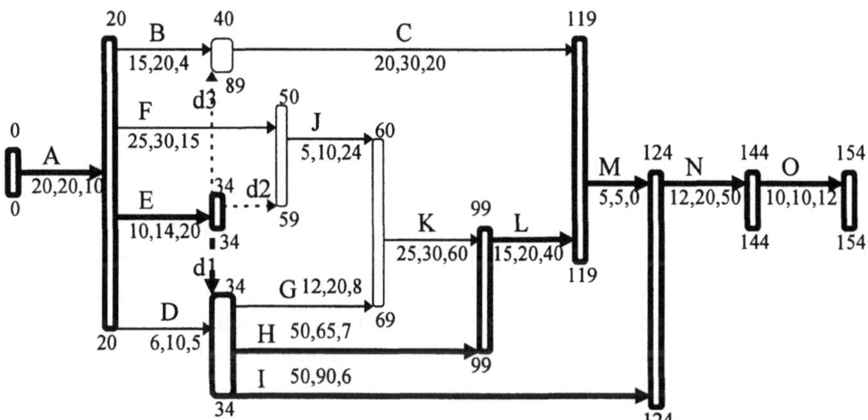

Figure 3-7 *Decreasing the project duration from 154 to 145 days*

Iteration IV.

Step 1. Figure 3-7 shows that there are still two critical paths to shorten. Shortening activities H and I together results in the most economical compression which costs 13 units a day.

Step 2. Activity I can be compressed by 40 days, activity H can be compressed by 15 days, so the Δ1 value is defined by this latter value. (Δ1=15days). The total float for B is 49 days, for C 49 days, for F, J and K 9 days, for G 15 days, for d2 25 days and for d3 55 days. The minimum of these values defines the Δ2 value. (Δ2=9 days) The measure of project compression is 9 days. (Δ=9 days)

Step 3. The activity duration for activity H is 56 days, for I 81 days. The increment of project direct cost is 9*(7+6)=117 units. The new project duration is 145 days. The results of this iteration phase can be seen in Figure 3-8.

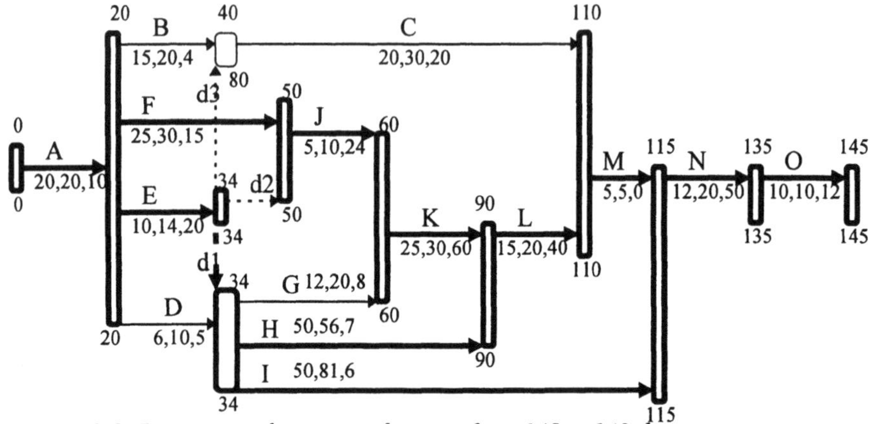

Figure 3-8 *Decreasing the project duration from 145 to 140 days.*

Iteration V.

Step 1. There are three critical paths to shorten. The new critical path is A-F-J-K-L-M-N-O. Activity O, M and A cannot be further shortened. The most economical compression of project duration is one where activity H, I and F are compressed. This activity combination results in 28 units of cost increment in a day.

Step 2. Activity H, I and F can be compressed by 6, 31 and 5 days, therefore $\Delta 1$=5days. All the non-critical activities but D have to be examined to determine $\Delta 2$. The total float for activity B and C is 40 days, for G 6 days, for d2 and d3 16 and 46 days, so $\Delta 2$=6 days and $\Delta=\min\{\Delta 1,\Delta 2\}$=5 days.

Step 3. The new duration for activity H, I and F are 51, 76 and 25 days. The project direct cost increment is 5*(15+7+6)=140 units. The new project duration is 140 days. The results of this iteration phase can be seen in Figure 3-9.

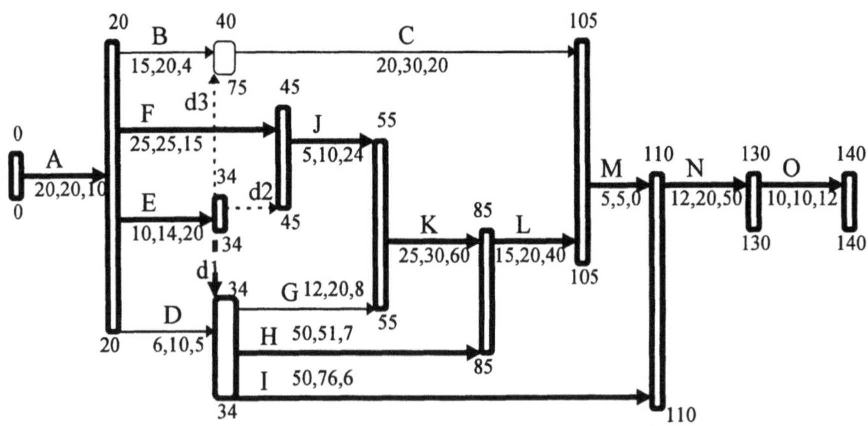

Figure 3-9 *Decreasing the project duration from 140 to 139 days.*

Iteration VI.

Step 1. There are three critical paths to shorten. Activity M, O, A, and F cannot be further shortened. The most economical combination for shortening is the one where activity H, I and J are compressed together.

Step 2. The possible compression for activity H, I and J is 1, 26 and 5 days, so Δ1=1. The total floats of activities B, C, G and d3 have to be examined. The total float of B, C, G and d3 is 35, 35, 1 and 41 days, so Δ2=1 day and Δ=1 day.

Step 3. The new activity durations for activities H, I and J are 50, 75 and 9 days. The project direct cost increment is 1*37=37 units. The project duration has changed from 140 days to 139 days. The results of this iteration can be seen in Figure 3-10.

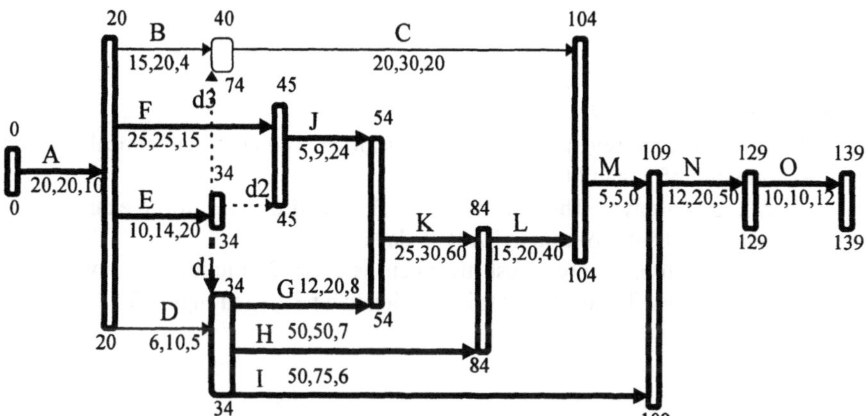

Figure 3-10 *Decreasing the project duration from 139 to 135 days.*

Iteration VII.

Step 1. There are four critical paths to shorten. Activities A, F, M and O cannot be further shortened. Shortening of J and E together is the most economical combination.

Step 2. Both activity J and E can be shortened by 4 days, so Δ1=4 days. The total float of activity B and D has to be examined, which is 34 and 4 days. (Δ2=4 days) Δ=min{Δ1,Δ2}=4 days.

Step 3. The new activity durations for J and E are 5 and 10 days. The project direct cost increment is 4*44=176 units. The new project duration is 135 days. The results of this iteration can be seen in Figure 3-11.

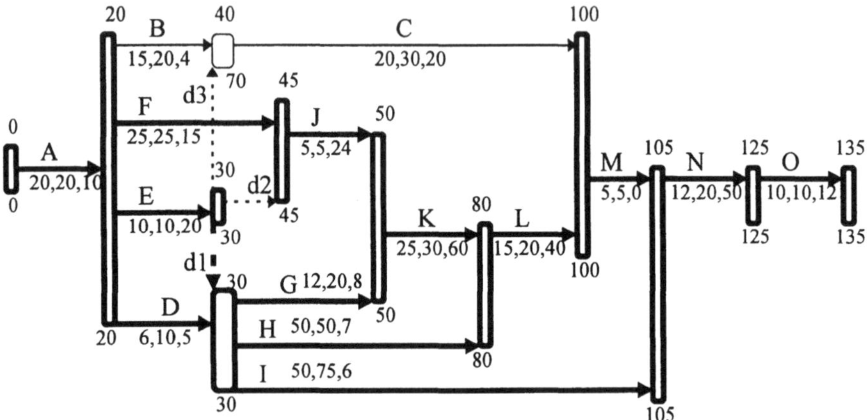

Figure 3-11 *Decreasing the project duration from 135 to 130 days.*

Iteration VIII.

Step 1. There are seven critical paths to shorten. The most economical solution is the shortening of L and I together.

Step 2. Δ1=5 days. The total floats of B, C and d3 have to be examined. It derives that Δ2=30 days. Δ=min{Δ1,Δ2}=5 days.

Step 3. The project direct cost increment is 5*46=230 units. The new project duration is 130 days. The results of this iteration can be seen in Figure 3-12.

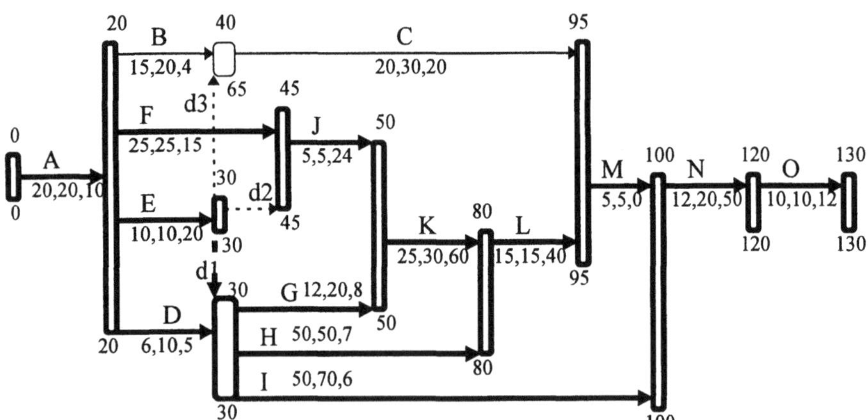

Figure 3-12 *Decreasing the project duration from 130 to 122 days.*

Iteration IX.

Step I. In this iteration there are seven critical paths to shorten. The most economical shortening is the compression of activity N.

Step 2. Activity N can be shortened by 8 days, and no non-critical activities have to be examined.

Step 3. The project direct cost increment is 8*50=400 units. The new project duration is 122 days. The results of this iteration can be seen in Figure 3-13.

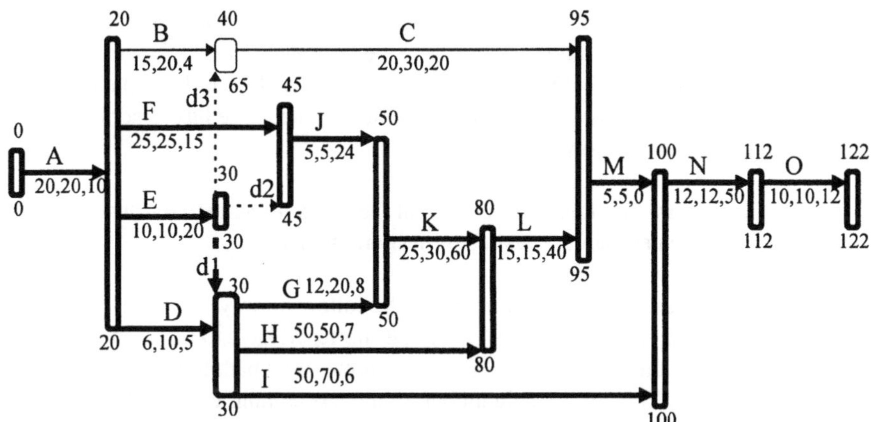

Figure 3-13 *The project duration cannot be further decreased*

There is a critical path for which no activities can be further shortened. (A,E,d1,H,L,M,N,O) This means that no more compression can be done in the project duration, so this is the minimal project duration. The optimal cost curve for the sample project can be seen in Figure 3-14.

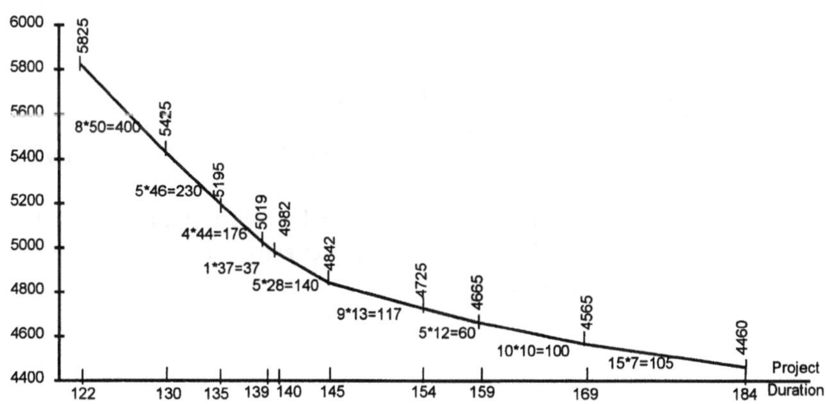

Figure 3-14 *The optimal direct cost curve for the sample project.*

Siemens' Method

This method also starts from the trivial optimum solution. The difference compared to Fondahl's method is that we reach the desired solution in one step, that is, this procedure does not provide the optimal solution curve in the p_b and desired project duration interval. The idea of the procedure can be summarized as follows. Each path that has a greater length than the desired project duration has to be shortened. To choose activities for shortening, the value of the so called '*effective cost slope*' is used. The effective cost slope is derived as the cost slope of the activity divided by the number of paths to be shortened containing the activity. During the process the effective cost slope will change, as less and less paths have to be compressed. Among the activities to be shortened, the procedure described below chooses the one with the lowest effective cost slope.

Step 1. Prepare the time analysis and all the paths in the network whose length is greater than the desired project duration, p.

Step 2. Prepare the time-cost matrix in the following way:

- Activities lying on a path to be shortened have to be listed in rows.
- All the columns but the last two represent a path to be shortened.
- The cells of the column before the last column contain the cost slope of the activities.
- The cells of the last column contain the value the activity can be compressed with. (This is the difference between the normal and the crash durations.)
- The last row contains the time reduction required, i.e., the path length minus p.
- Cells contain the effective cost slopes. The value of compression can be found below the slopes.

Step 3. Prepare the effective cost slopes as follows:

- Determine the paths that have to be shortened.
- Divide the cost slopes with the number of paths the activity is included in.
- If a path has been reduced adequately, then it does not have to be dealt with further, so some of the effective cost slopes have to be newly determined.

Step 4. Choose the activities to shorten. For this:

- Determine the longest path. If there are more paths with the same length, then choose the path which contains the lowest effective cost slope. If there are more paths, choose a path at random.
- Select the activity on this path with the lowest effective cost slope and determine the possible shortenings of it. (It comes as a difference of maximal possible shortenings -- last column -- and the compression already done. These latter data can be found in the cells below the effective cost slopes.) If there are more activities on the path with the same effective cost slope, then choose by using the following list.
  - Choose the activity which lies on the largest number of inadequately shortened paths.
  - If it is still not enough for decision making, then choose the activity which allows for the largest amount of compression.

Step 5. Determine the amount of compression, calculations.

- Compression of the chosen activity can changes in the length of many paths, therefore the amount of compression is determined by the smaller of the following two factors:
  - the smallest demand for shortening the inadequately shortened paths containing the chosen activity.
    (This value is derived as a difference of the original demand for reduction, which can be found in the cells of the last row, and the sum of the shortenings that have already been done. This latter value can be found in the cells of the column.)
  - the possible amount of shortening for the selected activity. (This value comes as a difference of the maximum possible shortening of an activity, which value can be found in the cells of the last column, and the shortening that has already been done for the activity. This latter value can be found in the cells.)
- The amount of the activity compression has to be displayed in the cells of the inadequately shortened paths. If an activity cannot be shortened any further, then it can be omitted from the next iteration phases.

Step 6. Last step of the iteration.

- If all the paths have been adequately shortened, then STOP.
- In the opposite case, we must go back to Step 3.

To illustrate the algorithm, we perform a calculation on our sample project to achieve the minimal cost solution for p=135 days. The procedure can be followed in Table 3-2. The columns, except for the last two, represent the path necessary to shorten. The column before the last shows the activity cost slopes while, the last column shows the possible maximum reduction of the activities. The cells of the paths contain the effective cost slopes, and in the bottom line of the cells the activity reductions which already have taken place can be found. These values have been printed in italics format. As the reader can see, activity B and activity C are not displayed in the matrix. The reason for this is, that they are not part of any paths that are longer than the desired 135 days. It is recommended to erase or cross out superseded values as the procedure goes forward. For the sake of lucidity we will not do this, but will indicate in the text, which ones are the superseded values .

Step 1: There are seven paths to shorten. The first column of bracketed numbers shows the lengths of these paths. The last column shows the time requirement for compression.

I.	A-F-J-K-L-M-N-O	(160 days)	(25 days)
II.	A-E-G-K-L-M-N-O	(154 days)	(19 days)
III.	A-E-H- L-M-N-O	(184 days)	(49 days)
IV.	A-E-I- N-O	(169 days)	(34 days)
V.	A-D-G-K-L-M-N-O	(150 days)	(15 days)
VI.	A-D-H- L-M-N-O	(180 days)	(45 days)
VII.	A-D-I- N-O	(165 days)	(30 days)

Step 2. Preparation of the time-cost matrix. The matrix can be seen in Table 3-2. The bottom line of the cells shows the reductions already made. At this stage, all of the performed reductions are zero.

Iteration 1.(Steps 3,4,5,6)

Step 3. The effective cost slopes are the first values of the cells.

Step 4. Path *III* has the largest demand for shortening, which is 49 days, and activity A has the lowest effective cost slope in this path, which is 10/7. Activity A has to be chosen for reduction.

Step 5. The possible reduction of A is 10 days. Activity A is included in all the paths. The reduction requirements for these path are: for path *I* 25 days, for path *II* 19 days, for path *III* 49 days, for path *IV* 34 days, for path *V* 15 days, for path *VI* 45 days, for path *VII* 30 days. The amount of reduction comes as a minimum of these values, that is, Δ=min{A(10), *I*(25), *II*(19), *III*(49), *IV*(34), *V*(15), *VI*(45), *VII*(30)}=10days. The time compression has been displayed in italics format at the bottom of the cells, so the original zero values have been superseded. The possible time reduction for activity A has been reached, so activity A can be omitted from further calculations.

Step 6. Inadequately shortened paths have remained, therefore new iteration is necessary.

Activit	Paths requiring compression							c_{ij}	b_{ij}-a_{ij}
	I	*II*	*III*	*IV*	*V*	*VI*	*VII*		
A	10/7 *0, 10*	10/7 *0, 10*	10/7 *0, 10*	10/7 *0, 10*	10/7 *0, 10*	10/7 *0, 10*	10/7 *0, 10*	10	10
F	15 *0, 5*							15	5
E		20/3 *0*	20/3 *0*	20/3 *0*				20	4
D					5/3 *0*	5/3, 5/2 *0*	5/3, 5/2, *0*	5	4
J	24 *0, 1*							24	5
G		4, 8 *0*			4 *0*			8	8
H			7/2 *0, 30*			7/2 *0, 30*		7	30
I				3, 6 *0, 15*			3 *0*	6	40
K	20, 30,60 *0*	20, 30 *0*			20 *0*			60	5
L	8,10,40/3, 40 *0*	8,10, 40/3 *0*	8, 10, 40/3 *0*		8, 40/3 *0*	8,10,40/3 *0*		40	5
M								0	0
N	50/7, 50/6, 10, 50/4 50 *0, 4*	50/7,50/6 10, 50/4 *0*	50/7,50/6, 10, 50/4 *0*	50/7, 50/6, 10, 50/4 *0*	50/7 *0*	50/7,50/6 *0*	50/7,50/6 10 *0*	50	8
O	12/7 *0, 5*	12/7 *0, 5*	12/7 *0, 5*	12/7 *0, 5*	12/7 *0, 5*	12/7 *0, 5*	12/7 *0, 5*	12	5
	Reduction demands for paths								
	25	**19**	**49**	**34**	**15**	**45**	**30**		

Table 3-2 *Reducing project duration of the sample project with Siemens' method*

Iteration II. (Steps 3,4,5,6)

Step 3. No effective cost slopes have changed.

Step 4. Path *III* has the largest demand for shortening, 39 days. Activity O has the lowest effective cost slopes, 12/7.

Step 5. The possible reduction for O is 5 days. The amount of reduction is Δ=min{O(5), *I*(15), *II*(9), *III*(39), *IV*(24), *V*(5), *VI*(35), *VII*(20)}=5 days. The reduction has been displayed in the cells of activity O, so the original zero values have been superseded. As no further reduction is available for activity O, it can be omitted from the calculation process. As the compression demand for path *V* has been satisfied, path *V* also must also be eliminated from the calculations.

Step 6. New iteration is necessary.

Iteration III. (Steps 3,4,5,6)

Step 3. Activities D, G, K, L and N are the elements of path *V*, therefore their effective cost slopes will change. The new values are: for D 5/2, for G 8, for K 30, for L 10, for N 50/6. The new values are displayed in the cells, the old values can be deleted.

Step 4. Path *III* has the largest demand for shortenings, 34 days. Activity H has the lowest effective cost slope, 7/2.

Step 5. The amount of reduction is Δ=min{H(30), III(34), VI(30)}= 30 days. The shortening has been displayed in the cells of the row H, near the zero values. As no further reduction is available for H and the compression demand of path *VI* has been satisfied, the row of H and the column of path *VI* can be canceled.

Step 6. New iteration is necessary.

Iteration IV. (Steps 3,4,5,6)

Step 3. The effective cost slopes of activities D, L and N have to change as they are lye on path *VI.* The new effective cost slope for D is 5, for L is 40/3 and for N is 10.

Step 4. Path IV has the largest demand for compression, 19 days. (Demands for the paths are: *I* 10, *II* 4, *III* 4, *IV* 19 *VII* 15.) Activity I has the lowest effective cost slope in this path, which is 3.

Step 5. The amount of reduction is Δ=min{I(40), IV(19), VII(15)}=15 days. The shortening has been displayed in the cells of I, near the old zero values, in italics. Activity I has remained active as it is still not at its crash duration. Path *VII* has been reduced to 135 days, so it can be eliminated from the further calculations.

Step 6. New iteration is necessary.

Iteration V. (Steps 3,4,5,6)

Step 3. The effective cost slopes of activities I and N have to be changed. The new values are 6 and 50/4 for I and N.

Step 4. Path *I* has the largest demand for shortening, 10 days (*I* 10, *II* 4, *III* 4, *IV* 4). Activity N has the smallest effective cost slope, which is 50/4.

Step 5. The amount of reduction is Δ=min{N(8), *I*(10), *II*(4), *III*(4), *IV*(4)}=4 days. The shortening has been displayed in the row of N, near the zero values, in italics. The compression demand of paths *II*, *III* and *IV* has been satisfied, so they must be eliminated from the further calculations.

Step 6. New iteration is necessary.

Iteration VI. (Steps 3,4,5,6)

Step 3. The effective cost slopes of activities N, K and L have changed as they lye on the latest canceled paths. The new value for N is 50, for K is 60, for L is

40. These are exactly the same as the real cost slopes because only one path has remained to be shortened.

Step 4. Path *I* has to be shortened by 6 days. Activity F has the lowest effective cost slope on this path, so this is the activity we have to shorten.

Step 5. The amount of reduction is Δ=min{F(5), *I* (6)}=5 days. The shortening has been displayed in the appropriate cells. No further compression is available for F, so it can be omitted from the further calculations.

Step 6. New iteration is necessary.

Iteration VII. (Steps 3,4,5,6)

Step 3. No effective cost slopes have been changed.

Step 4. Path *I* has to be shortened by 1 day. Activity J, which has the lowest effective cost slope, 24, has to be shortened.

Step 5. The amount of reduction is Δ=min{J(5), I(1)}=1 day. The shortening has been displayed in the active cells in the row of F.

Step 6. No path has remained with inadequately satisfied demand for shortening. The procedure has finished.

The procedure has ended as the compression demand for all the paths has been satisfied. The activity durations can be determined as the difference of the normal durations and the amount of activity shortenings which can be seen in the cells in italics. The results are shown on Figure 3-15.

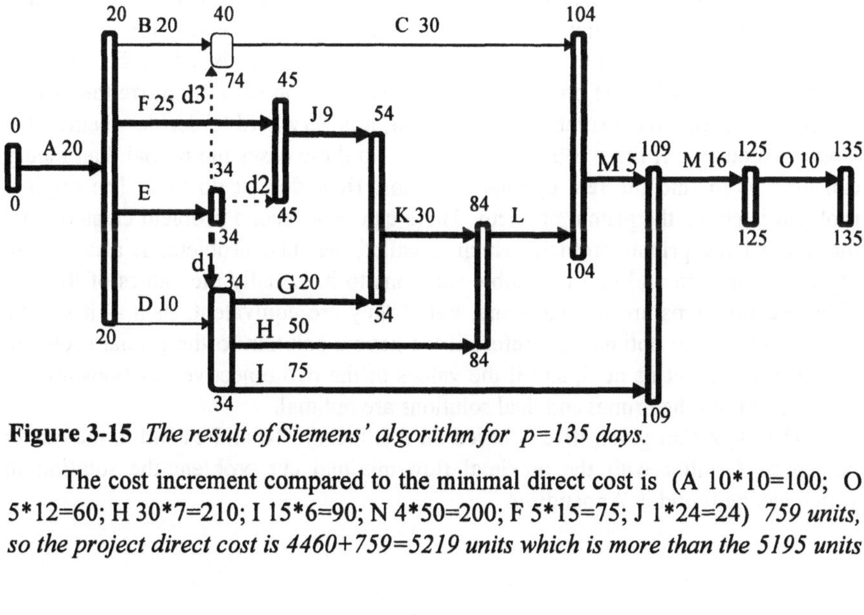

Figure 3-15 *The result of Siemens' algorithm for p=135 days.*

The cost increment compared to the minimal direct cost is (A 10*10=100; O 5*12=60; H 30*7=210; I 15*6=90; N 4*50=200; F 5*15=75; J 1*24=24) *759 units, so the project direct cost is 4460+759=5219 units which is more than the 5195 units* *from Fondahl's method.* In this case the Siemens method resulted in a higher direct cost solution than the Fondahl method, which provided the real optimal solution for the whole p_b, p_a interval.

Remarks to the algorithm:

- *The most time consuming step in the procedure is finding all the paths necessary to be shortened. Above a relatively small number of activities, it takes more time than to find the real optimal cost curve in the p_b, p_a interval. To systematically detect all the paths, there is a procedure, which can be found in Appendix A.*
- *Our experience is that the Siemens algorithm results in optimal solutions more often than the Fondahl method.*
- *The Siemens algorithm is recommended for use, if one does not need the direct cost curve, but only the direct cost related to the desired project duration.*

3.3 EXACT SOLUTION

In this subchapter we give an exact solution to the CPM cost problem. The solution we give is based on the primal-dual theory of operations research. What is the primal-dual theory and what is it good for? In operations research we usually look for a solution to a problem that is better from some point of view than any other one. There can be many solutions, but we want the best, e.g., minimizing the project indirect cost. The function which describes this goal is called the objective function. In the CPM/cost problem the objective function is to minimize the project direct cost. If we have a solution to a problem, it is very hard to decide whether it is a real optimum or if there is a better solution. In these cases the primal dual theory can help us to find the real optimal solutions. How does it do this? The original problem is called the primal problem. There can be a second problem created from the data of the primal problem which is called the dual problem. It also has an objective function and many possible solutions to it. Usually the values of the two objective functions are not the same, but if they are equivalent, then -- it can be proven -- they are optimal. Therefore if we have a solution to the primal problem, we have to look at its dual, and if the values of the two objective functions are the same, then both the primal and dual solutions are optimal.

The algorithm given here is based on this theory. To understand it, the reader has to be familiar with the maximal flow-minimal cut problem the solution of which can be found in Appendix A.

First, we are going to formulate the mathematical model of the problem which is called the primal problem.

The following, well known notations are used in this section:

N	- set of events
S	- subset of **N** which contains the initial event
T	- subset of **N** which contains the terminal event
A	- set of activities (included dummies)
s, t	- initial and terminal events
i	- optional event
(i,j)	- activity from event i to j
τ_{ij}	- duration of activity (ij)
E_i	- early event time of event i.
L_i	- late event time of event i
π_i	- arbitrary event time of i between E_i and L_i
ES_{ij}	- earliest start of activity (ij)
EF_{ij}	- earliest finish of activity (ij)
LS_{ij}	- latest start of activity (ij)
LF_{ij}	- latest finish of activity (ij)
p	- project duration
p_b	- project duration calculated from normal duration times
p_a	- project duration calculated from crash duration times
b_{ij}	normal duration for activity (ij)
a_{ij}	- crash duration time for activity (ij)
Cn_{ij}	- normal direct cost for activity (ij)
Cc_{ij}	- crash direct cost for activity (ij)
c_{ij}	- cost slope for activity (ij)

Let a directed loopless graph with one initial and one terminal node be given. Nodes serve as events, arrows serve as activities. Activities have to be carried out continuously with the same intensity. For activity identification we use the tail and head event, therefore multiple activities are not allowed. Activity durations have a lower and a higher bound, the crash and the normal durations. The assumption of the model is that the activity performance cost is higher at its crash than its normal duration, and the change of this cost is linear in the b_{ij} a_{ij} interval. With this assumption the cost slope can be determined as $c_{ij}=(Cc_{ij}-Cn_{ij})/(b_{ij}-a_{ij})$. From the definition of activities and events (see Definitions 2-1, 2-2) the following constraints can be given for all activities.

$$\pi_j-\pi_i\geq\tau_{ij} \qquad \forall\ (ij)\in\mathbf{A} \qquad \mathbf{3.1}$$

The activity durations have to be within their boundaries.

$$b_{ij} \geq \tau_{ij} \geq a_{ij} \qquad \forall\ (ij) \in \mathbf{A} \qquad \mathbf{3.2}$$

There are many different systems π_i exist for a fixed system τ_{ij}, which satisfy **3.1**. The system π_i, which is the minimal, but still satisfies **3.1** is called the minimal time policy[i]. In this system the π_t value is called the project duration and denoted by p, assuming that $\pi_s=0$.

$$\pi_s=0 \qquad \mathbf{3.3}$$

$$\pi_t=p \qquad \mathbf{3.4}$$

In a minimal time policy π_i there is at least one path along which **3.1** satisfies equality. This path is called the critical path.

On a network where τ_{ij} is not fixed but -- corresponding to **3.2** -- can change between its lower and upper bound, different project durations are possible. What is more the same project duration p can be achieved in several different ways. The minimal project duration is the one when we use crash durations, and the maximal project duration comes from the calculation using normal durations. The desired project duration has to be in this range ($p_a \leq p \leq p_b$).

It is evident that in case of different cost slopes c_{ij}, different project direct costs will correspond to the same project duration. Our goal is to find a system τ_{ij} and a system π_i which minimize the project direct cost that corresponds to a given p project duration and also satisfy **3.1** and **3.2**. The objective function can be formulated as follows:

$$min \sum_{A} Ca_{ij} = \sum_{A} \{Cn_{ij} + (b_{ij} - \tau_{ij}) c_{ij}\} \qquad \mathbf{3.5}$$

As Cn_{ij} and $(b_{ij}*c_{ij})$ are constant, the above objective function is equivalent to the following:

$$\max(\sum_{A} c_{ij}\, \tau_{ij}) \qquad \mathbf{3.6}$$

So, our mathematical model, hereinafter referred to as the CPM/cost primal problem, is the following:

---

[i] This is nothing else but the early event times for a schedule.

$\pi_j-\pi_i\geq\tau_{ij}$ $\forall\ (ij)\in\mathbf{A}$ **3.1**

$b_{ij}\geq\tau_{ij}\geq a_{ij}$ $\forall\ (ij)\in\mathbf{A}$ **3.2**

$\pi_s=0$ **3.3**

$\pi_t=p$ **3.4**

$$\max(\sum_{A} c_{ij}\,\tau_{ij}) \qquad \textbf{3.6}$$

A dual problem is established herein corresponding to the primal problem. These problems will be solved together. In the dual problem we are looking for a flow through the network. The flow goes into the network at the initial node and the same amount leaves the network at the terminal node. The value of the flow that goes in and goes out is denoted by Θ. A flow in an arrow is denoted by φ_{ij}. The amount of flow that goes into an internal node is the same as the amount of flow that goes out from the node. Knowing these, we can formulate the following constraints.

$\varphi_{ij}\geq0$ $\forall\ (ij)\in\mathbf{A}$ **3.7**

$$\sum_{j}\varphi_{ij}-\sum_{j}\varphi_{ji}=\begin{cases}0 & \forall i\in N-s-t\\ \Theta & i=s\\ -\Theta & i=t\end{cases} \qquad \textbf{3.8}$$

The goal in this model is to minimize the following objective function:

$$min\{\Theta\,p+\sum_{\substack{c_{ij}>\varphi_{ij}\\ A}}(c_{ij}-\varphi_{ij})\,b_{ij}-\sum_{\substack{\varphi_{ij}>c_{ij}\\ A}}(\varphi_{ij}-c_{ij})\,a_{ij}\} \qquad \textbf{3.9}$$

As the reader can see, the dual problem is based on the data of the original problem, only the variables are new.

We transform the first part of 3.9 as the later theorems will be easier to prove. Using **3.3** and **3.4,** the expression $\Theta\,p$ will change as follows:

$$\Theta\,p=(\pi\ s(-\Theta))+(\pi_2\,0)+\ldots+(\pi_j\,0)+\ldots+(\pi_t\,\Theta)=$$

$$\sum_{j\in N}\{\pi_j[\sum_{\substack{i\\(ij)\in A}}\varphi_{ij}-\sum_{\substack{i\\(ji)\in A}}\varphi_{ji}]\}=\sum_{j\in N}\{\pi_j\sum_{\substack{i\\(ij)\in A}}\varphi_{ij}\}-\sum_{i\in N}\{\pi_i\sum_{\substack{j\\(ij)\in A}}\varphi_{ij}\}=$$

$$=\sum_{j\in N}\pi_j*\sum_{i}\sum_{j}\varphi_{ij}-\sum_{i\in N}\pi_i*\sum_{i}\sum_{j}\varphi_{ij}=\sum_{A}(\pi_j-\pi_i)*\varphi_{ij}$$

Instead of 3.9 we can use the following equivalent form:

$$\min\{\sum_{A}(\pi_j-\pi_i)\varphi_{ij}+\sum_{\substack{c_{ij}>\varphi_{ij}\\A}}(c_{ij}-\varphi_{ij})b_{ij}-\sum_{\substack{\varphi_{ij}>c_{ij}\\A}}(\varphi_{ij}-c_{ij})a_{ij}\ \} \qquad \mathbf{3.10}$$

In the following we are going to prove that the second model is really the dual of the original cost optimization model.

In a lemma we are going to prove that the value of the objective function of the CPM/cost problem is usually less than the value of the objective function of the second model, but under some circumstances they can be equal.

Following this, we prove in a theorem that if the value of the objective functions are the same, then both solutions are optimal. We thus prove that these two models make a primal-dual pair.

After this, in a second theorem, we will prove that if there are feasible solutions to the problems, then there are solutions when the value of the two objective functions are equal, that is, there exist optimal solutions to the problems. The proof of this theorem is constructive, that is, it gives the algorithm, too.

In the last theorem we will prove that no optimal solutions exists for which the values of the objective functions are not the same.

Lemma 3-1 If there exist policy τ and policy π to a given project duration that satisfy the primal conditions and flow φ that satisfies the dual conditions, then the value of the primal objective function is less than or equal to the value of the dual objective function, that is,

$$\sum_{A}c_{ij}\tau_{ij}\le\sum_{A}(\pi_j-\pi_i)\varphi_{ij}+\sum_{\substack{c_{ij}>\varphi_{ij}\\A}}(c_{ij}-\varphi_{ij})b_{ij}-\sum_{\substack{\varphi_{ij}>c_{ij}\\A}}(\varphi_{ij}-c_{ij})a_{ij}$$

and equality holds if and only if the followings are satisfied:

if $\pi_j-\pi_i>\tau_{ij}$	then	$\varphi_{ij}=0$	1^0
if $b_{ij}>\tau_{ij}$	then	$\varphi_{ij}\ge c_{ij}$	2^0
if $a_{ij}<\tau_{ij}$	then	$\varphi_{ij}\le c_{ij}$	3^0

Proof :

Replacing b_{ij} with τ_{ij}, and a_{ij} with τ_{ij}, the value of the right side of the lemma will decrease or remain the same. After substitution we come to the following expression:

$$\sum_{A}c_{ij}\tau_{ij}\le\sum_{A}(\pi_j-\pi_i)\varphi_{ij}+\sum_{A}(c_{ij}-\varphi_{ij})\tau_{ij}$$

After simplification we get the following:

$$\sum_A c_{ij}\tau_{ij} \le \sum_A (\pi_j - \pi_i)\varphi_{ij} + \sum_A (c_{ij} - \varphi_{ij})\tau_{ij}$$

$$0 \le \sum_A (\pi_j - \pi_i)\varphi_{ij} - \sum_A \varphi_{ij}\tau_{ij}$$

If we replace π_j-π_i by τ_{ij}, the value of the right side of the lemma will decrease or remain the same.

$$0 \le \sum_A \tau_{ij}\varphi_{ij} - \sum_A \varphi_{ij}\tau_{ij}$$

$$0=0$$

With this, the original statement is proven, and equilibrium holds if and only if

$\varphi_{ij} > 0$	then	$\pi_j-\pi_i=\tau_{ij}$	1^{oo}
$c_{ij} > \varphi_{ij}$	then	$b_{ij}=\tau_{ij}$	2^{oo}
$c_{ij} < \varphi_{ij}$	then	$a_{ij}=\tau_{ij}$	3^{oo}

Reversing these,

if $\pi_j-\pi_i > \tau_{ij}$	then	$\varphi_{ij}= 0$	1°
if $b_{ij} > \tau_{ij}$	then	$\varphi_{ij} \ge c_{ij}$	2°
if $a_{ij} < \tau_{ij}$	then	$\varphi_{ij} \le c_{ij}$	3°

These are exactly the equilibrium conditions of the lemma.

The lemma is thus proved.

An important consequence of the lemma will be presented as follows:

Theorem 3-1 (*weak form of equilibrium*): If there exists a policy (π, τ) that satisfies the primal conditions to a given project duration p and a flow φ_{ij} that satisfies the dual conditions and the values of both objective functions are equal, then the solutions of both problems are optimal.

Proof (*indirect way*). Let us denote the value of the primal objective function by P and that of the dual by D.

Let us assume that P=D, but there exists a better primal solution P* where P*>P. In this case, P*>P=D but this is a contradiction according to the lemma.

Let us assume that P=D, but there exists a better dual solution D* where D*<D. In this case, D*<D=P but this is a contradiction according to the lemma.

The theorem is thus proved.

Now it is proved that the two models make a primal-dual pair. In the next theorem we prove that there exists a solution where the value of the objective functions are the same, that is, both solutions are optimal.

Theorem 3-2 (*duality theorem*): If a solution exists for a given project duration p that satisfies both the primal and dual conditions (p is bounded $p_a \leq p \leq p_b$,), then there is a solution so that the values of both objective functions are equal, that is, the solution is optimal.

Proof: To start a trivial solution let π_i be the time policy calculated from $\tau_{ij}=b_{ij}$ values, let $p=p_b$ and $\varphi_{ij}=0 \;\; \forall \; (ij)\in \mathbf{A}$. This is an optimal solution as it satisfies the primal conditions 3.1-3.4 and the dual conditions 3.7-3.8, and (3.6)=(3.10).
So in the initial step (Step 0.) we have to make a simple time analysis, and with this we will have the optimal solution of the primal-dual model which corresponds to the project duration p_b.

If we know an optimal solution (π, τ, φ) corresponding to any project duration p, then we can give a system (p^*,π^*, τ^* and φ^*), which also satisfies the primal and dual, even more the equilibrium conditions, i.e., it is also optimal and $p^*<p$.

This states that if there exists an optimal solution to a project duration p, we can move on to an another optimal solution which corresponds to a smaller project duration. As we know the optimal solution (Step 0.) corresponding to the project duration p_b we can give the optimal solution to all the project durations, in the p_a, p_b interval.

We comprehend this through a two step construction.

In the first step we increase the flow from φ to φ^* so that the duality conditions formulated in the lemma continue to be true, that is, the solution remains optimal. In the second step we decrease the project duration p such a way that the duality conditions remain fulfilled, that is, the solution is still optimal. After finishing Step 2., we have to go back to Step 1. and start the increasing of the flow again. We have to repeat Step 1. and Step 2. continuously until we reach the desired project duration or the minimum project duration.

Step 1. (increasing the flow):

Let us examine an arrow from an optimal solution. Which duality conditions can be accomplished in what combination? These are shown in the first three columns of Table 3-3. The '+' indicates that the first part of the equilibrium condition is fulfilled, the '-' indicates that it is not. The '+,+,+' and '+,+,-' cases are special. The first indicates an activity which is not critical and $b_{ij}>\tau_{ij}>a_{ij}$. An activity like this can only be a part of an optimal solution if its shortenings does not result in any cost increment, that is, its cost slope is zero. The '+,+,-' combination indicates a non-critical activity for which $\tau_{ij}=a_{ij}$, An activity like this can be a part of an optimal solution if its cost slope is zero. We will prove later that the algorithm excludes the possibility of these combinations arising, so they were indicated just for the sake of completeness.

Depending on which conditions are satisfied for a certain activity from an optimal solution we can get some information on the flow φ passing through on the activities. These pieces of information can be found in the fourth column of Table 3-3.

We have classified the activities of an optimal solution for further reference. These notations can be found in the fifth column.

In the first step the flow has to be increased so that the solution remains optimal, i.e., the dual conditions and the equilibrium conditions 1°, 2°, 3° are satisfied for all activities. For this, we have to increase the flow in a way that activities remain in their own class, that is, the solution remains optimal. With the help of the data on flows, which can be found in the fourth column, we can give a value of how much the flow φ_{ij} can be increased or decreased. The 6th and 7th columns of Table 3-3 show the capacities which have come about as a result of such an argument. These capacities ensure that the equilibrium conditions of the lemma remain fulfilled, that is, the solution remains optimal after the increment of the flow.

Equilibrium conditions			Information on flows	Classing of activities	κ_{ij}	κ_{ji}
1°	2°	3°				
+	-	+	$\varphi_{ij}=0$	A1	0	0
+	-	-	$\varphi_{ij}=0$	A1	0	0
-	+	+	$\varphi_{ij}=c_{ij}$	A2	0	0
-	+	-	$\varphi_{ij}\geq c_{ij}$	A3	∞	$\varphi_{ij}-c_{ij}$
-	-	+	$\varphi_{ij}\leq c_{ij}$	A4	$c_{ij}-\varphi_{ij}$	φ_{ij}
-	-	-	$\varphi_{ij}\geq 0$	A5	∞	φ_{ij}
+	+	+	$\varphi_{ij}=c_{ij}=0$	A6	0	0
+	+	-	$\varphi_{ij}=c_{ij}=0$	A6	0	0

Table 3-3 *Activities in an optimal solution.*

After the initial step, as all the activities are at their normal durations, activities can be in class A1 and A4. Searching for maximal flow in the network with the given capacities, we get a flow ψ_{ij}. Adding this to the flow φ, the new flow will be:

$$\varphi_{ij}^{*}=\varphi_{ij}+\psi_{ij} \quad \forall\,(i,j)\in \mathbf{A}$$

During this step activities have remained in their own classes due to the applied capacities. The equilibrium conditions and the conditions of the primal and the dual task are still satisfied, that is, the solution is still optimal.

Step 2: In the network we are to find a system ($p^*<p$, π^*, τ^*) corresponding to the increased flow φ^* where the duality and the primal conditions are accomplished, that is, the solution remains optimal. During this step some of the activities will change their class, but these changes will correspond to their flows. In the first step we were looking for a maximal flow. In this case there exists an (**S**,**T**) cut[i] in the network the activities of which are saturated, and there are no flows on the activities which are going backward in the cut. ($\mathbf{S}\cap\mathbf{T}=0$, $\mathbf{S}\cup\mathbf{T}=\mathbf{N}$) Saturated means that the flow on the arrow is equal to the capacity of the arrow, that is, $\psi_{ij}=\kappa_{ij}$. In the cut there can be activities from A1, A2, A4, and A6 classes. A3 and A5 cannot be in the cut as they cannot be saturated, due to their infinite capacity. In the cut backwards there can be activities from all classes (A1-A6). To determine the new system (p^*, π^*, τ^*) we give a δ value. System π will be decreased in set **T** by δ, which result in the decreasing of p by δ as well. We determine δ in such a way that all the activities that are in or backward in the cut have to remain the part of the optimal solution. The value of δ can be determined as:

$\delta:=\min\{\delta_{A1}, \delta_{A2}, \delta_{A4}, \delta_{A6}, \delta_{A1*}, \delta_{A2*}, \delta_{A3*}, \delta_{A4*}, \delta_{A5*}, \delta_{A6*}\}$
where

In case of activities in the cut

$\delta_{A1}:=\min\{\pi_j-\pi_i-b_{ij} \quad : \forall \quad (ij)\in(\mathbf{S},\mathbf{T}), (ij)\in A1\}$
$\delta_{A2}:=\min\{\pi_j-\pi_i-a_{ij} \quad : \forall \quad (ij)\in(\mathbf{S},\mathbf{T}), (ij)\in A2\}$
$\delta_{A4}:=\min\{\pi_j-\pi_i-a_{ij} \quad : \forall \quad (ij)\in(\mathbf{S},\mathbf{T}), (ij)\in A4\}$
$\delta_{A6}:=\min\{\pi_j-\pi_i-a_{ij} \quad : \forall \quad (ij)\in(\mathbf{S},\mathbf{T}), (ij)\in A6\}$

In case of activities backward in the cut

$\delta_{A1*}=\infty \quad : \forall \quad (ij)\in(\mathbf{T},\mathbf{S}), (ij)\in A1\}$
$\delta_{A2*}=\min\{b_{ij}-(\pi_j-\pi_i) \quad : \forall \quad (ij)\in(\mathbf{T},\mathbf{S}), (ij)\in A2\}$
$\delta_{A3*}=\min\{b_{ij}-(\pi_j-\pi_i) \quad : \forall \quad (ij)\in(\mathbf{T},\mathbf{S}), (ij)\in A3\}$
$\delta_{A4*}=\infty \quad : \forall \quad (ij)\in(\mathbf{T},\mathbf{S}), (ij)\in A4\}$
$\delta_{A5*}=\infty \quad : \forall \quad (ij)\in(\mathbf{T},\mathbf{S}), (ij)\in A5\}$
$\delta_{A6*}=\infty \quad : \forall \quad (ij)\in(\mathbf{T},\mathbf{S}), (ij)\in A6\}$

The δ defined in this way is definitely greater than zero. If the remaining project compression demand of a given project duration is less than δ, then it has to be changed to this value.

Let the new system π^* be determined in the following way:

---

[i] **S** and **T** are two subsets of the **N** with no common node, but all the nodes have to be an element of one of them The initial node has to be the element of **S** and the terminal node has to be the element of **T**. $s\in\mathbf{S}$ and $t\in\mathbf{T}$. The (**S**,**T**) symbol denotes all the activities that are originated in **S** and ended in **T**. More about this can be found in Appendix A.

$$\pi_i^* := \begin{cases} \pi_i & \text{if} \quad i \in S \\ \pi_i - \delta & \text{if} \quad i \in T \end{cases}$$

Thus p will become $p^*=p-\delta$ and the duration of an activity (ij) can be determined from the following formula:

$$\tau_{ij}^* = \min \{ \pi^*_j - \pi^*_i; b_{ij} \} \quad \forall (i,j) \in \mathbf{A}$$

The δ had been constructed so that the duality conditions on the arrows continue to be accomplished.

The cost increment can be determined as the sum of the product of cost slopes and the changes in activity durations.

After these modifications the below described changes take place.

Activities of which both nodes fall into the same set of nodes (**S** or **T**) remain in the same class.

In the case of activities in the cut the following changes take place corresponding to the flow.

Activities from class A1 remain in class A1 or get into class A4.
Activities from class A2 remain in class A2 or get into class A3.
Activities from class A4 get into class A3 or class A2.
Activities from class A6 get into classes A1, A2, A3 or A4.

In the case of activities backward in the cut the following changes take place, corresponding to the flow.

Activities from class A1 remain in class A1.
Activities from class A2 remain in class A2 or get into class A4.
Activities from class A3 get into classes A2 or A4.
Activities from class A4 get into class A1.
Activities from class A5 get into class A1.
Activities from class A6 get into classes A1, A2, A3 or A4.

After the initial step activities can be in classes A1 or A4. None of the changes in the second step can result in an activity getting into class A6. In other words, the algorithm will never result in activities with duration times smaller than their normal durations if they are not critical, even if their cost slopes are zero. Therefore all the notes on activities in class A6 are redundant from the algorithm point of view.

With this we have finished Step 2.

Now we have to go back to step one and increase the flow again and then in the second step the project duration can be decreased. These steps must be alternately repeated until we reach the desired project duration or the flow becomes infinitely large. Let us examine the case of infinite flow. In this case there exists a $P(s \rightarrow t)$ path along which all the capacities are infinite. This means that activities in this path can be of classes A3 or A5. The durations of the activities in class A3, which is a_{ij}, cannot be further decreased. In the case of activities in class A5, $a_{ij}=b_{ij}$ so these activities cannot be decreased either.

In this way the length of the critical path cannot be further decreased, that is, the p project duration is minimal, $p=p_a$. The curve of optimal solutions is now given in the p_b, p_a interval.

Thus the theorem is constructively proven.

An important consequence of this theorem is presented hereby.

Theorem 3-3 (*strong form of equilibrium*). If there exist an optimal primal solution the value of the objective function of which is (P), and an optimal dual solution the value of the objective function of which is (D), then these two values are equal.

Proof: According to the duality theorem, there exist a maximal primal solution (P*) and a minimal dual solution (D*), which are optimal, that is, (P*)=(D*). As (P) and (D) are also optimal, so (P*)=(P) and (D*)=(D), but then (P)=(D).

Remarks to the algorithm:

Several methods are available to solve the CPM/cost problem. We give more details on this at the end of this chapter, in Historical Review.

Any kind of maximal flow-minimal cut algorithm can be used in the flow increasing step. In Appendix A we have given the so called augmenting path procedure. There are much faster algorithms for the problem, the so called preflow-push algorithms.

We have implemented both the augmenting path and the preflow push algorithms in a computer software and generated extra large random CPM problems. The running time using the augmenting path algorithm was less than three minutes for a network consisting of 3000 activities and 800 events, with 149 crash points in the project cost curve. (It means that step 1 and step 2 was repeated 149 times.) The running time with preflow- push flow algorithm was 30 seconds on a current generation personal computer (IBM 486).

This algorithm is faster than the algorithm based on LP or integer programming, or out-of-kilter method both theoretically and practically. We do not prove this as it would take to much mathematical consideration and the knowledge of the rest of the algorithms.

To follow the algorithm on our sample project would be too big a task, so we will illustrate it on a very small five activities network. The network can be seen in Figure 3-16, the input data can be found in Table 3-4. We have to determine the minimal cost curve in the interval p_b, p_a.

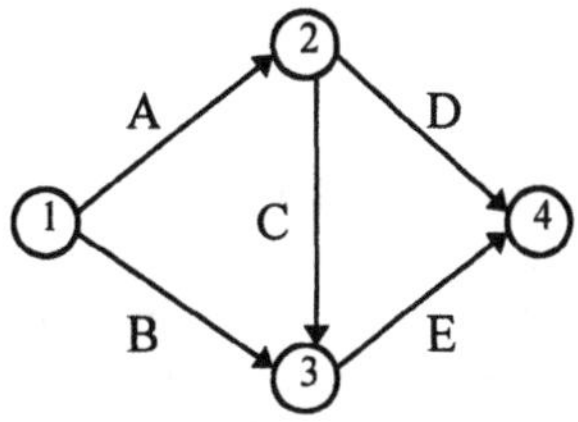

Figure 3-16 *Small sample project to illustrate the CPM/cost algorithm.*

Activity	A	B	C	D	E
Crash duration	3	5	3	8	5
Normal duration	4	7	5	10	7
Cost slope	10	12	8	6	14

Table 3-4 *Input data for the small sample project.*

Initial step. (Determining the trivial optimum solution)

The trivial optimum solution is the solution where all the activities are considered at their normal duration, and $p=p_b$. The result of this calculation can be seen in Figure 3-17. The project duration is $p_b=16$ days. We have not performed the backward calculations as we need only one system π for the further calculations.

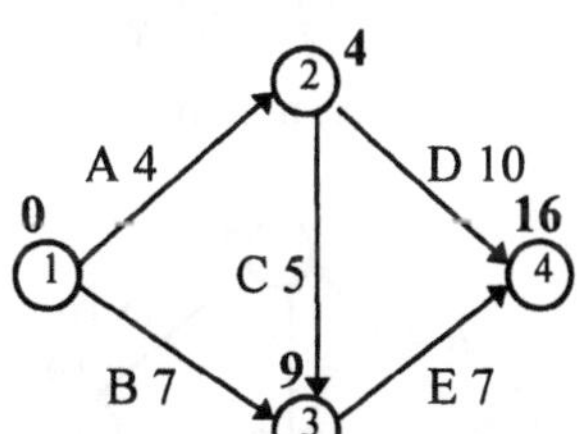

Figure 3-17 *The result of the initial step for the small sample project.*

The iteration starts with flow increments. Each iteration can be followed in a figure consisting of five parts. The first part shows what equilibrium conditions are fulfilled in what combinations, that is, it shows the classifying of activities. The second part shows the capacities the flows can be increased or decreased with. The first number shows the possible increment, the last the possible decrement on an arrow. The third part shows the maximal flow in the network, and also shows the minimal cut which defines this flow. The nodes after the cut form set **T**, and the nodes before the cut form set **S**. The fourth part shows the new flow after the increment. The last part shows the new project duration p*, the new system π and the new system τ.

Iteration I.

The steps of the first iteration can be followed in Figure 3-18. The critical activities are in class A4, the non-criticals are in class A1 (see Figure 3-18.a). The flow on critical activities can be increased with the cost slopes, and cannot be decreased. No flow increment and decrement can be taken place in non critical activities (see Figure 3-18.b). The maximal flow is shown in Figure 3-18.c). Activities B, C and D are in the cut. The value of δ comes as the minimum of δ_{ij} which has been determined for B, C and D (δ=min{δ_B, δ_C, δ_D}=min{2,2,2}=2). The new flow can be seen in Figure 3-18.d). For determining the new π^*, τ^* and p^* values we have to decrease the event times with δ in the set **T**. The activity durations come about the following way: $\tau_{ij}=\min\{\pi_j-\pi_i, \tau_{ij}\}$ The results can be seen in Figure 3-18e). The duration of activity C has changed. The cost increment is $c_C*\Delta\tau_C=16$, that is, 8 units per day.

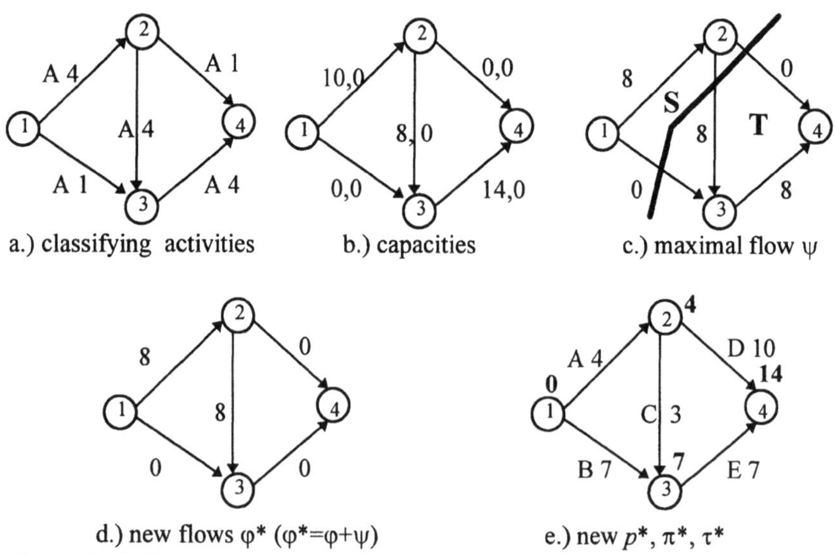

Figure 3-18 *Iteration I. Moving from p_b=16 days to p=14 days.*

Iteration II.

All the activities have become critical, but only the duration of C decreased. The iteration can be followed in Figure 3-19. The value of δ comes as a minimum of values δ of the activities in the cut and backward in the cut. As the reader can see, activity C goes from **T** to **S**, that is, backward in the cut and after Step 2 its duration will increase. The value of δ for A, as it is in class A4, is $\delta_A=1$, for E is $\delta_E=2$ and for C, which is backward in the cut, is $\delta_C=2$. It follows that $\delta=\min\{\delta_A,\delta_C,\delta_E\}=1$. The new project duration is 13 days. The cost increment is $(c_A\tau_A)$-$(c_C\Delta\tau_C)+(c_E\Delta\tau_E)=10-8+14=16$ units. The cumulative cost increment is 16+16=32 units.

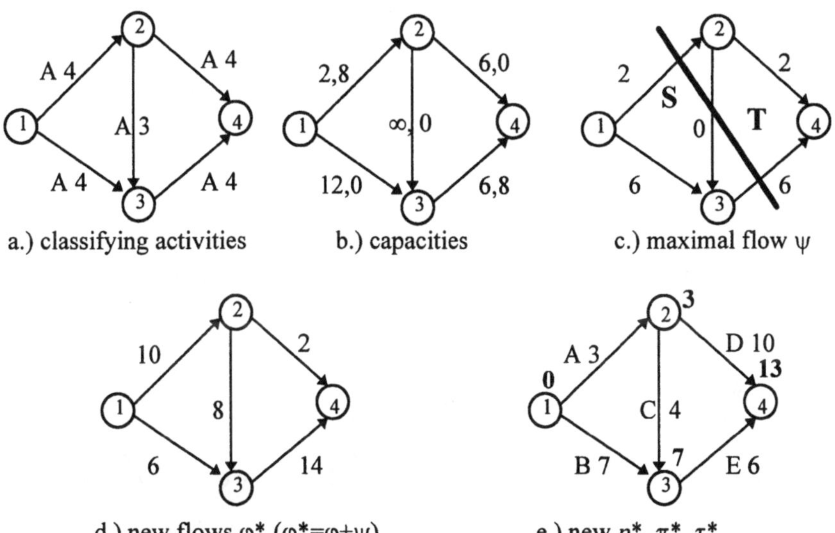

Figure 3-19 *Iteration II. Moving from p=14 days to p=13 days.*

Iteration III.

All the activities are critical, B and D are at their normal duration, A is at its crash duration. The iteration can be followed in Figure 3-20. The δ value comes as a minimum of δ of activities in the cut. Activities D and A are in the cut, δ for D is $\delta_D=2$, the value of δ for E is $\delta_E=1$, that is, $\delta=1$. The new project duration is 12 days. The duration of the activities in the cut has changed.

The project cost increment compared to $p=14$ days project duration is $(c_D\Delta\tau_D)+(c_E\Delta\tau_E)=6+14=20$ units. The cumulative cost increment is 52 units.

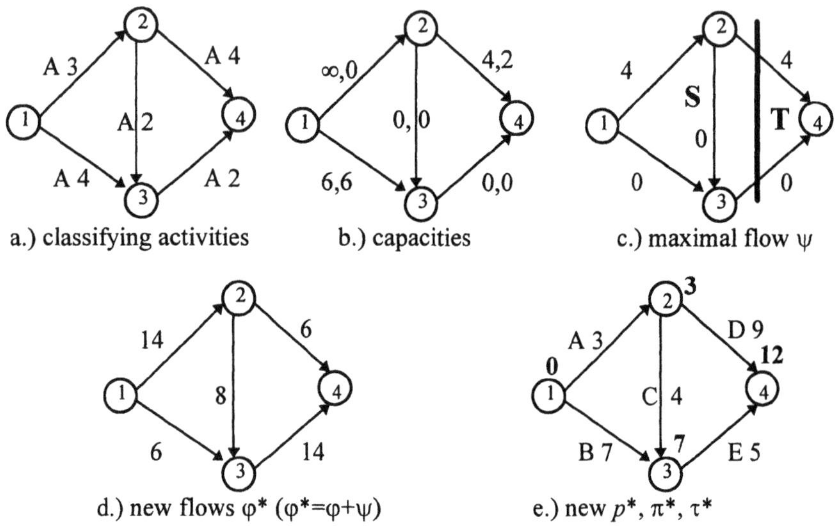

Figure 3-20 *Iteration III. Moving from p=13days to p=12 days.*

Iteration IV.

The steps of the iteration can be followed in Figure 3-21. The duration of B, C and D have decreased by one day. The cost increment compared to the p=12 days project duration is: $(c_B\Delta\tau_B)+(c_C\Delta\tau_C)+(c_D\Delta\tau_D)=12+8+6=26$ units. The cumulative cost increment is 52+26=78 units.

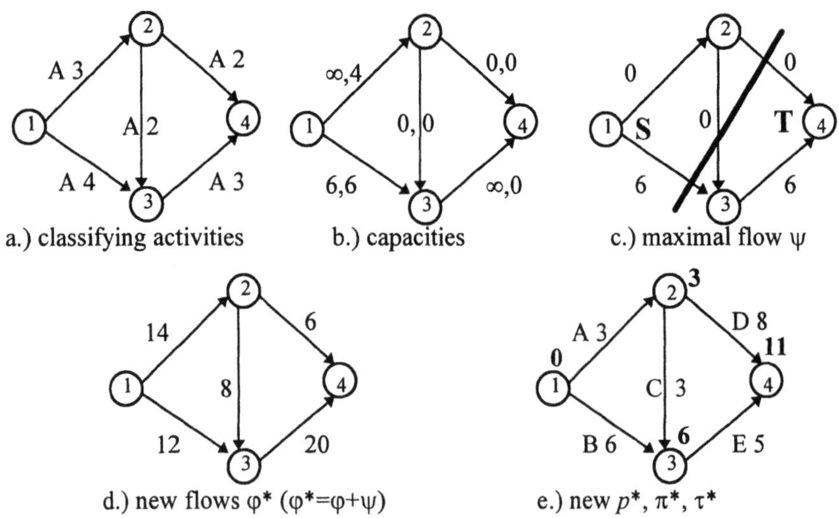

Figure 3-21 *Iteration IV. Moving from p=12 days to p=11 days.*

Iteration V.

There are paths along which the capacities are infinite, so the flow is infinite, too. This means that activities lying on these path are at their crash durations, that is, they cannot be shortened anymore. The minimal project duration is p_a=11 days, and the procedure has finished (see Figure 3-22).

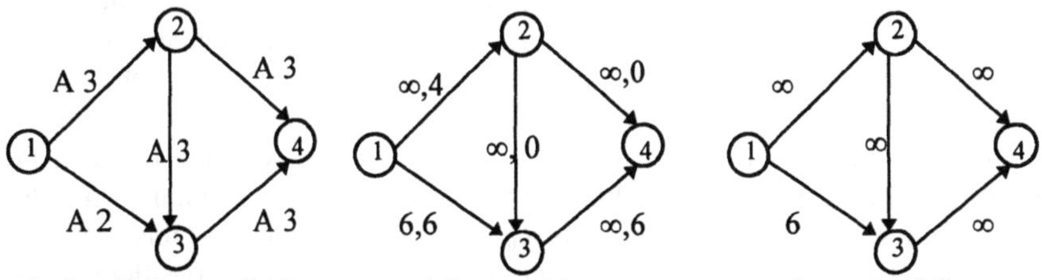

a.) classifying activities b.) capacities c.) maximal flow ψ

Figure 3-22 *Iteration V. Last iteration.*

Finally we mention that the optimal cost curve for the original, that is, the depot building sample, is the same we got from the Fondahl method.

In the next subchapter we give an algorithm to determine the maximal cost curve which is the upper envelope of the feasible solution set. We will calculate the maximal cost curve for the original sample project. The difference between the minimal and maximal cost curve can convince the skeptics about the usefulness of time-cost trade-off methods.

3.4 MAXIMAL COST SOLUTION

We have shown already that there are many different solutions to arrive at a desired project duration. As the cost slopes of the activities differ, the different solutions usually result in different project direct costs. Using the Fondahl and Siemens methods, different project costs belong to the solutions for the same project durations. These are heuristic methods which serve finding the optimal or close-to-optimal solutions. If an inexperienced project manager tries to achieve the desired project duration by randomly changing the activity durations, he can get results far behind the optimal solutions. In the worst case, he might schedule the plan the possibly most expensive way. In this chapter we present an algorithm to find the "worst solution" curve in the interval of p_b, p_a. The significance of this section is that the deviation between the "worst" and the "best" solution justifies the grounds for using the original CPM/cost problem.

We use the following new notations during the chapter.

l_b	- length of an arbitrary s→t path calculated from norma durations.
l_a	- length of an arbitrary s→t path calculated from cras durations
Cmax	- direct cost increment of the "all crash" solution to the "al normal" solution

The mathematical model of the CPM/cost problem (see Fig. 3-2, A-B curve) according to a given project duration p has been stated in the following manner:

$$\pi_j - \pi_i \geq \tau_{ij} \qquad \forall\ (ij) \in \mathbf{A} \qquad \mathbf{3.11} = 3.1$$

$$b_{ij} \geq \tau_{ij} \geq a_{ij} \qquad \forall\ (ij) \in \mathbf{A} \qquad \mathbf{3.12} = 3.2$$

$$\pi_s = 0 \qquad \mathbf{3.13} = 3.3$$

$$\pi_t = p \qquad \mathbf{3.14} = 3.4$$

$$\max\left(\sum_{A} c_{ij}\, \tau_{ij}\right) \qquad \mathbf{3.15} = 3.6$$

To find the curve of the "worst" solution it seems obvious to modify the objective function (3.15) of the model to search for the minimum instead of the maximum. However, for any p between p_a and p_b, this modification gives us the absolute "worst" solution, which is the "all crash" point and belongs to p_a, and not the "worst" solution to the desired project duration p. This can be easily verified by using the problem's LP dual and the LP's equilibrium conditions. However, this book is aimed mainly at engineers and not mathematicians, so we provide an easier, clearer explanation instead.

Let the following system (π, τ) be created to a given project duration $p>p_a$. Let $\tau_{ij}=a_{ij}$ for every activity, and let system π_i be calculated using this τ system, except π_t where $\pi_t=p$ instead of p_a coming from this calculation. This system (τ, π) satisfies conditions 3.11-3.14, however, it does not give the cost of the "worst" solution belonging to p, but rather the cost of the absolute "worst" solution (Figure 3-2 point D).

The reason for this is that the CPM/cost model with the originally stated conditions (3.1-3.4) is deficient. The condition that there always have to be at least one p-length path in the network, that is, a critical path, is lacking. This condition is automatically realized if we are trying to find the minimal cost; thus its statement is superfluous. To find the maximum cost for a given p, however, it is necessary to adopt this condition to the model. Therefore, the model of finding the maximum cost solutions is the following:

$\pi_j-\pi_i\geq\tau_{ij}$	$\forall\ (ij)\in \mathbf{A}$	**3.16**
$b_{ij}\geq\tau_{ij}\geq a_{ij}$	$\forall\ (ij)\in \mathbf{A}$	**3.17**
$\pi_s=0$		**3.18**
$\pi_t=p$		**3.19**
There should be at least one path with length p		**3.20**
$\min(\sum_{A} c_{ij}\,\tau_{ij})$		**3.21**

Thus the above stated model is not solvable by either linear programming or related methods -- like flow algorithms --, because of condition 3.20. In the following, we provide an algorithm to the solution of the above mentioned problem.

The algorithm for finding the solution to the maximum cost of project duration p^* is based on the concept that we can move from the optimal solution belonging to p^* to the optimal solution belonging to p if $p>p^*$. Since the optimal, that is, the maximal cost solution to $p^*=p_a$ is trivial, the "worst" solution to an individual project duration is definable within the entire p_a-p_b interval. The maximum cost solution to a given project duration p is characterized by two things. Firstly, it is characterized by a single critical path (P) which has the highest direct cost at this length. Secondly, all the activity durations are at their crash duration for activities that are not elements of the P path, that is, $\tau_{ij}=a_{ij}$ if $(ij)\notin$P. (Project duration $p=p_a$ is an exception, for which several critical paths are possible.) The critical path is the one (among the possible p-length paths) that causes the least possible cost reduction as compared to p^*. The following algorithm is based upon these observations.

Step 1. Define the absolute "worst" solution (Figure 3-2, point D). The shortest project duration (p_a) can be calculated from the activities' crash durations. The cost deviation from the absolute "best" solution (Figure 3-2, point B) is given by the sum $Cmax=\Sigma c_{ij}*(b_{ij}-a_{ij})$.

Step 2. Determine all those paths whose lengths calculated by the normal durations are greater than or equal to p_a.

Step 3. Determine the lengths of the paths defined in the previous step using the crash durations.

Step 4. Determine the cost decrement (ΔC) of increasing a path from l_a to p for every path which were defined in Step 2. The path with the least cost reduction will be the critical one; for all other activities $\tau_{ij}=a_{ij}$. Along the paths, the cost reduction should be defined in such a way that it allows for only the smallest possible reduction. Firstly, activities with the smallest cost slopes must be increased, and when any further increases in durations are impossible, then switch over to the next activity having the smallest cost slope. The "worst" solution to project duration p is expressed by $Cmax\text{-min}\{\Delta C\}$.

Step 5: Repetition of the previous step for any other project duration.

Remarks to the algorithm:

This algorithm is not heuristic, that is, it gives the curve of the optimal solutions.

The critical point is to find all the paths the lengths of which are greater than the minimal project duration. The algorithm which is used here is a kind of enumeration.

We produced a software based on the algorithm and tested on large networks. For a network which consisted of 3000 activities and 600 events, it took more than 5 hours to produce the maximal cost curve. Finding the more than 10.000 paths took 280 minutes.

If one is not only looking for a solution corresponding to a given project duration, but rather would like to find the maximal cost curve in the whole interval (p_a, p_b)*, then it is recommended to start from* p_a *and increase p step by step by one day.*

Let us find the curve of the "worst" solutions in the sample network.

Step 1.

The minimal project duration is p_a=122 days. The project direct cost increment to the "all normal" solution is $Cmax=\Sigma c_{ij}(b_{ij}-a_{ij})$=2089 units.

Step 2.

The paths with greater length than p_a are: I. A-F-J-K-L-M-N-O=160; II. A-E-G-K-L-M-N-O=154; III. A-E-H-L-M-N-O=184; IV. A-E-I-N-O=169; V. A-D-G-K-L-M-N-O=150; VI. A-D-H-L-M-N-O=180; VII. A-D-I-N-O=165.

Step 3.

The crash length of these paths, that is, $\tau_{ij}=a_{ij}$, are: I. l_a=117; II. l_a=109; III. l_a=122; IV. l_a=102; V. l_a=105; VI. l_a=118; VII. l_a=93.

Step 4.

The solutions for the entire interval p_a-p_b can be seen in Table 3-5. The table is divided into two parts. In the upper part, the values in the cells indicate the remaining possible lengthening for the activities.

At the beginning all the activities can be increased by $b_{ij}-a_{ij}$. The first numbers of the cells show these values. During the calculation as the activities are shortened these values are superseded, and new values will be valid. In the lower part of the table, cells show when an activity duration has to change in order to achieve the desired length for the path in question.

For example, in the column of Path I, the value near p=137 days has to be read in the following way: To lengthen this path to 137 days, we have to lengthen activity A by ten days, which causes 100 units cost decrement, then activity O by five days, which causes 60 units, but cumulative 160 units of cost decrement, and activity F by 5 days, which causes 75, but cumulative 235 units of cost decrement. The cell of the first column near p=136 days has be read as : To lengthen this path to 136 days, we have to increase A by 10 days, then O by 5 days, and F by 4 days, which causes 160+4*15=220 units of cost increment. As the reader can see, only those rows are displayed in which some of the activities have reached their possible duration increments, but non-displayed rows can be calculated for each column by linear interpolation. To achieve the desired project duration at its maximal cost, we have to choose the path with the smallest cost decrement. For example, to achieve p=137 days, it would take 235 units for Path I., 304+40=344 units for Path II., 15*7=105 units for path III., 120+15*6=200 units for Path IV, 424 units for Path V., 120+15*7=2225 units for Path VI. and 260 units for path VII. Path III. causes the minimal cost decrement, so this is going to be the critical path. For all the remaining activities not on this path, $\tau_{ij}=a_{ij}$.

The definitive path for the project duration in question has been shaded. There are crash points in the maximal cost curve when the definitive path will change, or a new activity has to be increased on this path.

Activity	c_{ij}	I.	II.	III.	IV.	V.	VI.	VII.
A	10	10,5,0	10,5,0	10,0	10,0	10,5,0	10,0	10,0
F	15	5,0						
E	20		4,0	4,0	4,0			
D	5					4,0	4,0	4,0
J	24	5,0						
G	8		8,0			8,0		
H	7			30,0			30,0	
I	6				40,20			40,20
K	60	5,0	5,0			5,0		
L	40	5,0	5,0	5,0		5,0	5,0	
M	0	0	0	0		0	0	
N	50	8,0	8,0	8,0	8,0	8,0	8,0	8,0
O	12	5,0	5,0	5,0	5,0	5,0	5,0	5,0
crash length l_a		117	109	122	102	105	118	93
normal length l_b		160	154	184	169	150	180	160
122		A5 55	G8 114 A5	0	I20 *120*	D4, A5, G8 134	D4 120	I25, D4 *170*
127		A10 100	A10 164			A10 184		
132		O5 160	O5 224			O5 244		
136			E4 304					
137		F5 235				L5 444		I40 260
141			L5 504					
142		J5 355			I40 240			
145						N8 844		
147		L5 555						A10 360
149			N8 904					
150						K5 1144		
152				H30 210	A10 340		H30 230	O5 420
154			K5 1204					
155		N8 955						
157					O5 400			
160		K5 1255						N8 820
161					E4 480			
162				A10 310			A10 330	
167				O5 370			O5 390	
169					N8 880			
171				E4 450				
172							L5 490	
176				L5 650				
180							N8 890	
184				N8 1050				

Table 3-5 *Maximal cost calculations for the sample network*

The result of the algorithm can be seen in Figure 3-23. The optimal cost curve has also been displayed, so the reader can see the huge difference between the minimal and maximal cost curve. The difference is usually the largest in the middle of the interval, that is, the chance for making a much more expensive schedule for a given project duration is the highest. While the minimal cost curve is always convex, assuming linear activity time-cost curves where $Cc_{ij} \geq Cn_{ij}$, the maximal cost curve is not necessarily concave, as different paths can form the critical path in different phases.

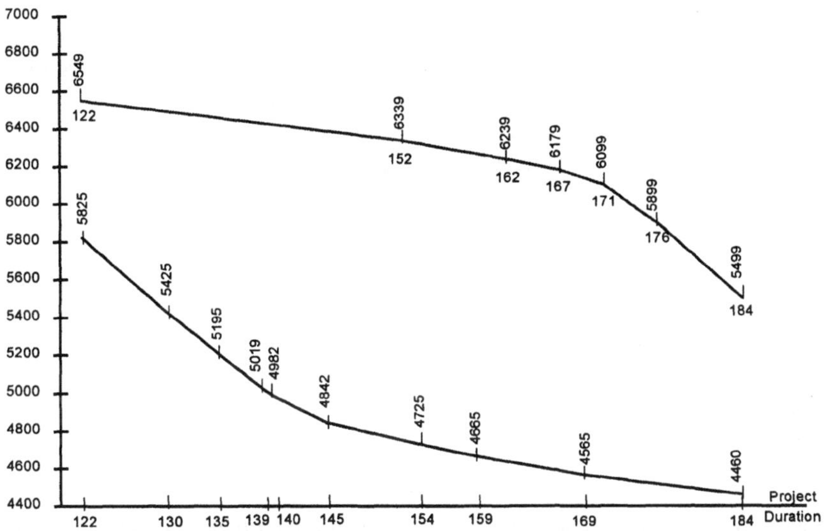

Figure 3-23 *Minimal and maximal direct cost curves for the sample project.*

3.5 WHICH IS BETTER?

In this part we compare the Siemens, the Fondahl and the exact methods. We make the examinations from the following point of view: In case of different size networks, which method is advisable to use.

First we have to make it clear that both Fondahl's and Siemens' methods are heuristic methods, that is, the solution may not be optimal. We have shown this for the Siemens method, it has resulted in higher direct costs for the same project duration than Fondahl's method. Fondahl's method has resulted in optimal solutions for the sample project in the entire interval p_a-p_b. Now we give a small sample to prove that the Fondahl method is heuristic too.

We solve the example problem used to illustrate the exact method by Fondahl's method. The input data of the sample project can be seen in Figure 3-24 and Table 3-6. These are identical to Figure 3-16 and Table 3-4.

The minimal project direct cost curve, and the "minimal" cost curve that have been derived from using Fondahl's method can be seen in Figure 3-25.

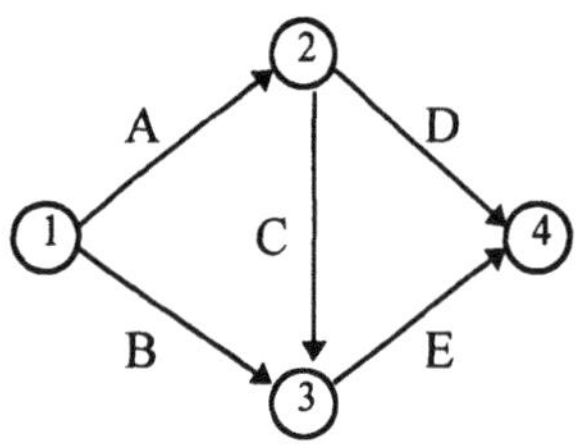

Figure 3-24 *A small example to illustrate the unreliability of Fondahl's method..*

Activity	A	B	C	D	E
Crash duration	3	5	3	8	5
Normal duration	4	7	5	10	7
Cost slope	10	12	8	6	14

Table 3-6 *Input data for the small sample project.*

The normal duration is 16 days. In Fondahl's method we start from this point. The critical path is A-C-E, we shorten C by two days. The new project duration is 14 days, the cost increment is 16 units. Now we have three critical paths. It seems to be the most economical to shorten D and E by two days. This will result in 12 days project duration and 2*20=40 units increment, that is 56 units cumulative cost increment. In the next step we have to shorten A and B by one day, which will result in 11 days project duration and 22 units, that is 78 units cumulative cost increment.

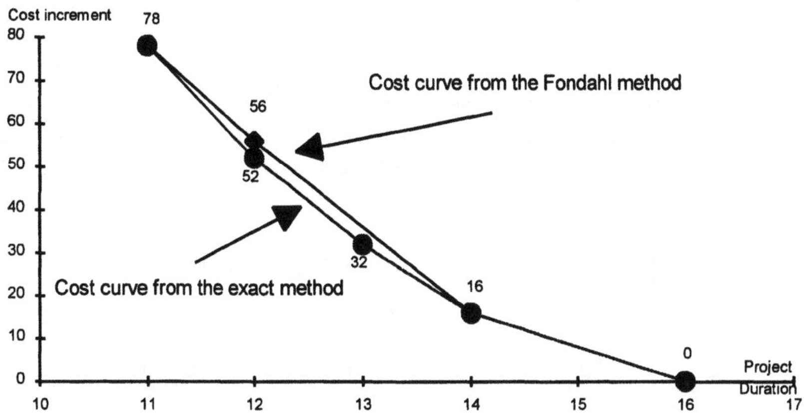

Figure 3-25 *Optimal and non-optimal cost curves for the small sample project.*

The reader can see, that even in the case of very small networks both of these heuristic methods can lead to a wrong solution.

What methods shall we use for CPM least cost scheduling? There is no question, if we need optimal solution and not just close-to-optimal, then we have to use one of the exact methods. If we want to perform a computer implementation of CPM time-cost trade-off, then we also have to use the exact method for two different reasons. First, an optimal solution is better than a non-optimal. Second, the exact method is far more faster than these heuristics methods. We have mentioned, that in case of big networks the computational time took five hours while with exact method it has taken just couple of minutes.

If we have to perform hand calculations, and the network is very small, the heuristics methods can be used. In this case use the Siemens method, as it is easier as the determination of the paths involved has finished.

3.6 SPECIAL CASES

In the original CPM/cost method we use the assumption that the change of activity direct cost is linear in the duration interval $[a_{ij}\text{-}b_{ij}]$, as a close approximation of the usual convex activity time-cost curve (Figure 3-1). The activity cost can sometimes be so large that itself is a very serious factor in the project cost. In this case we use the so called piece-wise linear approximation. Each piece is being treated as a separate, in other words a pseudo-activity. An activity can be divided into any number of pieces. One has to consider how appropriate the original activity time-cost estimation has been. If there is uncertainty in the estimation, then it does not make sense to create a lot of pseudo-activities from the activity in question. The way of creating pseudo activities from an arbitrary activity can be seen in Fig. 3-26.

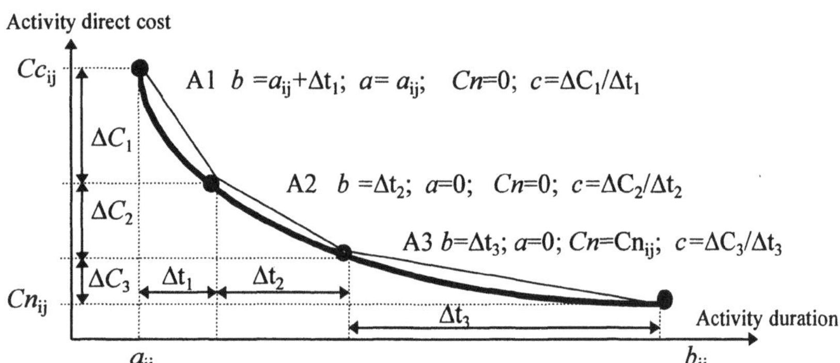

Figure 3-26 *Piece-wise linear approximation to activity time-cost curves.*

The network representation of pseudo-activities is shown in Figure 3-27.

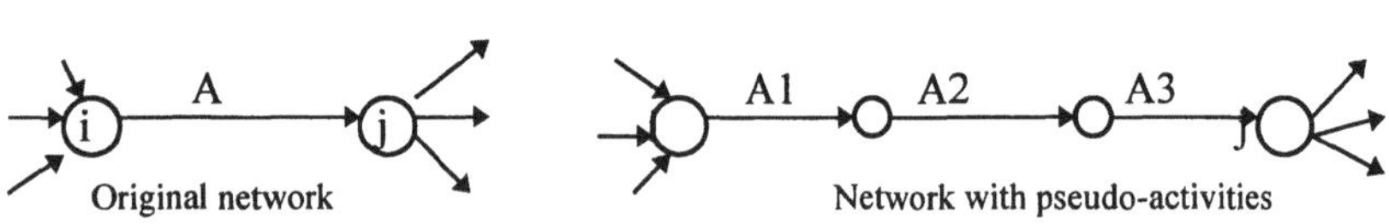

Figure 3-27 *Pseudo activities in the network.*

There are other types of activity time-cost curves too. They can be concave, discontinuous, discrete, or the combination of these. It can be increasing, but this is usually a result of a technically inadequate time-cost estimation. Possible activity time-cost curves can be seen in Figure 3-28.

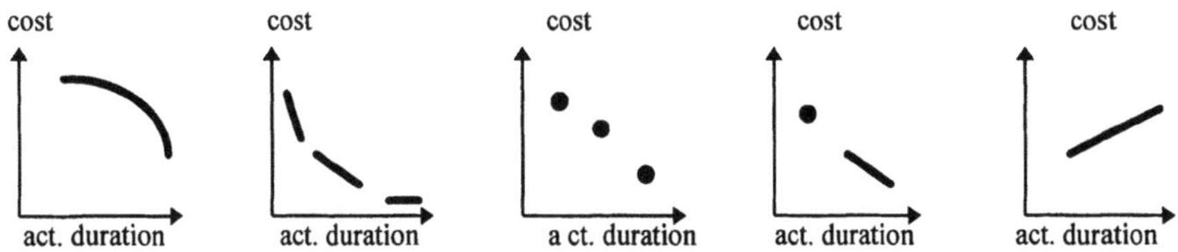

Figure 3-28 *Different activity time-cost curves*

The exact method which has been described in this chapter cannot handle all these activity cost curves. In case of a concave activity time-cost curve, the linear approximation can be used as it was used in the convex case as well. It can also handle the discrete time-cost function if there are feasible points for each day in the activity time-cost curve. The reason for this is that our procedure will result in integer activity durations if the normal and crash durations are integer. For example, if the normal duration is 5 days and the crash duration is 2 days, the possible result of the calculation can be 5, 4, 3 or 2 days but never 3.5 days. The problem of how to handle the last time-cost curve is the most difficult in mathematical sense, but the simplest from the engineers point of view. If the cost of an activity is the smallest at its crash accomplishing time, then we have to fix its duration at this value and leave the time-cost curve out of consideration.

Our method is not suitable to handle the discontinuous or arbitrary discrete time-cost functions. Different methods have been developed for solving the problem when different types of time-cost functions are in the network. These are usually based on some of the well known mathematical programming methods like linear programming for the discontinuous time-cost functions, or integer programming for the discrete time-cost functions. It is not too hard to combine these models for the case of both discrete and continuous curves.

An another interesting approach to handle different time-cost functions is the Decision CPM method. In this method all possibilities for an activity are represented by individual fixed time activities. These possibilities are originated from a so called decision node and the algorithm will decide which is the best choice from the point of view of objective function. The only shortcoming of this approach is that the size of the network can dramatically increased, and the computational time will take much longer than in the case of continuous linear time-cost functions.

In most cases the simple linear time-cost approximation is sufficient.

3.7 PRACTICAL PROBLEMS AND SOLUTIONS

Problems

1. Solve the below given CPM/cost problem in the p_a-p_b interval using the Fondahl method!

	A	B	C	D	E	F	G	H
b_{ij}	4	7	3	5	2	10	7	2
a_{ij}	3	5	2	3	2	8	5	1
c_{ij}	10	12	5	8	0	6	14	10

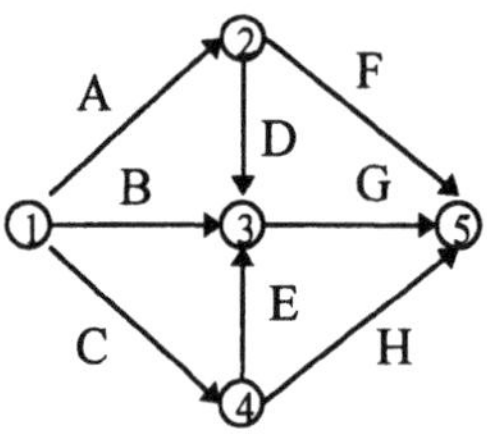

2. Solve the CPM/cost problem given in **1**. using the Siemens method!

3. Solve the CPM/cost problem given in **1**. using the maximal flow algorithm!

4. Solve the CPM/cost problem given in **1**. for the maximal cost curve!

5. Solve the below given CPM/cost problem using the Fondahl method!

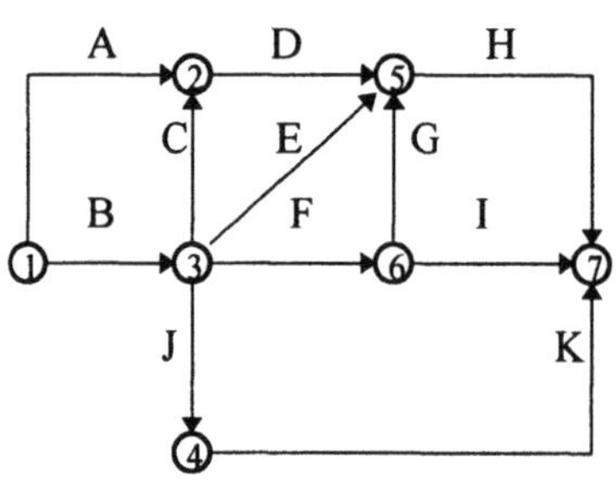

	A	B	C	D	E	F	G	H	I	J	K
b_{ij}	4	5	2	3	6	2	1	7	5	6	3
a_{ij}	2	4	1	1	2	2	1	1	2	5	2
c_{ij}	8	3	5	2	4	1	3	6	4	3	5

6. Solve the CPM/cost problem given in **5**. using the Siemens method!

7. Solve the CPM/cost problem given **5**. using the maximal flow algorithm!

8. Solve the CPM/cost problem given in **1**. for the maximal cost curve!

9. Solve the below given CPM/cost problem using the maximal flow algorithm!

	A	B	C	D	E	F	G	H
b	3	4	2	5	5	3	2	4
a	1	3	2	2	4	1	1	3
c	6	3	4	7	8	5	2	1

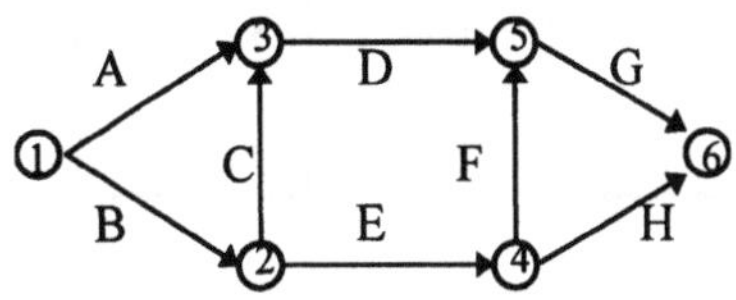

10. Solve the CPM/cost problem given in **9**. for the maximal cost curve!

11. Solve the below given CPM/cost problem using the maximal flow algorithm!

	A	B	C	D	E	F	G
b	2	3	4	5	4	3	2
a	1	2	3	2	3	2	1
c	1	2	3	4	3	2	3

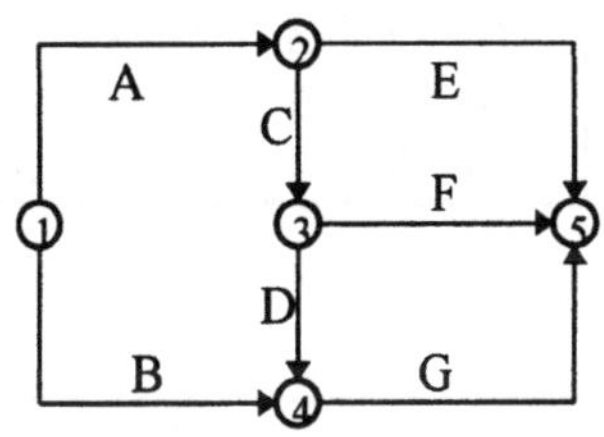

12. Solve the CPM/cost problem given in **11**. for the maximal cost curve!

Solutions

1. The results can be followed in the table below. The duration and event times which have been changed in the iteration in question have been written in bold face.

Project duration	Cost increment	A	B	C	D	E	F	G	H	1	2	3	4	5
		Activity duration								Event time				
16	0	4	7	3	5	2	10	7	2	0	4	3	9	16
14	+8=16	4	7	3	**3**	2	10	7	2	0	4	3	7	**14**
12	+40=56	4	7	3	3	2	**8**	**5**	2	0	4	3	7	**12**
11	+22=78	3	6	3	3	2	8	5	2	0	**3**	3	**6**	**11**

2.

Project	Cost	Activity duration								Event time				
duration	increment	A	B	C	D	E	F	G	H	1	2	3	4	5
16	0	4	7	3	5	2	10	7	2	0	4	3	9	16
15	+8=8	4	7	3	**4**	2	10	7	2	0	4	3	8	**15**
14	+8=16	4	7	3	**3**	2	10	7	2	0	4	3	7	**14**
13	+16=32	**3**	7	3	**4**	2	10	**6**	2	0	**3**	3	7	**13**
12	+20=52	3	7	3	4	2	**9**	**5**	2	0	3	3	7	**12**
11	+26=78	3	**6**	3	**3**	2	**8**	5	2	0	3	3	**6**	**11**

3.

Project	Cost	Activity duration								Event time				
duration	increment	A	B	C	D	E	F	G	H	1	2	3	4	5
16	0	4	7	3	5	2	10	7	2	0	4	3	9	16
14	+16=16	4	7	3	**3**	2	10	7	2	0	4	3	7	**14**
13	+16=32	**3**	7	3	**4**	2	10	**6**	2	0	**3**	3	7	**13**
12	+20=52	3	7	3	4	2	**9**	**5**	2	0	3	3	7	**12**
11	+26=78	3	**6**	3	**3**	2	**8**	5	2	0	3	3	**6**	**11**

4.

Project	Cost	Activity duration								Event time				
duration	decrement	A	B	C	D	E	F	G	H	1	2	3	4	5
11	105	3	5	2	3	2	8	5	1	0	3	2	6	11
13	-12=93	3	5	2	3	2	**10**	5	1	0	3	2	6	**13**
14	-10=83	**4**	5	2	3	2	10	5	1	0	**4**	2	**7**	**14**
15	-18=65	4	5	2	**5**	2	**8**	**6**	1	0	4	2	**9**	**15**
16	-14=51	4	5	2	5	2	8	**7**	1	0	4	2	9	**16**

5.

Project	Cost	Activity duration											Event time						
duration	increment	A	B	C	D	E	F	G	H	I	J	K	1	2	3	4	5	6	7
18	0	4	5	2	3	6	2	1	7	5	6	3	0	7	5	11	7	**18**	11
17	+3=3	4	**4**	2	3	6	2	1	7	5	6	3	0	6	**4**	10	6	**17**	**10**
16	+4=7	4	4	2	3	**5**	2	1	7	5	6	3	0	6	4	**9**	6	**16**	10
14	+12=19	4	4	2	**1**	**3**	2	1	7	5	6	3	0	6	4	7	6	**14**	10
13	+6=25	4	4	2	1	3	2	1	**6**	5	6	3	0	6	4	7	6	**13**	10
12	+9=34	4	4	2	1	3	2	1	**5**	5	**5**	3	0	6	4	7	6	**12**	**9**
11	+11=45	4	4	2	1	3	2	1	**4**	5	5	**2**	0	6	4	7	6	**11**	9

6.

Project	Cost	Activity duration											Event time						
duration	increment	A	B	C	D	E	F	G	H	I	J	K	1	2	3	4	5	6	7
18	0	4	5	2	3	6	2	1	7	5	6	3	0	7	5	11	7	**18**	11
17	+3=3	4	**4**	2	3	6	2	1	7	5	6	3	0	6	**4**	10	6	**17**	**10**
16	+4=7	4	4	2	3	**5**	2	1	7	5	6	3	0	6	4	**9**	6	**16**	10
15	+6=13	4	4	2	3	**4**	2	1	7	5	6	3	0	6	4	**8**	6	**15**	10
14	+6=19	4	4	2	**1**	**3**	2	1	7	5	6	3	0	6	4	**7**	6	**14**	10
13	+6=25	4	4	2	1	3	2	1	**6**	5	6	3	0	6	4	7	6	**13**	10
12	+9=34	4	4	2	1	3	2	1	**5**	5	**5**	3	0	6	4	7	6	**12**	**9**
11	+11=45	4	4	2	1	3	2	1	**4**	5	5	**2**	0	6	4	7	6	**11**	9

7. Observe that the result are the same as in 6!

Project	Cost	Activity duration											Event time						
duration	increment	A	B	C	D	E	F	G	H	I	J	K	1	2	3	4	5	6	7
18	0	4	5	2	3	6	2	1	7	5	6	3	0	7	5	11	7	**18**	11
17	+3=3	4	**4**	2	3	6	2	1	7	5	6	3	0	6	**4**	10	6	**17**	**10**
16	+4=7	4	4	2	3	**5**	2	1	7	5	6	3	0	6	4	**9**	6	**16**	10
15	+6=13	4	4	2	3	**4**	2	1	7	5	6	3	0	6	4	**8**	6	**15**	10
14	+6=19	4	4	2	**1**	**3**	2	1	7	5	6	3	0	6	4	**7**	6	**14**	10
13	+6=25	4	4	2	1	3	2	1	**6**	5	6	3	0	6	4	7	6	**13**	10
12	+9=34	4	4	2	1	3	2	1	**5**	5	**5**	3	0	6	4	7	6	**12**	**9**
11	+11=45	4	4	2	1	3	2	1	**4**	5	5	**2**	0	6	4	7	6	**11**	9

8. The results can be seen in the table below. The cost increment of the "all crash" solution is 100 units compared to the all normal solution.

Project	Cost	Activity duration											Event time						
duration	decremen	A	B	C	D	E	F	G	H	I	J	K	1	2	3	4	5	6	7
11	100	2	4	1	1	2	2	1	1	2	5	2	0	5	4	7	6	11	9
13	94	2	**5**	1	1	2	2	1	1	2	6	2	0	**6**	**5**	**8**	**7**	**13**	**11**
14	89	2	5	1	1	2	2	1	1	2	6	**3**	0	6	5	8	7	**14**	11
15	64	2	5	**2**	**3**	2	2	1	5	2	**5**	3	0	**7**	5	**10**	7	**15**	**10**
17	52	2	5	2	3	2	2	1	**7**	2	5	**2**	0	7	5	10	7	**17**	10
18	45	2	5	**1**	**1**	**6**	2	1	7	2	5	2	0	**6**	5	**11**	7	**18**	10

9.

Project duration	Cost increment	Activity duration								Event time					
		A	B	C	D	E	F	G	H	1	2	3	4	5	6
14	0	3	4	2	5	5	3	2	4	0	4	6	9	12	14
13	2	3	4	2	5	5	3	**1**	4	0	4	6	9	12	**13**
12	5	3	**3**	2	5	5	3	1	4	0	**3**	**5**	**8**	**11**	**12**
11	11	3	3	2	5	5	**2**	1	**3**	0	3	5	8	**10**	**11**
10	26	3	3	2	**4**	**4**	2	1	3	0	3	5	7	**9**	**10**

10. The direct cost increment of the "all crash" solution is 57 units comparing to the "all normal" solution. The results can be seen in the table below.

Project duration	Cost decrement	Activity duration								Event time					
		A	B	C	D	E	F	G	H	1	2	3	4	5	6
10	57	1	3	2	2	4	1	1	3	0	3	5	7	8	10
11	56	1	3	2	2	4	1	1	**4**	0	3	5	7	8	**11**
12	53	1	**4**	2	2	4	1	1	4	0	**4**	**6**	**8**	**9**	**12**
13	45	1	4	2	2	**5**	1	1	4	0	4	6	**9**	**10**	**13**
14	34	1	4	2	2	5	**3**	**2**	**3**	0	4	6	9	**12**	**14**

11.

Project duration	Cost increment	Activity duration							Event time				
		A	B	C	D	E	F	G	1	2	3	4	5
13	0	2	3	4	5	4	3	2	0	2	6	11	**13**
12	1	**1**	3	4	5	4	3	2	0	**1**	**5**	**10**	**12**
11	4	1	3	**3**	5	4	3	2	0	1	**4**	**9**	**11**
10	7	1	3	3	5	4	3	**1**	0	1	4	9	**10**
7	19	1	3	3	**2**	4	3	1	0	1	4	**6**	**7**

12.

Project duration	Cost decrement	Activity duration							Event time				
		A	B	C	D	E	F	G	1	2	3	4	5
7	26	1	2	3	2	3	2	1	0	1	4	6	7
8	25	**2**	2	3	2	3	2	1	0	**2**	**5**	7	**8**
10	19	2	2	**4**	2	3	2	**2**	0	2	**6**	**8**	**10**
13	7	2	2	4	**5**	3	2	2	0	2	6	**11**	**13**

3.8 HISTORICAL REVIEW

In the Historical Review of Chapter 2, we have cited the original CPM papers, and those which are connected to the time analysis phase or to the CPM network-based methodology. Here we merely concentrate on the papers that are dealing with time-cost trade-off. The CPM/cost method originates from James E. Kelley and Morgan Walker [41],[42]. Their method was so sophisticated that construction industry assimilated this technique as two, as it was mentioned in the introductory chapter.

To make the CPM time-cost trade-off technique understandable for construction industry experts, in the early 1960's heuristic algorithms were developed by the pioneers of network techniques. The aim of these works were to develop easy-to-understand techniques which can be used in hand computations. A classical work in this field is John Fondahl's technical report [18]. In this work he gave three different heuristic methods. One when the calculation starts from the "all normal" point, one when the calculation starts from the "all crash" point, and one when the calculation starts from an arbitrary but feasible point. He also discussed which of these methods and when have to be used, and gave a systematic procedure to detect all the possible combinations in order to choose the most economical combination for shortening. In this chapter we have presented only the first of his methods without the procedure used for detecting the possible combinations. The reason for this is that in the world of computers it is not worth dealing with heuristic algorithms if they are much slower then the exact ones.

The heuristic method of Siemens [61] was developed a decade later. Our experience is that it gives optimal results more often than the Fondahl method. While the main shortcoming of Fondahl's method is to determine all the possible activity combinations in order to choose the most economical shortening, the drawback to the Siemens method is, that one has to find all the possible paths for which the "normal" length is greater than the minimal project duration. To develop these path we used the algorithm that was developed by Yen [68] in 1971. He developed a procedure for determining the first "*k*" *shortest* paths in the network. A slightly modified algorithm can be found in Appendix A.

The algorithm for the maximal cost solution is a result of a talk between Keith Crandall, professor emeritus of construction management at University of California at Berkeley and the author [10] was taken place in 1993. The algorithm has only been published in Hungarian before. Probably this is not the first algorithm for this problem, but we could not traced antecedent works. Moder et al. [50]give a sample problem for which the maximal and minimal curve are determined, but they do not give the algorithm.

The algorithm of this chapter based on maximal flow algorithm that was first published in Klafszky's [43] book in 1969. A modified and improved version of it is Hajdu's and Klafszky's[29] work.

Another algorithm, which is based on network flow theory can be found in Fulkerson's [25] paper from 1961.

Several embellishments of the original CPM method have been presented over the years. One of the most remarkable extensions is the possibility of handling special activity time-cost curves. Meyer and Shaffer [47],[48] have developed a model, which can handle any combinations of continuos curves and discrete points. A dynamic programming approach tool was developed by Hindelang and Muth [38] whose approach is called the decision CPM network. In this network all the possible durations for an activity are represented by arbitrary activities which are originated from the same decision node. This is really a powerful technique and it can be used for resource leveling as well. Unfortunately the size of the network can dramatically increase which can cause many more computational steps to appear.

An interesting extension is Elmegrabhy's [16] work, who has developed an algorithm, which minimizes the project total cost, plus the penalties for tardy completion.

A comprehensive discussion of the CPM time-cost trade-off problem can be found in different books, e.g., Moder et al [50].

Chapter 4

PRECEDENCE DIAGRAMMING

As it has become evident from the preceding chapters, the CPM technique has several shortcomings. One of these imperfections causes a lot of unnecessary work, as we could see it in the time consuming network drawing process. The most serious problem with CPM is that we cannot describe activity interdependencies in an adequate way. According to the CPM definitions (see Def. 2-1 and 2-2) an activity can start only when all the preceding activities have finished. If we want to describe activity overlapping in CPM, we will face several difficulties which cannot be solved perfectly. In this chapter we introduce a new approach to scheduling. This method is called the Precedence Diagramming Method. In the non-English-speaking world, e.g., Germany, France, Spain, Hungary, etc., this method is called the Metra Potential Method (MPM) or Method of Potential. The reader can find more about the history of PDM at the end of this chapter. This technique is much more flexible, it is more appropriate to model different activity interdependencies.

First, we are going to establish a sample project which is going to be used through the chapter, for illustration.

In the second subchapter the basic definitions are discussed, which are necessary to understand the model.

In the third section we deal with the preceding activity lists.

In the next subchapter we discuss how to draw a PDM network.

In the fifth section we give the rules which govern the PDM time analysis.

A lot of otherwise excellent texts contain a small step the use of which can be questioned. We cover this topic in the sixth section, 'Notes on Calculation Rules.'

In the fifth section we assumed that activities are non-splittable, that is, they have to be carried out continuously in time, without interruption. However, in a project there can be activities which are interruptable. An algorithm which can handle both types of activities can be found in the seventh subchapter, 'PDM Calculation: Splitting Allowed.'

In the subsequent subchapter, 'Loops in PDM Network,' we deal with the different interpretations of loops.

In the next part, 'Notes on Negative Lag and Dangling Activities,' we pay attention to some wrong practices one should better avoid.

In the following section, 'The Paradox Behavior of PDM,' we discuss an interesting scheduling phenomenon which can be described with PDM but cannot be modeled with CPM.

Floats in PDM are covered in the part 'Information Obtained from Calculations'. Conclusions follow after this.

At the end of this chapter we give some problems to practice the new technique and their solutions. Finally, we discuss the history of this technique.

4.1 INTRODUCING THE SAMPLE PROJECT

Let us consider the sample project that has been described in the first part of Chapter 2, Introducing a Sample Project, on page 18. A covered depot has to be built from prefabricated standardized steel elements in order to protect a piece of large industrial equipment used for engine production in heavy industry. The project has been divided into fifteen processes from A to N. We have also kept the original time estimations for the activities.

A	-	Preparations	30 days
B	-	Ordering machine	20 days
C	-	Delivering machine	30 days
D	-	Ordering building materials	10 days
E	-	Preparatory works on site	14 days
F	-	Conveyance of bulldozers	30 days
G	-	Delivery of gravel and cement	20 days
H	-	Delivery of mount structure	80 days
I	-	Delivery of prefabricated steel structure	90 days
J	-	Groundwork	10 days

K	-	Foundation for the depot	30 days
L	-	Placing the mount for the machine	20 days
M	-	Placing the machine on the mount	5 days
N	-	Assembling the building structure	20 days
O	-	Finishing works, project delivery	15 days

We, the project managers in charge, were directed to make a faster schedule for this project than it was done in Chapter 2. As the activity durations are given, the only way to speed up the completion time of the project is to overlap some of the activities.

For a new schedule, we have to establish the activity relationships.

We do this in such a way that we take each activity from A to O and answer the following question: "*What activities precede logically the activity in question and how?*" Remember, in CPM scheduling we have asked, "*What activities have to be finished before the activity in question can start?*"

Answering the question, we can make the following list.

- There are no activities that have to be done before preparations (A).

- Ordering of machine (B) is possible after choosing it (A).

- Before the delivery of the machine (C), we have to order the machine (B) and finish the preparatory works on site (E).

- The ordering of building materials (D) can take place after preparations (A) have finished.

- Preparatory work on site (E) can start after finishing preparations (A).

- Delivery of bulldozers (F) can start at the same time as preparatory work on site begins.

- Delivery of gravel and cement (G) can start after the ordering of building materials (D) and three day's work is being accomplished from the preparatory work on site (E). (3 days are necessary to prepare the storage conditions for materials.)

- Delivery of mount structure (H) can start after the ordering of materials (D) and as the preparatory work on site (E) has started.

- Delivery of prefabricated steel structure (I) can start as the preparatory work (E) has started and after the ordering of materials (D).

- Groundwork (J) can start after the delivery of bulldozers (F) and after the preparatory work has finished (E).

- Foundation (K) can be done during the groundwork (J), but groundwork has to lead at least by 5 days at any stage of the activity. Foundation can start as 3 days' amount of gravel and cement has been delivered, and this safety reserve must be kept during the whole process of delivery.

- At least 10 days have to be between the placing of mount structure (L) and the foundations (K), in any stage of L and K. This time is necessary to cure the concrete. There should be enough mount structure (H) on the site for at least ten days of placement at any stage of L.

- The machine can be placed on mount (M) after the delivery of machine (C) and after the placement of mount structure (L). The concrete has to be cured at least 20 days before the machine is placed.

- The assembly of prefabricated steel structure (N) can start when the placement of machine (M) has finished. It can be done during the delivery of steel structure (I), but it has to two start at least two days later, and it can finish as I has finished.

- The project delivery (O) can finish at the same time as the assembly of the building (N) has finished, and can start after N starts.

One can argue that these activity interdependencies are fictitious and some of them are not real. It may be true, but this is just an example, and the main emphasis is on how we can translate these verbal statements to formulas which can then be handled in a mathematical way.

As the reader can see, the activity interdependencies can be different, and at this stage it is hard to imagine what kind of formulas can describe this variety of relationships. It is clear that the usual CPM activity relationship, which is 'An activity can start if all the preceding activities have finished,' does not seem to be suitable for this logic. In PDM, different kinds of precedence relationships can be used to adequately describe these activity relationships.

These relationships and the basic definitions are being discussed in the following subchapter.

4.2 DEFINITIONS. PRECEDENCE RELATIONSHIPS.

The mathematical structure of a PDM network is a directed loopless graph. The set of nodes is denoted by **N**, where **N**=$\{1,2,...i...j...n\}$, the set of arrows is denoted by **A**. In PDM the nodes stand for the activities and arrows stand for the precedence relationships between activities. An arrow between two nodes (activities) means that the activity at the head of the arrow logically depends on the activity at the tail of the arrow. Any number of arrows can be between two activities. We denote the precedence relationships by the $\{(ij)k\}$ formula where k denotes the kth precedence relationship between i and j. The difference between the PDM and CPM structure is obvious. We do not have events in this structure, and the nodes serve as activities, whereas in CPM arrows stand for activities.

Activities can be defined in the following way:

Definition 4-1 *Activities are individual processes that can be well separated in space and time, and have to carry out continuously with the same intensity.*[i] *The duration of the activity i is denoted by* τ_i.

In the PDM network technique two types of precedence relationships can be used. The first is called the minimal type of precedence relationships, the second is called the maximal type of precedence relationships or negative or backward acting relationships. The maximal type of precedence relationships are rarely used and not very well known, but - no doubt - they have importance as they can help to produce networks which are closer to the reality. The maximal type of precedence relationships will be discussed in Chapter 5: Advanced Precedence Diagramming.

Minimal type of precedence relationships

Definition 4-2 $SSz_{(ij)k}$ *(Start-to-Start-*$z_{(ij)k}$*)*
An $SSz_{(ij)k}$ *precedence relationship means that activity j can start as* $z_{(ij)k}$ *lag*[ii] *time passed after the start of activity i.*

In Figure 4-1 the graphical portrayal of the precedence relationship and its interpretation by Linear Scheduling Method can be seen. The half-line, which

---

[i]A model when discontinuity is allowed is discussed in Chapter 4.7.

[ii]The lead and lag time can be any positive, zero or negative value. Originally the term of lead was used for the SS and SF relationships while the term of lag time was used for the FF and FS relationships. Today both terms are freely applied for any kind of precedence relationships. In the following, when we use the term lag or lead we usually think about positive values. In case of special considerations when difference has to be made between the positive and negative factors, we will note it accordingly.

originates from point O, shows the set of the possible starts for activity B. Three possible accomplishments are portrayed on the LSM chart, the earliest is denoted by a continuous line, the later ones by dotted lines.

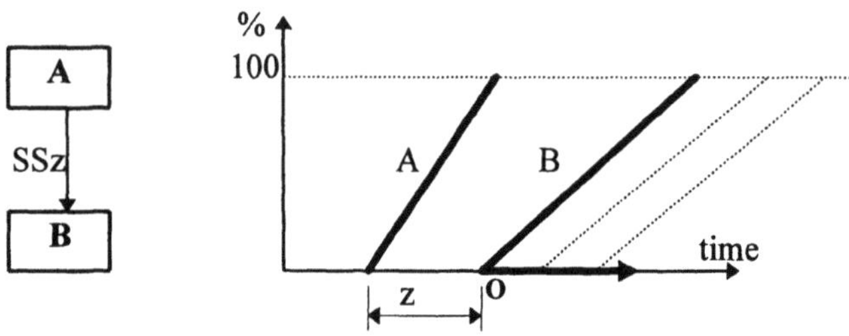

Figure 4-1 *LSM interpretation of Start-to-Start precedence relationship.*

Definition 4-3 *$FFz_{(ij)k}$ (Finish-to-Finish-$z_{(ij)k}$)*
An $FFz_{(ij)k}$ precedence relationship means that activity j can finish as $z_{(ij)k}$ lag time passed after the finish of activity i.

In Figure 4-2 the graphical portrayal of the precedence relationship and its interpretation by Linear Scheduling Method can be seen. The half-line, which originates from point O, shows the set of the possible finishes of activity B. Three possible accomplishments are portrayed on the LSM chart, the earliest is denoted by a continuous line, the later ones by dotted lines.

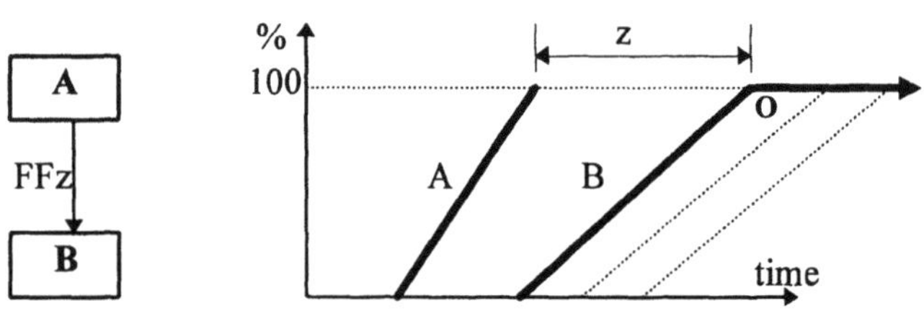

Figure 4-2 *LSM interpretation of Finish-to-Finish precedence relationship.*

Definition 4-4 *$FSz_{(ij)k}$ (Finish-to-Start-$z_{(ij)k}$)*
An $FSz_{(ij)k}$ precedence relationship means that activity j can start as $z_{(ij)k}$ lag time passed after the finish of activity i.

In Figure 4-3 the graphical portrayal of the precedence relationship and its interpretation by Linear Scheduling Method can be seen. The half-line, which originates from point O, shows the set of the possible finishes of activity B. Three possible accomplishments are portrayed on the LSM chart, the earliest is denoted by a continuous line, the later ones by dotted lines. *Notice that the CPM activity interdependency is nothing else but a FS0 relationship between the activities!*

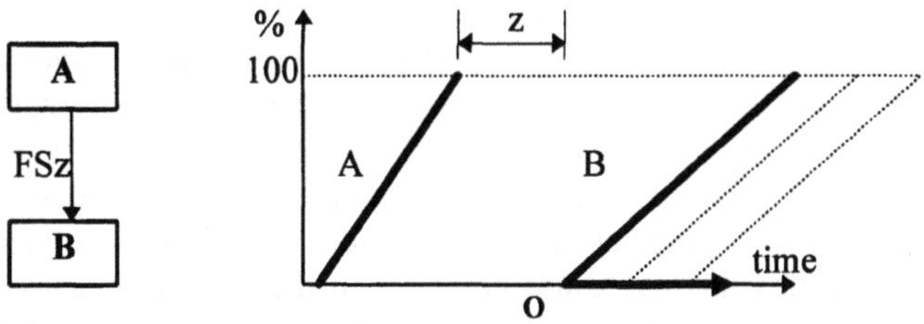

Figure 4-3 *LSM interpretation of Finish-to-Start precedence relationship.*

Definition 4-5 *$SFz_{(ij)k}$ (Start-to-Finish-$z_{(ij)k}$)*
An $SFz_{(ij)k}$ precedence relationship means that activity j can finish as $z_{(ij)k}$ lag time passed after the start of activity i.

In Figure 4-4 the graphical portrayal of the precedence relationship and its interpretation by Linear Scheduling Method can be seen. The half-line, which originates from point O, shows the set of the possible finishes of activity B. Three possible accomplishments are portrayed on the LSM chart, the earliest is denoted by a continuous line the later ones by dotted lines.

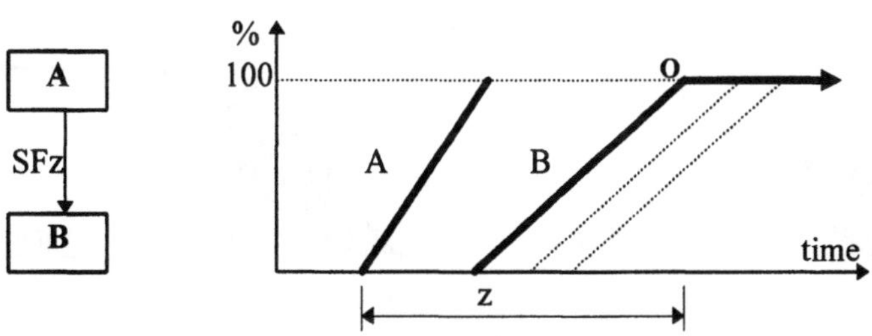

Figure 4-4 *LSM interpretation of Start-to-Finish precedence relationship.*

Definition 4-6 *$ZZz_{(ij)k}$ (A frequently used combination)*
A $ZZz_{(ij)k}$ precedence relationship means that there should be at least z or more lag times between activity j and activity i at any stages of i and j. This is nothing else but the application of SSz and FFz precedence relationships together.

In Figure 4-5 the graphical portrayal of the precedence relationship and its interpretation by Linear Scheduling Method can be seen. The half-line, which originates from point O, shows the set of the possible finishes of activity B. If $\tau_i > \tau_j$, then the application of ZZz gives the same result as the application of only the FFz precedence relationship. If $\tau_i < \tau_j$, the use of ZZz gives the same result as we get by using an SSz precedence relationship only. One can argue that we should simply examine the activity durations and then add SSz or FFz links to the schedule instead of using ZZ relationship. We disagree with this argument. Precedence relationships always have to reflect the real activity interdependencies and not some substitutions. If the computer program does not allow the use of the ZZ combination, then one have to add both SS and ZZ relationships to the network. If the software does not allow either the use of ZZ combinations or the use of more precedence relationships between two activities, then this can cause problems in the modification and the updating of the project. In this case, the scheduler who is aware of the limitations and abilities of the used software can apply SS or FF relationship. Afterwards he or she has to indicate in the schedule that the applied relationship serves only as a half solution.

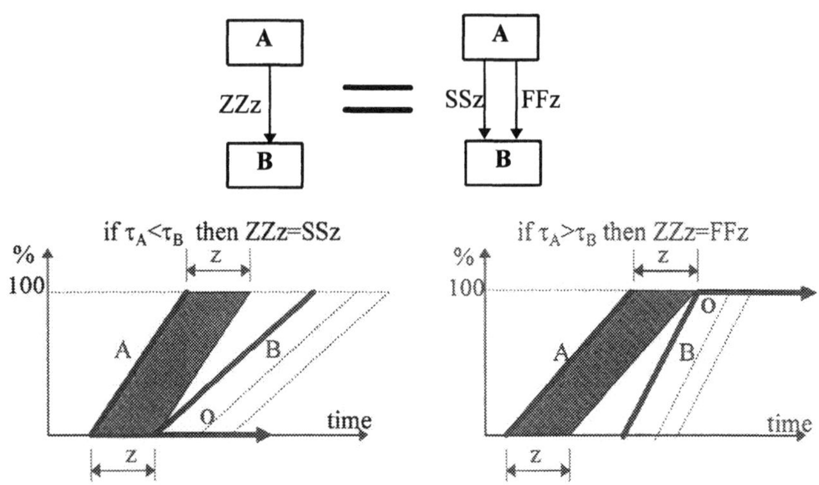

Figure 4-5 *LSM representation of ZZ combination.*

These precedence relationships are called as *minimal precedence relationships*, because the *z* values stand for the minimal allowable distance between the two distinguished points -- start or finish -- of the activities. This means that the lag time between these points can be more than *z*, but never less than it.

Now we can transform the verbal precedence relationships to PDM precedence relationships. We strongly recommend the reader to follow this process in order to understand this methodology. The preceding activity list can be seen in Table 4-1.

Activity	Duration in days	Preceding activities with precedence relationships
A	30	-
B	20	A FS0;
C	30	B FS0; E FS0;
D	10	A FS0;
E	14	A FS0;
F	30	E SS0;
G	20	D FS0; E SS3;
H	80	D FS0; E SS0;
I	90	D FS0; E SS0;
J	10	F FS0; E FS0;
K	30	J ZZ5; G ZZ3;
L	20	K ZZ10; H ZZ10;
M	5	L FS0; C FS0; K ZZ20;
N	20	M FS0; I SS2; I FF0;
O	15	N FF0; SS0

Table 4-1 *Input data for the sample project using PDM technique.*

In CPM special emphasis was put on producing an immediate preceding activity (IPA) list from a given preceding activity (PA) list. IPA lists were considered the best tool of comparing different versions of the same project. Let us see how the role of the preceding activity list changes in PDM.

4.3 PRECEDING ACTIVITY LIST

In the PDM network-based methodology one must very carefully reduce the preceding activity list in order to develop an IPA list. Due to the different available precedence relationships a wrong step can result in the loss of valuable information. To prove this, let us examine the following part of the preceding activity list (see Table 4-1).

L	20	K ZZ10; H ZZ10;
M	5	L FS0; C FS0; K ZZ20;

This part of the preceding activity list can be seen in Figure 4-6 in PDM and LSM chart format. Let us examine K, L and M only. Following the CPM logic, one could say that the ZZ20 precedence relationship between M and K is redundant, so it can be canceled. The results we have after the cancellation can be seen in Figure 4-7. This cancellation results in a schedule in which activity M can start immediately after L has finished, that is, during the placement of the machine, a part of the foundation will not be cured for 20 days.

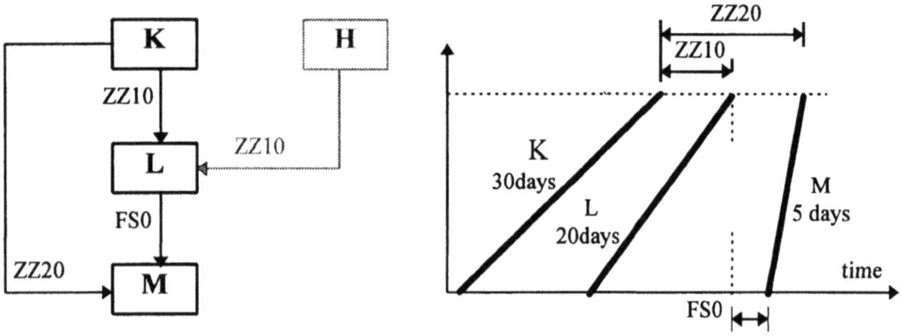

Figure 4-6 *PDM and LSM representation of a small part of the sample project*

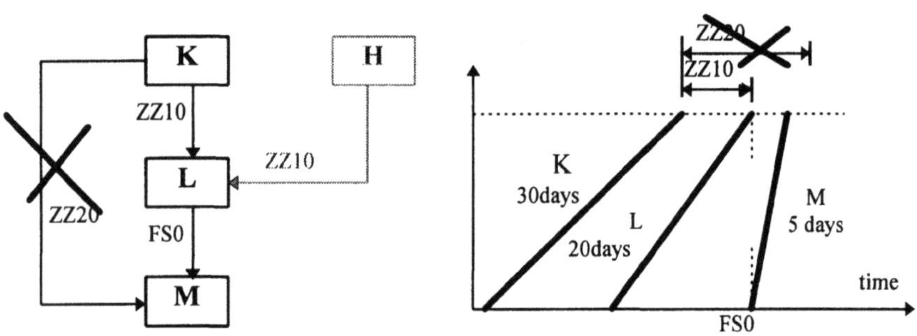

Figure 4-7 *Results after the cancellation of a precedence relationship.*

In PDM one does not deal with producing an IPA list because it makes no sense. It can be true that after deletion the results remain the same. It depends on the types and the lag times of the applied precedence relationships. *Despite this, do not erase them if they are real logical relationships! A network could be modified dozens of times during the project. Even a simple modification of an activity durations can make non-determinant relationships determinant!*

4.4 Drawing of a Network

Remember that this step was the most time consuming in CPM. In PDM one can draw a logically perfect network diagram from a logically good preceding activity list in the first try. In PDM, we do not have dummies, or the problem of multiple activities. We simply put the activities down as nodes and connect them with precedence relationships according to the preceding activity list. Of course, it can take a lot of time after this to draw a well-organized and easy-to-follow network, but this problem exists in CPM as well. The PDM network representation of the sample project can be seen in Figure 4-8.

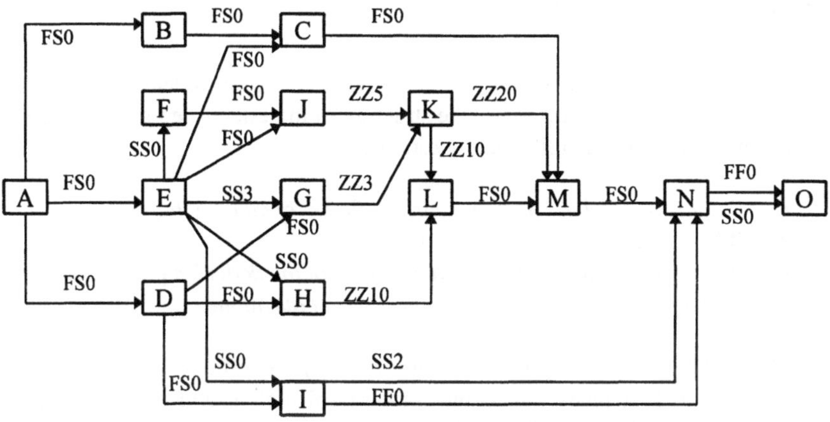

Figure 4-8 *PDM network representation of the sample project*

Another frequently used drawing format is where only the lag times are displayed on the arrows. In this format, the left side of a node stands for the start of the activity and the right side of the node stands for the finish of the activity.

An SS precedence relationship starts from the left side of the predecessor activity and ends at the left side of the activity in question. This drawing format can be followed in Figure 4-9.

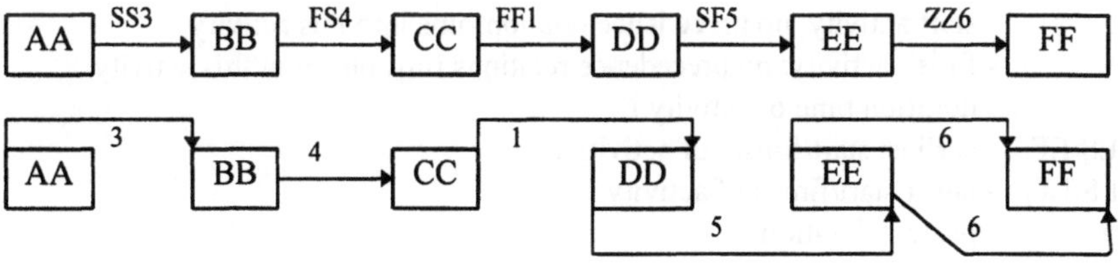

Figure 4-9 *Two different PDM drawing formats.*

So far, PDM seems to be more friendly than CPM. Due to its different precedence relationships, it is more suitable for describing activity interdependencies. In PDM we do not have the problem of transforming different activity lists into another lists and back. Last but not least, we do not have the time consuming complication of network drawing.

The next step in the network-based project methodology is the time analysis. This algorithm will be a bit more difficult than it was in CPM network.

4.5 PDM CALCULATION

If we know the duration of the activities and the precedence relationships between the activities, what is more, the structure of the established graph is correct, we can get answers for the following questions:

- *What is the earliest start and earliest finish of an activity?*
- *What is the project total duration?*
- *What are the latest start and latest finish of the activities if we want to accomplish the project within the project duration?*

The calculation that gives answers to these questions is a two phase calculation. The first phase called the forward calculation and gives the earliest starts and finishes of the activities and the project duration. The second phase called the backward calculation gives the latest starts and finishes of the activities.

During the calculation, we assume that there is only one initial and one terminal node in the network, and Definition 4-1is valid, that is, activities must carry out continuously, with the same intensity.

During the algorithm, we use the following notations:

n - number of activities in the network
i - notation of an activity, i=1..n
s - start activity; no precedence relation runs into this activity
t - finish activity; no precedence relations runs out from this activity
τ_i - duration time of activity i
ES/EF_i - earliest start/finish of activity i
LS/LF_i - latest start/finish of activity i
p - project duration

The algorithm is basically the same as the CPM algorithm. It is based on the observation that the early start and early finish of an activity can be determined if the ES and EF values are known for all the predecessor activities. In the beginning, it is known for only the initial activity, for which $ES_s=0$, $EF_s=\tau_i$. The late start and the late finish of an activity can be determined if these values are known for all the succeeding activities. Knowing these, the following rules govern the calculation of the early and late activity times.

PDM Calculation: No Splitting Allowed

Forward Calculation

Step1. Let $ES_s=0$; $EF_s=ES_s+\tau_s$; $ES_i= -\infty$ = INITIAL TIME for all the remaining activities[i], f:=0

Step 2.

REPEAT

f:=f+1

Take an activity j , which is not calculated yet, but ES and EF are known for all its predecessors.

IF there is no such an activity, then *GO TO* step 3.

$$ES_j=\max\{\text{INITIAL TIME};\; ES_i+SSz_{(ij)k};\; EF_i+FSz_{(ij)k};\; ES_i+SFz_{(ij)k}-\tau_j;\; EF_i+FFz_{(ij)k}-\tau_j\} \quad \forall\ \{(ij)k\}\in \mathbf{A}$$ [ii]

$EF_j=ES_j+\tau_j$

UNTIL $f=n$

Step 3

IF $f<n$ *THEN* STOP! (There is a loop in the network)

ELSE $p=EF_t$ [iii]

The backward pass is completed by working from the terminal activity to the initial activity in the reverse direction of arrows. The steps of the backward pass rules can be summarized as follows:

---

[i] The INITIAL TIME is not used in the algorithm, as it has an infinitely small value, thus it is practically redundant. Some authors argue then some finite value has to be used as an initial value. See remarks on the initial values after the algorithm.

[ii] The symbol should read : 'for all the relationships which are elements of the set of the arrows and finish in j,' that is, for all the predecessor relationships of j

[iii] See remarks on project duration.

Backward Calculations

Step1. Let $LF_t=p$; $LS_t=LS_t-\tau_t$; $Lf_i= \infty$ =TERMINAL TIME[i] $\forall\ i \in \mathbf{N}\text{-}t$

Step 2.

REPEAT

f:=f-1

Take an activity *i*, which is not calculated yet, but LS and LF are known for all its successors.

IF there is no such an activity, THEN GO TO Step 3.

$$LF_i=\min\{\text{TERMINAL TIME}; LF_j-FFz_{(ij)k}, LS_j-FSz_{(ij)k}; LF_j-SFz_{(ij)k}+\tau_i; LS_j-SSz_{(ij)k}+\tau_i\} \quad \forall\ \{(ij)k\} \in \mathbf{A}$$

$LS_i=LF_i-\tau_i$

UNTIL f=0

Step 3. STOP! The calculation has finished.

Remarks to the algorithm:

- *During both the forward and the backward calculation process we can get less than zero start times and greater than* EF_t *finish times for some activities. This means that these activities may start before the project start date or they may finish after the project ends. To avoid this, some authors argue that in the forward calculation process the* $ES_i=0$ *initial value has to be used for each activity instead of the applied* $ES_i= -\infty$*, and in the backward calculation process, the* $LF_i=EF_t$ *or* $LF_i=max\{EF_i\}$ *terminal values have to be used. This is, however, a wrong approach for two reasons, which will be discussed in the next subchapter.*
- *The backward pass can be made faster, if the order in which the activities have been taken into account during the forward pass is being kept. The backward pass can follow the reverse order.*

To illustrate the algorithm, the solution of the sample project will be presented here. The initial value for the start activity is $ES_s=0$, for the rest of the activities, $ESi=-\infty$. The results can be seen in Figure 4-10.

---

[i] As the TERMINAL TIME is being set to infinite large, it is redundant in this algorithm. Some authors argue that to prevent an activity finish time being greater than the project duration, a finite TERMINAL TIME is has to be used, which usually is set to the equal to the project duration. See remarks to the algorithm

Forward pass calculations: (ES_i, EF_i values)

$ES_A=0$	$EF_A=30;$
$ES_B=\max\{EF_A+0\}=30$	$EF_B=50;$
$ES_D=\max\{EF_A+0\}=30$	$EF_D=40;$
$ES_E=\max\{EF_A+0\}=30$	$EF_E=44;$
$ES_F=\max\{ES_E+0\}=30$	$EF_F=60;$
$ES_C=\max\{EF_E+0;\ EF_B+0\}=\max\{44,50\}=50$	$EF_C=80;$
$ES_J=\max\{EF_E+0;\ ES_F+0\}=\max\{44,60\}=60$	$EF_J=70;$
$ES_G=\max\{ES_E+3;\ EF_D+0\}=\max\{33,40\}=40$	$EF_G=60;$
$ES_H=\max\{ES_E+0;\ EF_D+0\}=\max\{30,40\}=40$	$EF_H=120;$
$ES_I=\max\{ES_E+0;\ EF_D+0\}=\max\{30,40\}=40$	$EF_I=130;$
$ES_K=\max\{ES_J+5;EF_J+5-\tau_K;ES_G+3;EF_G+3-\tau_K\}=65$	$EF_K=95;$
$ES_L=\max\{ES_K+10;EF_K+10-\tau_L;ES_H+10;\ EF_H+10-\tau_L\}=110$	$EF_L=130;$
$ES_M=\max\{EF_C+0;EF_L+0;ES_K+20;EF_K+20-\tau_M\}=130$	$EF_M=135;$
$ES_N=\max\{EF_M+0;\ ES_I+2;\ EF_I+0-\tau_N\}=\max\{135,42,110\}=135$	$EF_N=155;$
$ES_O=\max\{EF_N+0-\tau_O;\ ES_N+0)=140;$	$EF_O=155;$

The project duration is equal to the early finish time of the finishing activity, that is, $p=EF_{t=O}=155$ days.

The backward calculation starts with $LF_O=p=155$ days. For the rest of the activities, $LF_i=\infty$.

Backward pass calculations: (LS_i, LF_i values)

$LF_O=EF_O=155;$	$LS_O=140;$
$LF_N=\min\{LF_O-0\}=155$	$LS_N=135;$
$LF_M=\min\{LS_N-0\}=135$	$LS_M=130;$
$LF_L=\min\{LS_M-0\}=130$	$LS_L=110;$
$LF_K=\min\{LF_L-10;\ LS_L-10+\tau_H;\ LF_M-20;\ LS_M-20+\tau_H\}=115$	$LS_K=85;$
$LF_I=\min\{LS_N-2+\tau_I;\ LF_N-0\}=\min\{223,155\}=155$	$LS_I=65;$
$LF_H=\min\{LF_L-10;\ LS_L-10+\tau_H\}=\min\{120,180\}=120$	$LS_H=40;$
$LF_G=\min\{LF_K-3;\ LS_K-3+\tau_G\}=\min\{112,102\}=102$	$LS_G=82;$
$LF_J=\min\{LF_K-5;\ LS_K-5+\tau_J\}=\min\{110,90\}=90$	$LS_J=80;$
$LF_C=\min\{LS_M-0\}=130$	$LS_C=100;$
$LF_F=\min\{LS_J-0\}=80$	$LS_F=50;$
$LF_E=\min\{LS_C-0;LS_F-0+\tau_E;LS_J-0;LS_G-3+\tau_E;LS_H-0+\tau_E;\ LS_I-0+\tau_E;\}=$ $=\min\{100,74,80,93,54,79\}=54$	$LS_E=40;$
$LF_D=\min\{LS_G-0;\ LS_H-0;\ LS_I-0;\}=\min\{82,40,65\}=40$	$LS_D=30;$
$LF_B=\min\{LS_C-0\}=100$	$LS_B=80;$
$LF_A=\min\{LS_B-0;\ LS_E-0;\ LS_D-0\}=\min\{80,40,30\}=30$	$LS_A=0;$

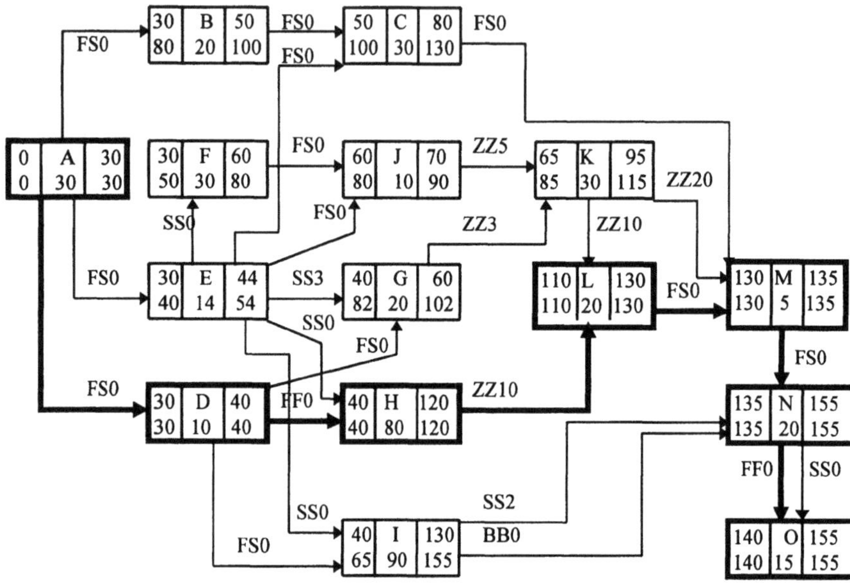

Figure 4-10 *PDM calculation of the sample project*

4.6 NOTES ON CALCULATION RULES.

Lots of fine texts are available on network techniques, and lots of them give a comprehensive treatment to PDM. A great deal of them use the INITIAL and the TERMINAL TIME in the algorithm. The introduction of these initial values are not suggested in the pure time analysis stage. Their use can be questioned, therefore it is advisable to replace them with simple precedence relationships.

For better understanding what behind this argument is we give the definitions of the dangling activities.

Definition 4-7 *(Dangling start activity)*
An activity without a relationship designating to the start point of the activity is called dangling start activity. That means that no FS or SS relationships go to the activity.

Definition 4-8 *(Dangling finish activity)*
An activity without a relationship originating from the finish point of the activity is called dangling finish activity. That means that no FS or FF relationsip goes out from the activity.

The calculation usually starts using the assumption that the early start time for the initial activity has zero value. We have also used the assumption that the early finish of the terminal activity gives the project duration, and this value will be used as the late finish time for the terminal activity. Using these basic assumptions, the rules that govern the calculation process can result in activities starting before the project begins and finishing after the project duration. This can be seen in Figure 4-11.a). Some authors say that the start of C is 'an erroneous negative value,' and 'to prevent the occurrence of this error,' an additional time, called the INITIAL TIME is used as an initial value of the early start dates, which is usually set equal to zero. To prevent the occurrence of the finish times being greater then the project duration the so called TERMINAL TIME is used as the late finish time of the activities. The TERMINAL TIME is usually set equal to the project duration, that is, equal to EF_t or it is set to $\max\{EF_i\}$ i=1..n. The introduction of INITIAL TIME and TERMINAL TIME in this way is a result of a wrong theoretical approach for two reasons.

First, dangling activities[i] can start before the project starts because they really do not have any connection to the project start, and dandling activities can finish after EF_t because they really do not have any connection to the project delivery. After the calculation that has been accomplished with infinite initial and terminal values one must consider if an activity can really start before the project starts. If not, then this is a logical relationship, i.e., a SS0 relationship has to be used to avoid negative start times. If an activity cannot finish after the terminal activity's finish, that is, beyond the project duration, then a FF0 constraint has to be used between the terminal activity and the activity in question. In short, dangling activities have to be made non-dangling activities.

Second, by using the ES_i=0 and LF_i=EF_t initial and terminal values, one can get meaningless results *without a critical path* and *with late start values that are smaller than the early start* values. To prove this, let us consider the following PDM network (Figure 4-11).

In Figure 4-11a.) we have not used finite initial and terminal values. The project duration is 19 days, activity B finishes after the project completion, activity C starts before the project, i.e., activity A starts. To prevent this situation from occuring, the ES_i=0 INITIAL TIME is suggested to be used.

---

[i]The activities with activity times before or beyond the project start or finish in case of positive lag times are always dangling activities . We deal with dangling activities in detail in Chapter 4: Notes on Negative Lag and Dangling Activities.

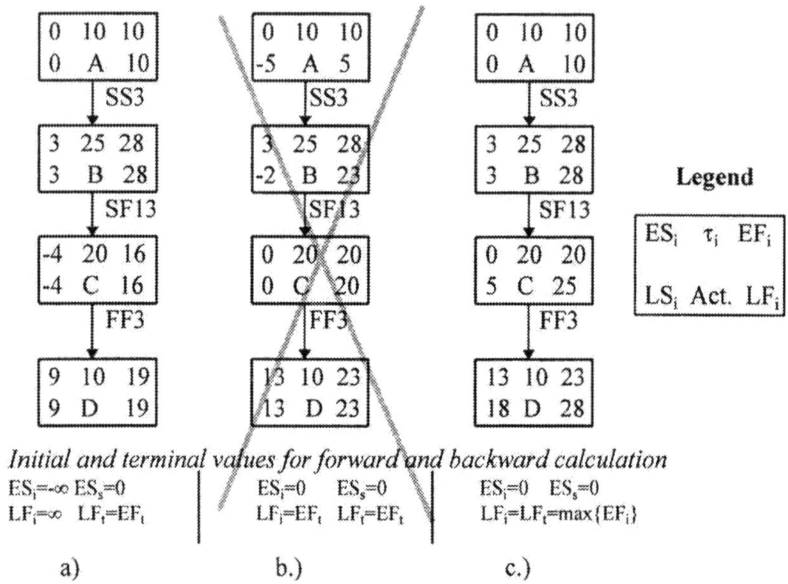

Figure 4-11 *A small example to prove the remarks.*

After the forward pass, we get the project duration which is p=23 days. Now, we set the TERMINAL TIME equal to this value, that is, LF_i=TERMINAL TIME=23. The result of the calculation can be seen in Figure 4-11b.). The project duration has changed and there is no critical path from s to t. The reader at this point does not know much about the floats and the critical path in PDM, but obviously there must be something wrong, if the latest possible start of A is smaller than its earliest start, and the latest possible start of B is smaller than its earliest start. The situation is also strange if we use the max{EFi} value as a TERMINAL TIME. The results of this calculation can be seen in Figure 4-11c). It is hard to believe that the terminal activity, which is usually the project delivery, is not critical.

What is the reason for these erroneous results? We simply tried to impose constraints on the network, which are opposite to the network logic. The INITIAL TIME is also known as 'Must start on or later' constraint. The TERMINAL TIME is also known as 'Must finish on or before' constraint.

Constraints in PDM are the subject of Chapter 5: Advanced Precedence Diagramming. Let us just say here that constraints are effective but dangerous tools. Since they are independent from the network logic, they can easily confront with it, which can result in suspicious activity start and finish times.

What is the solution if one wants to set all the activity starts after the project's, that is, the first activity's start? What is the solution if one wants to finish every activity before the project, that is, the last activity finishes?

The answer is simple. Use SS0 relationship between the start activity and the activity in question, and FF0 relationship between the activity and the finishing activity, to make dangling activities non-dangling, if these relationships logically exist. If they do not exist, do not use them, but in this case some activities can start before the project starts and some activities can finish after the project finishes because their start/finish have no connection to the project.

The correct solution to the problem can be seen in Figure 4-12. A SS0 precedence relationship has been added from activity A to activity C. To prevent the occurrence of an activity time going beyond the project duration, a FF0 precedence relationship has been added between activity B and the terminal activity, that is, activity D. Now the network logic is perfect. It includes the logical relationships that C has to start after A starts, and B cannot finish before the project duration, i.e., the project can finish after B has finished. The calculations can be done without the INITIAL TIME and TERMINAL TIME values. The project duration is 28 days, and all the activities are within the project time limits.

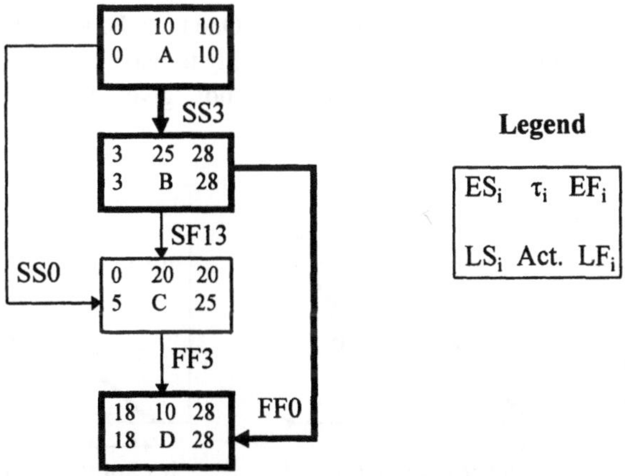

Figure 4-12 *The correct way of avoiding less than 0 or greater than p activity times*

Again, logic problems in scheduling should be solved by using precedence relationships if possible, and not by using non-network tools which can collide with the network logic.

4.7 PDM CALCULATION: SPLITTING ALLOWED

In construction some activities have to be carried on continuously, but some activities can be interrupted. Some schedulers say that this fact has to be reflected in the model itself. They say, if activities are interruptable then they have to be treated as such and not as though they were non-interruptable activities. In this section, we relax the definition of activities and we allow the activities to be interrupted (see Definition 4-1). This assumption may cause some problems in modeling a real situation, and also some complications in the algorithm.

What are these problems?

First, in PDM we can make logical relationships between activities through the application of precedence relationships, despite that precedence relationships make logical contact between only the extreme (start, finish) points of activities and not between all stages of the processes. As activities are assumed to be non-interruptable, by using these relationships one can make logical relations between whole activities. For example, if one wants to make two activities partially concurrent, then he or she should apply a ZZ relationship. As it can be seen in Figure 4-13, between excavation (A) and pipe laying down (B), there has to be 100m safety distance at every stage of these tasks, which is , say, z days lag time. If the activities are assumed to be non-interruptable, then the ZZz (=SSz+FFz) relationship will ensure this safety distance at each stage of the project. *If interruption is allowed, then pipe laying down (in part) can go before excavation, while the SSz and FFz relationships are kept.* One must be very much aware of the s fact, that if interruption is allowed, the possibility to make contact between all stages of activities is lost. The network will not give details about the exact accomplishment of these activities.

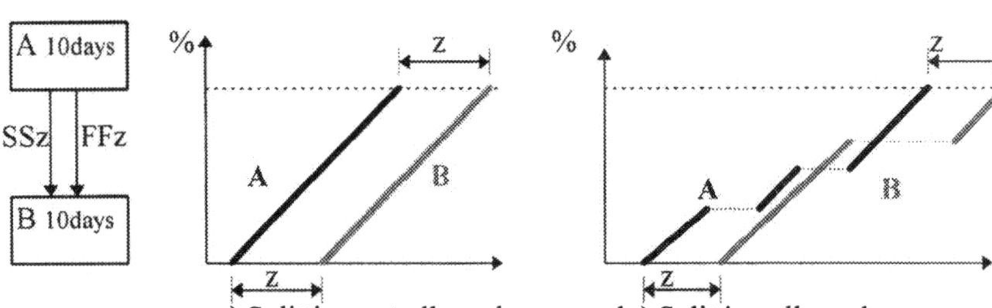

Figure 4-13 *Partially concurrent activities in PDM and LSM interpretation*

The second problem is the problem of dangling activities, which manifests itself more significantly. To prove this, let us see Figure 4-14 where all the activities are assumed to be interruptable. Activity B is a start dangling activity which means that no precedence relationship leads to its start point. In this case, its early start is at a minus infinite point in time. What is worse is that in this case activity C will early start and early finish in minus infinite time as well. One can argue that it is high time to apply some finite INITIAL TIME and TERMINAL TIME in the algorithm. No, it is not. If we get wrong results, it is the results of a wrong network logic, which has to be corrected. PDM gives a result based on our logic, nothing else.

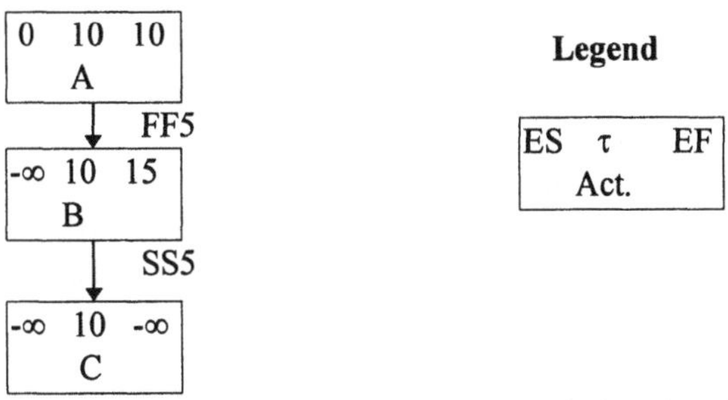

Figure 4-14 *Dangling activities when splitting is allowed: Forward pass*

The third problem is concerns the algorithm. In the non-splitting case we used the $ES_i+\tau_i=EF_i$ and $LS_i+\tau_i=LF_i$ formulas. Now these formulas will change as $ES_i+\tau_i\leq EF_i$ and $LS_i+\tau_i\leq LF_i$. These changes will cause some inevitable modifications in the algorithm.

Sometimes splitting is constrained by time. For example, we have a machine used on an activity, which has a 1 week duration and is splittable, but the machine will be taken away to another project after 4 weeks. In this case, the amount of possible splitting is 3 weeks.

We use the following new notation in the algorithm.

d_i - The time limit within which the activity has to be carried out minus the activity duration.

The algorithm is, in fact, the modified PDM algorithm discussed in the previous section. In the first phase, we define the early start and early finish times. This phase is known as the forward pass computations phase. The project duration will be the early finish time of the terminal activity. In the second phase the late start and late finish times will be defined. This phase is called the backward pass calculation phase. During the calculations, some of the activities are splittable and some are non-splittable.

Knowing these, the following steps govern the calculations:

PDM Calculations: Splitting Allowed

Forward Pass Computations

Step1. Let $ES_s=0$; $EF_s=ES_s+\tau_s$ f:=0

Step 2.

REPEAT

f:=f+1

Take an activity *j* , which is not calculated yet, but ES and EF are known for all its predecessors. (In the first iteration, this is act. *s*)

IF there is no such an activity, then *GO TO* step 3.

IF neither SS nor FS relationships run into *j AND j* is splittable

THEN $ES_j=-\infty$ (*this is the place for putting* INITIAL TIME)

ELSE $ES_j=\max\{ES_i+SSz_{ijk}, EF_i+FSz_{ijk}\}$ $\forall\{(ij)k\}\in\mathbf{A}$ and SS or FS.

$EF_j=\max\{ES_j+\tau_j, ES_i+SFz_{ijk}, EF_i+FFz_{ijk}\}$ $\forall\{(ij)k\}\in\mathbf{A}$ and SF or FF

IF j is non-splittable, then $ES_j=EF_j-\tau_j$

IF j is splittable, but splitting is limited, then $ES_j:=\max\{ES_j; EF_j-\tau_j-d_j\}$

UNTIL f=k

Step 3

IF f<k THEN

STOP! (There is a loop in the network)

ELSE p=EF_t

Backward Pass Computations

Step1. Let $LF_t=p$; $LS_t=LS_t-\tau_t$ *f=k*

Step 2.

REPEAT

f=f-1

Take an activity *i*, which is not calculated yet, but LS and LF are known for all its successors. (In the first iteration, this is act. *t*)

IF neither FF nor FS relations run out from *i AND i* is splittable

THEN $LF_i=\infty$ (*the place for putting* TERMINAL TIME)

ELSE $LF_i=\min\{LF_j-FFz_{ijk}, LS_j-FSz_{ijk}\}$ for all FF, FS immediate succeed *i*.

$LS_i=\min\{LF_i-\tau_i, LF_j-SFz_{ijk}, LS_j-SSz_{ijk}\}$ for all SF, SS immediate succeed *j*.

IF j is non-splittable, *THEN* $LF_i=LS_i+\tau_i$

IF j is splittable, but splitting is limited, *THEN* $LF_i:=\min\{LF_i; LS_i+\tau_i+d_i\}$

UNTIL f=0

Step 3. STOP! The calculation is completed.

As the reader can see, we included the possibility of using INITIAL and TERMINAL values, but one should be aware that these values can collide with the network logic, that is, with the schedule of dangling activities. If these dangling activities do not go before or beyond the project start or finish, or there are no dangling activities and negative lag times in the network, then the use of INITIAL/TERMINAL TIMES does not affect the schedule.

4.8 LOOPS IN PDM NETWORK

In CPM, the definition of the loop meant a circle along which one could go back to a node that has been left already. Two loops (A-D-B, and B-C-D) can be seen in the following CPM network (see Figure 4-15).

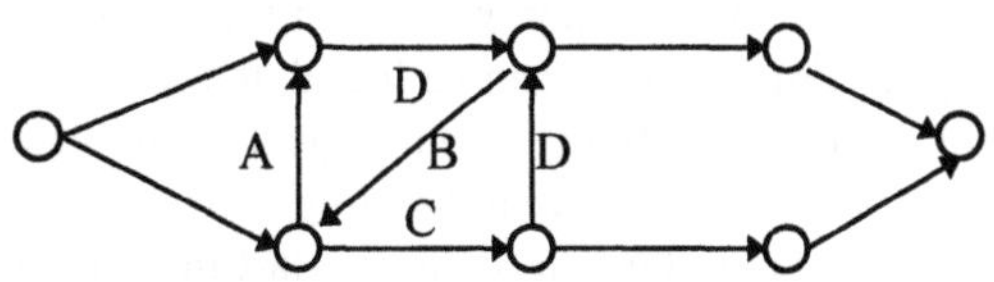

Figure 4-15 *Loops in CPM network*

In PDM we give the same definitions for loops. If there is a circle along which one can go back to a node (activity) one has left before, then this activity is in loop, which is a logical mistake. According to this definition, a loop can be seen in PDM network in Figure 4-16a. There is an another definition of loops in PDM, according to which a loop is a circle along which we can go back to the same point (start, finish) of the activity that has been left already. According to this definition, B C and D do not form a loop as we cannot go from the finish of C to the start of C. In this text, we follow the first definition, i.e., B, C and D form a loop. During the analysis of this network, the computation will stop in the first iteration of Step 2, as it will not find an activity for which all the predecessors are calculated.

In Chapter 5, we give an algorithm which can handle loops, but again, we consider activities B, D, C to be a loop.

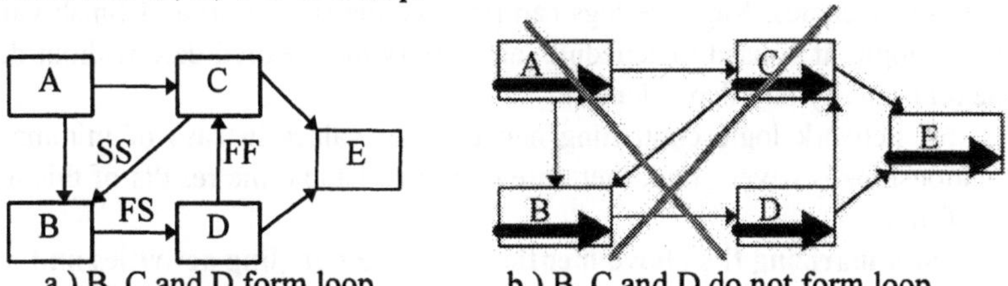

a.) B, C and D form loop b.) B, C and D do not form loop

Figure 4-16 *Loops in PDM network*

4.9 NOTES ON NEGATIVE LAG AND DANGLING ACTIVITIES

First, our notes on dangling activities. Dangling activities usually emerge as a result of a deficient network logic. First, this logic should be corrected. The best we can do is if during the process of creating the network logic, we answer the following two questions:

What activities precede the ***start*** *of the activity in question, and how?*
What activities succeed the ***finish*** *of the activity in question, and how?*

One should prepare a preceding activity list by answering the first question for all the activities. This can help to avoid the appearance of *dangling start activities*. It also helps to avoid having more start activities in the network.

By answering the second question, the formation of *dangling finish activities* can be avoided. This also helps to create only one finish node.

Summarizing the above mentioned, in a proper network usually only one dangling start activity and one dangling finish activity can be found, the start and the finish activity of the project. If the start of an activity has no logical relationship to any other activities, then this activity is allowed to be a dangling start activity. In this case it can start anytime, if it is splittable. Forcing its early start to come after the project start with initial time (which is a 'must start on or later' constraint) is usually possible, but sometimes brings about wrong results, as it can be seen in Figure 4-11.

Our notes on negative lag times. The calculation process can handle any positive, negative or zero lag times. Some schedulers like to use negative lags to make activities concurrent. They say, 'Activity B can start 3 days before activity A finishes, so the right relationship is FS-3.' However, this logic is false. Imagine that activity B has started before the scheduled finish of activity A, but due to some problems, activity A has stopped for a week. According to the network logic, we have a situation which is impossible. The lag time between the finish of A and the start of B is less than the prescribed value. This means that the network logic was false. The correct relationship would be that 'activity B can start after x days of the start of activity A,' or 'activity B can start after y percentage of A has been completed,' etc. This is the corrrect logic to express these kinds of activity interdependencies. Negative lags can produce the same start and finish values with wrong logic. It is used to schedule an activity to a desired date without describing the real activity interdependencies.

A network logic containing negative lag values in case of minimal type of relationships is never good, therefore one must handle the results of this logic very carefully.

Summarizing the above mentioned, avoid dangling activities and never use negative lag in minimal type of precedence relationships.

4.10 THE PARADOX BEHAVIOR OF PDM

A well known fact among production managers and engineers organizing assembly lines is that the production is faster if processes are synchronized to each other. This is true in construction as well. To prove this, let us examine the following very simple project.

A sewer has to be built in 1 km length. Activities have to carry out continuously, that is, without interruption. The project can be divided into three activities: A - excavation, 40 days; B - pipe laying down, 40 days; C - backfilling, 40 days. The minimal safety distance between the activities is 4 days. The project schedule can be followed in Figure 4-17.

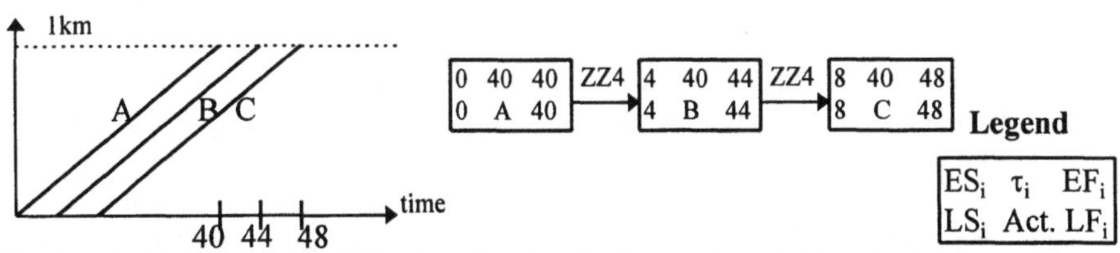

Figure 4-17 *LSM and PDM schedules: Perfectly synchronized activities.*

The project manager in charge would like to speed up the project, and thus hires a high capacity machine for automating the pipe laying down. By doing this, the duration of activity B can be shortened to 30 days. The project manager expects that the project duration will be shorter by 10 days. However, during the project this expectation will prove to be wrong. The project duration will increase instead of decreasing. The reason behind this phenomenon is simple. As activity B has to carry out continuously and there must be at least a 4 days' worth of safety distance between A and B, activity B can only start on the 14th day of the project. Hence the 4 days distance will only be ensured at the end of the ditch. In this way activity C is forced to start four days after the start of B, that is, on the 18th day. The results can be seen in Figure 4-18.

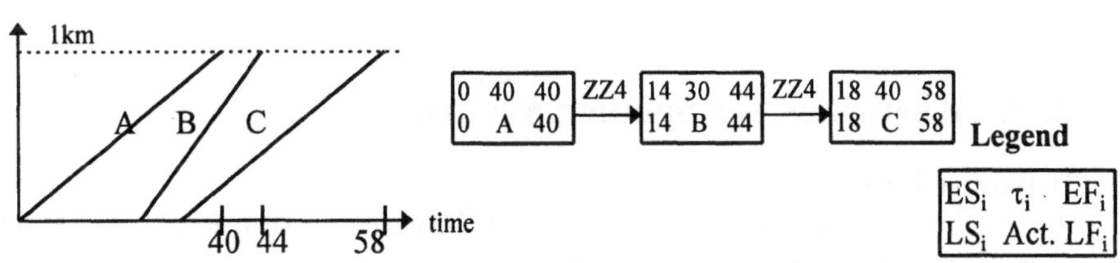

Figure 4-18 *LSM and PDM schedules: Activities are not synchronized.*

This is a really interesting phenomenon. In case of one critical path, the shortening of a critical activity in CPM always results in a decreased project duration. In PDM, the situation is completely different. Changing an activity duration can result in the change of the project duration in the same direction or the opposite direction. What is more either decreasing or increasing an activity duration can have the same effect on the project duration. What makes the situation very difficult is that an activity can change its behavior during the modification of the project. For example, when examining a network, we can come to the conclusion that shortening the duration of an activity in question will increase the project duration. If we make a minor change in the network, for example, decrease by one day a lag of a remote link, then the same change in the same activity may this time result in a decreasing project duration. During the preparation of the network, hundreds of changes may be made by the schedulers. It is almost impossible to follow the effect of these changes on the network, but it would nevertheless be of great value. It could show to the schedulers which activities have to be shortened or lengthened in order to adjust the schedule within the deadlines.

Generally, we can say that only non-splittable activities, or splittable activities if d_i is limited can behave in such an anomalous way.

In Chapter 7: PDM Least Cost Scheduling there is a method which is able to adjust automatically the schedule to the desired project duration at the least possible cost.

It seems that the criticality of an activity and of a path are not equivalent terms in CPM and PDM. This and the information obtained from the calculations are the topics of the following subchapter.

4.11 INFORMATION OBTAINED FROM CALCULATIONS

Below we give definitions for critical path, critical activities, and floats. Thanks to the possibilities of the PDM one can get more information out of it on critical activities and paths than out of the CPM. Unfortunately, splittable activities cause problems in the conventional definitions, therefore we have to discuss them separately.

Critical activities: No Splitting Allowed

Definition 4-9 *(Critical activity)*
An activity is considered to be critical if the difference between its late finish and early start time is equal to its activity time.

In PDM, the critical activities can be characterized by the following definitions:

Definition 4-10 *(Normal critical activity)*
An activity is considered to be normal critical if it is critical and the project duration shifts in the same direction as the shift in the duration of the activity in question.

Definition 4-11 *(Reverse critical activity)*
An activity is considered to be reverse critical if it is critical and the project duration shifts in the reverse direction as the shift in the duration of the activity in question.

Definition 4-12 *(Bi-critical activity)*
An activity is considered to be bi-critial if it is critical and the project duration ***increases*** *as the activity duration shifts in any direction.*

Definition 4-13 *(Start critical activity)*
An activity is considered to be start critical if it is critical and the the project duration shifts in the same direction as the shift of the start time, but any shift in the activity duration keeps the project duration the same.

Definition 4-14 *(Finish critical activity)*
An activity is considered to be finish critical if it is critical and the project duration shifts in the same direction as the shift of the finish time, but no shift in the activity duration has any effect on the project duration.

Activity Floats: No-Splitting Allowed

Definition 4-15 *(Activity total float)*
The total float of an activity denotes the amount of time by which the start of the activity in question can exceed its earliest start without affecting the latest times for the succeeding activities, that is without affecting the project duration. This value is given by the following formula:

$$TF_i = LF_i - ES_i - \tau_i \quad \text{or} \quad TF_i = LS_i - ES_i \quad \text{or} \quad TF_i = LF_i - EF_i\,.$$

Definition 4-16 *(Activity free float)*
The free float of an activity denotes the amount of time by which the start of the of the activity in question can exceed its earliest start without affecting the earliest time of the succeeding activities. This value is given by

$$FF_i = min\begin{pmatrix} ES_j - z_{ijk} - \tau_i & \forall\ FSz_{ijk} \\ ES_j - z_{ijk} & \forall\ SSz_{ijk} \\ EF_j - z_{ijk} - \tau_i & \forall\ FFz_{ijk} \\ EF_j - z_{ijk} & \forall\ SFz_{ijk} \end{pmatrix} - ES_i$$

Definition 4-17 *(Activity independent float, IF)*
Independent float of an activity denotes the time by which the start time of the activity in question can exceed its start time defined by the latest times of the preceding activities without affecting the early times of the succeeding activities.

$$IF_i = min\begin{pmatrix} ES_j - z_{ijk} - \tau_i & \forall\ FSz_{ijk} \\ ES_j - z_{ijk} & \forall\ SSz_{ijk} \\ EF_j - z_{ijk} - \tau_i & \forall\ FFz_{ijk} \\ EF_j - z_{ijk} & \forall\ SFz_{ijk} \end{pmatrix} - max\begin{pmatrix} LF_l + z_{lik} & \forall\ FSz_{lik} \\ LS_l + z_{lik} & \forall\ SSz_{lik} \\ LF_l + z_{lik} - \tau_i & \forall\ FFz_{lik} \\ LS_l + z_{lik} - \tau_i & \forall\ SFz_{lik} \end{pmatrix}$$

Definition 4-18 *(Activity conditional float, CF)*
The conditional float of an activity denotes the time by which the start of an activity can exceed its start defined by the latest times of the preceding activities without affecting the late times of the succeeding activities.

$$CF_i = LS_i - max\begin{pmatrix} LF_l + z_{lik} & \forall\ FSz_{lik} \\ LS_l + z_{lik} & \forall\ SSz_{lik} \\ LF_l + z_{lik} - \tau_i & \forall\ FFz_{lik} \\ LS_l + z_{lik} - \tau_i & \forall\ SFz_{lik} \end{pmatrix}$$

Floats: Precedence Relationships

In some sense precedence relationships can be handled as non-splittable activities, therefore all the four floats can be defined for them, what is more, all of them can have positive value for certain precedence relationships. The value of these floats comes about the following way:

Total Float	*TF=late time at the head - early time at the tail-lag time*
Free Float	*FF=early time at the head - early time at the tail - lag time.*
Independent Float	*IF=early time at the head - late time at the tail - lag time.*
Conditional Float	*CF=late time at the head - late time at the tail - lag time.*

We did not give definitions for these floats, -- one can figure them out from the formulas -- because they are not used. In practice, the term of determinant precedence relationship is used.

Definition 4-19 *(Determinant precedence relationship)*
A precedence relationship is called determinant if its FF value is zero, that i,s the early time at the head activity minus the early time of the tail activity is equal to the lag time.

If a definitive activity connects two critical activities, then the relationship is critical.

Definition 4-20 *(Critical precedence relationship)*
A precedence relationship is called critical if its determinant and connects two critical activities. In this case, the precedence relationship total and free floats are equal to zero.

The critical precedence relationships are always normal critical, that is, any shift in its lag value results in the same shift in the project duration ,in case of one critical path. Now we can give the definition of the critical path.

Definition 4-21 *(Critical Path)*
The critical path is the longest path in the network from s to t along which all the activities and precedence relationships are critical.

A small example can be seen in Figure 4-19 to illustrate the definitions. There are no dangling activities in the network, except the start which is always dangling start and except the finish, which is always dangling finish. There are no loops in the network and activity splitting is not allowed, based on the original PDM activity definition.

Activities A, C, D, G and the precedence relationships that connect them form a critical path, therefore all their floats are zero. The total float comes in a very simple way, it is the difference of EF and ES. It is displayed between the late finish and late start values. If we want to characterize the critical path, then activity A is start critical, activity C is normal critical, activity D is reverse critical and activity E is finish critical. Non-critical but determinant precedence relations are displayed in gray.

It is easy to determine the above labels. A critical activity is start critical if the preceding critical precedence relationships go into the start point , that is, FS or SS relationships, and the succeeding critical relationships leave from the start of the activity, that is, SS or SF relations. An activity is finish critical if the critical relationships are connected to the finish of the activity in question. An activity is normal critical if the critical predecessors go to the start and the critical successors leave from the end of the activity. An activity is reverse critical if the critical relationships go to the finish and leave from the start of the critical activity. An activity is bi-critical if it is normal and reverse critical at the same time.

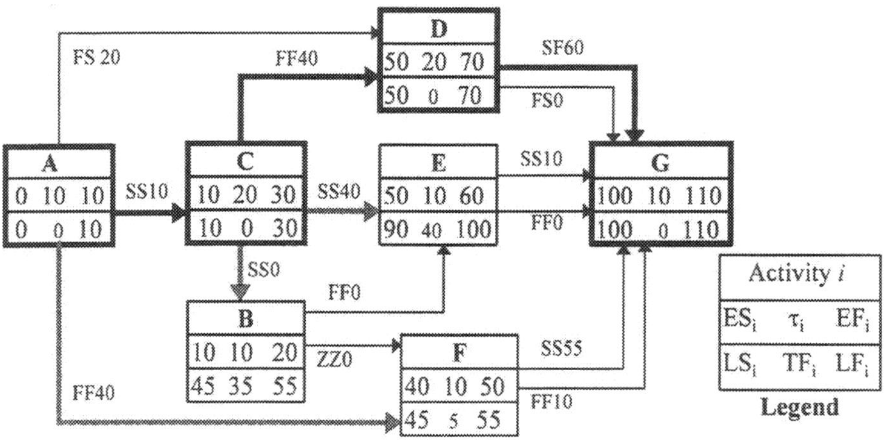

Figure 4-19 *Results and interpretation of calculations, no splitting allowed*

Critical Activities: Splitting Allowed

When activity splitting is allowed, the results of the computations can differ from the results when splitting is not allowed; what is more, the project durations are usually not the same. The solution of the network in Figure 4-19, with the splitting allowed assumption, can be seen in Figure 4-20. The results are completely different. The interpretation of the results are different too. For example, the difference between the late finish and the early start of activity D is 50 days. As its duration is only 20 days, following the logic when splitting is not allowed, we could say that we can delay the start of the activity by 20 days, without affecting the project duration. But this is not true. The latest possible start of activity D is the same as its early start, so we cannot delay it by even a single day without affecting the project duration. The reason behind this results is that the start and the finish of the activity "live separate lives," and we do not even have information on the earliest accomplishment of the activity. We only know its earliest start time and its earliest finish time. This means that if we want to accomplish the task at its earliest schedule, we have to start it exactly at the time of the early start, and we have to finish it exactly at its early finish. This can be a one minute work at the early start time and a one minute work at the early finish time, but we do not have any knowledge about the whole process. The definitions, when splitting is allowed, differs from those when splitting is not allowed.

Definition 4-22 *(Critical activity)*
An activity is considered to be critical if the difference between its late finish and early start time is equal to its activity time.

Activity Floats: Splitting Allowed

Definition 4-23 *(Activity total float)*
The activity total float shows, how much the activity duration can increase without affecting the project duration. It is derived as $TF=LF_i-ES_i-\tau_i$

The definition differs from the definition in non-splitting case, but the formula is the same.

Definition 4-24 *(Activity start float, ASF)*
An activity start float shows how much the early start of an activity can delay without affecting the project duration. It is derived as $ASF_i=LS_i-ES_i$

Definition 4-25 *(Activity finish float, AFF)*
An activity finish float shows how much the early finish of an activity can delay without affecting the project duration. Itis derived as $AFF_i=LF_i-EF_i$

The definitions of the determinant precedence relationships and there floats are the same, as they can be assumed to be, in a sense, non-splittable activities. The definition of the critical path, the longest path from point *s* to *t* is also the same. In the non-splittable case, the critical path can go through the start point of a non critical activity, (see Figure 4-20 activity D) and can go through in the finish point of a non critical activity (see Figure 4-20 activity G).

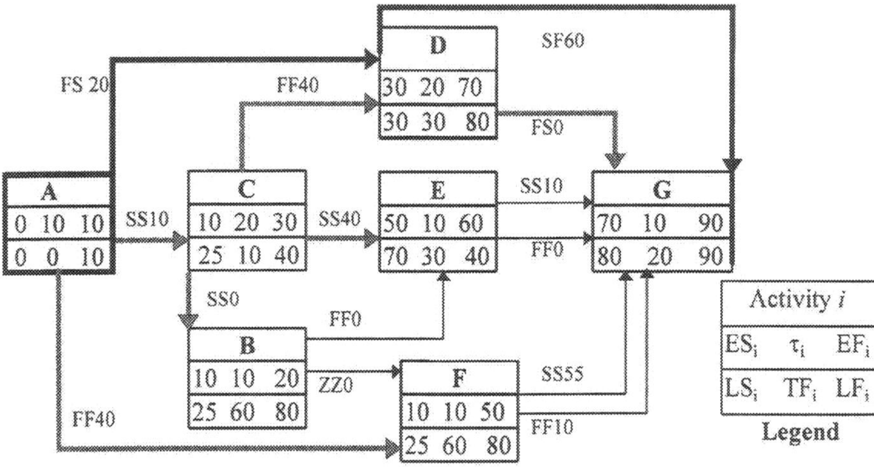

Figure 4-20 *Results and interpretation of calculations, splitting allowed*

4.12 CONCLUSIONS

At the beginning of this chapter, we started with some definitions, for example, that activities must carry out continuously, that there can be only one start and one terminal activity, that loops are not allowed, etc. We gave the definitions of the precedence relationships which can be used to describe interdependencies between activities, and not just between their two extreme points.

Then we introduced the concept of splittable activities. Making an activity splittable, we lost the possibility of making contact between whole activities. This concept results in a special problem, the problem of dangling activities, which can start at an infinite point of time in the past and can finish after an infinite point of time in the future. We have introduced the INITIAL and TERMINAL TIMES to force the activitz start and finish times into boundaries. We have also shown that this can result in earlier late start time than early start time, and can result in no critical path from s to t. In PDM, negative lag can also be used despite that it cannot describe real logical relationships. Further, we have also shown that according to some extensions, loops are not always loops, and in those cases the given algorithm will not work. The definitions of the criticality are different in the splittable and non-splittable case and will become more different if we use these two approaches in the same network.

There are very few schedulers able to work correctly with all these extensions. During the chapter we discussed which extension is advisable to use and which is not. Here we summarize the rules which are useful to keep, especially if the planner is not very much experienced.

The rules are the followings

1. Use one start and one finish activity. (If you use multiple terminal activities, the project duration will be the maximum among the latest starts of these activities, and the backward calculation will start from this value at all the terminal nodes).
2. Eliminate loops (see Loops in PDM network).
3. Do not use negative lags (see Notes on Negative Lag and Dangling Activities).
4. Convert dangling activities into non-dangling with a correct precedence list (see Notes on Negative Lag and Dangling Activities).
5. Do not use INITIAL and TERMINAL values. You do not need them if your network does not contain dangling activities (see Notes on Calculation Rules).
6. **Use the activity-splitting-not-allowed approach since this is the original PDM assumption.** The non-splitting approach provides you information on the early and late accomplishing of the activities, while the activity splitting approach usually does not.
7. If it is inevitable, then the activity splitting approach can be used, but remember that after this you make connections between the 'start event' and 'finish event' of the activities and not between the activities themselves.[i]
8. Be careful in case of disputes or when working with other schedulers. First you have to check whether your ideas about PDM are concordant.

These rules help to prevent the schedule from becoming confusing or incorrect. Unfortunately, not even the software developers tell us what rules they had implemented in their software, and the scheduler may get different results for the same network using different software packages. We are going to discuss this in Appendix B.

[i]When the start and finish points of the activities are set apart, and with this some 'events' emerge and certain loops are allowed, for example, A -SS -B -FF- A, then this network starts to become the so called ladder network which is an extension of the CPM with lead and lag factors, and has been discussed by many authors. It is even been standardized by the German Standards Institution (DIN).

4.13 PRACTICAL PROBLEMS AND SOLUTIONS

Problems

1. Perform PDM calculations on the below given network! (Activity splitting is not allowed.)

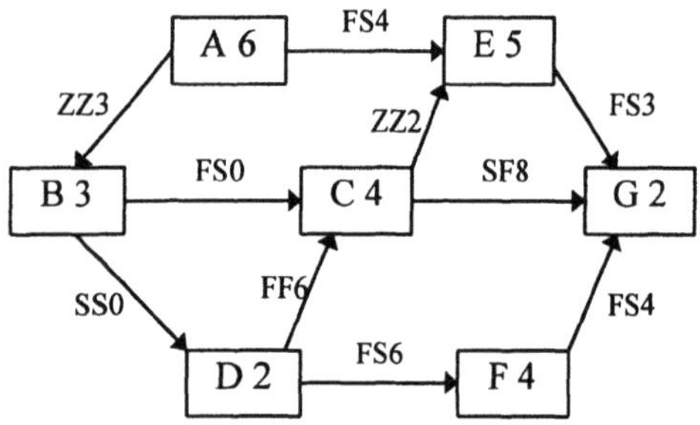

2. Perform PDM calculations on the below given network! (Activity splitting is not allowed.)

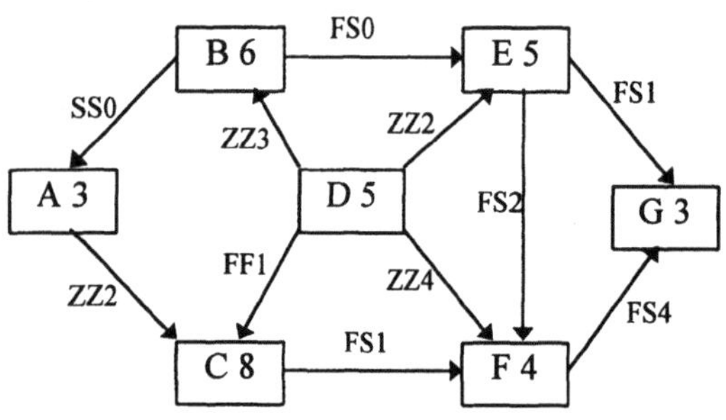

3. Perform PDM calculations on the below given network! (Activity splitting is not allowed.)

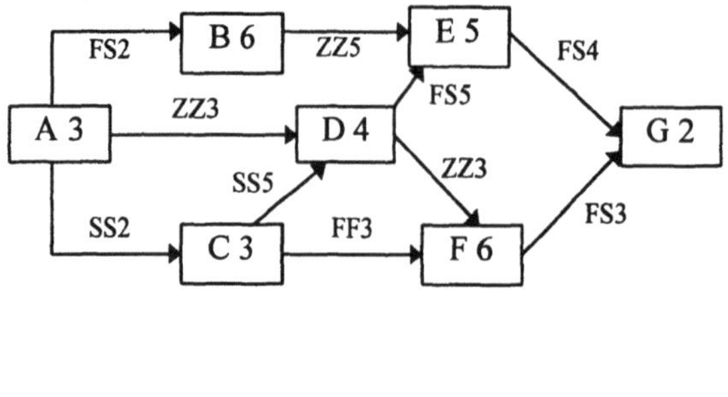

4. Perform PDM calculations on the below given network! (Activity splitting is not allowed.)

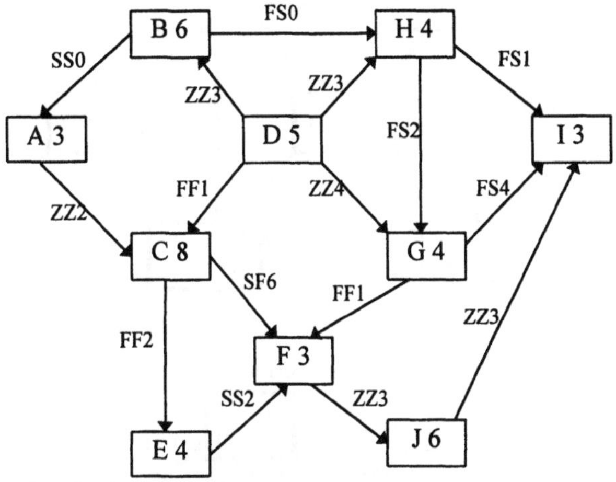

5. Solve the network given in 1. using the activity-splitting-allowed assumption!

6. Solve the network given in 2. using the activity-splitting-allowed assumption!

7. Solve the network given in 3. using the activity-splitting-allowed assumption!

Solutions

1. The results are given below. Observe that activity A is normal critical, B is reverse critical, and D, F, G are normal critical as well.

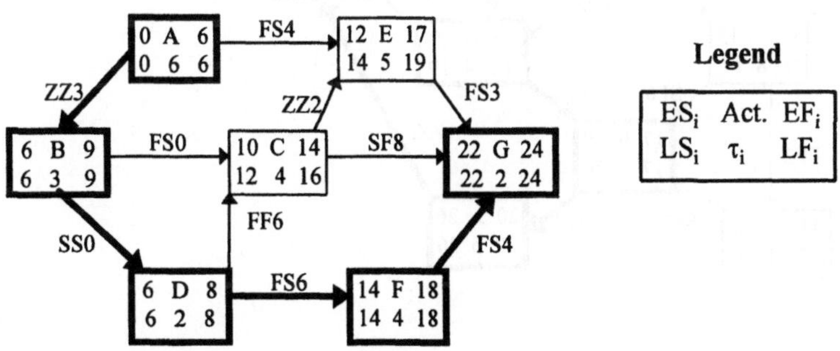

2. Activity D is start critical, activities B, E, F and G are normal critical.

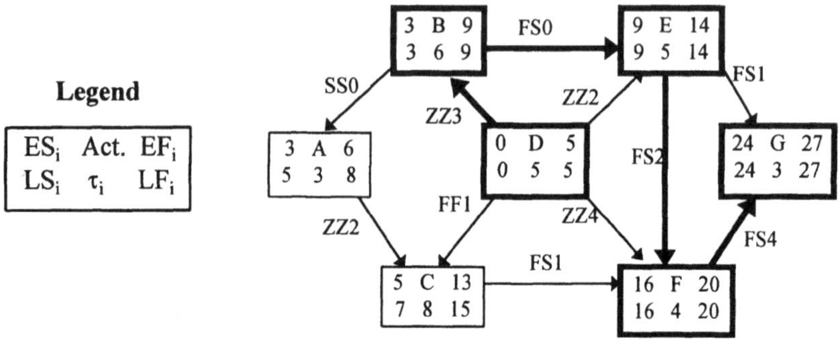

3. Activities A and C are start critical, D, E and G are normal critical.

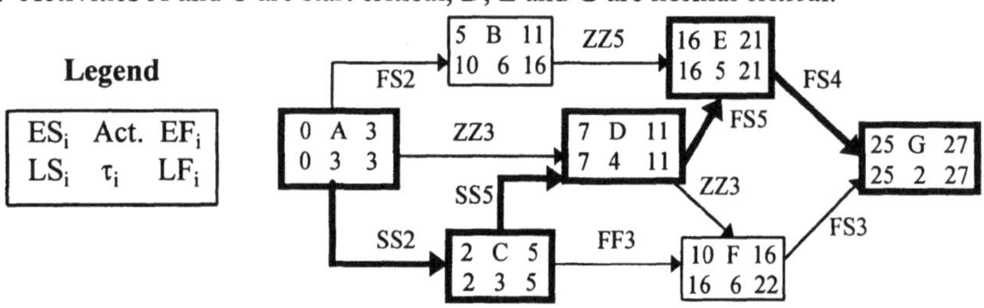

4. Activity D is start critical, B, H and G are normal critical, F is reverse critical, J is normal and I is reverse critical.

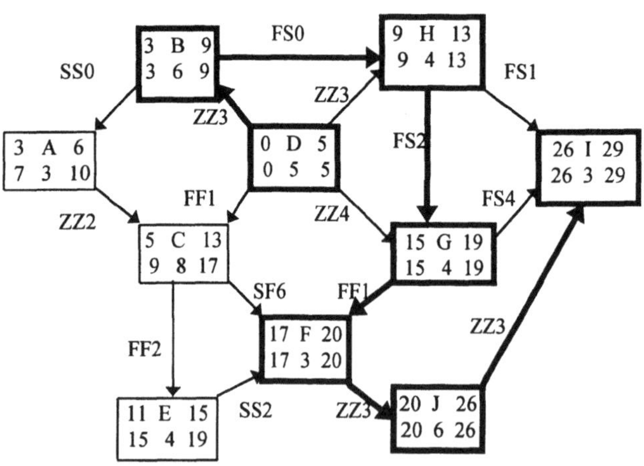

5. The results can be seen below. Observe that there are two critical paths in the network, both of them touch activity B which is not critical, and one of them goes through the start of C, which is a dangling finish activity, so it can be finished anytime. The critical paths touches all the activities.

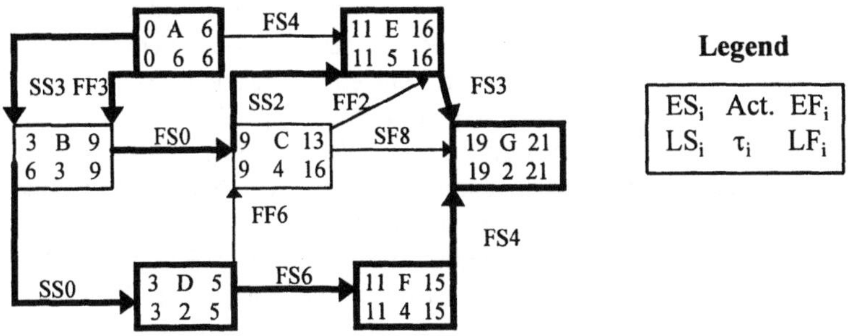

6. The project duration is the same as in the non-splitting case. Only the late finish of A has changed.

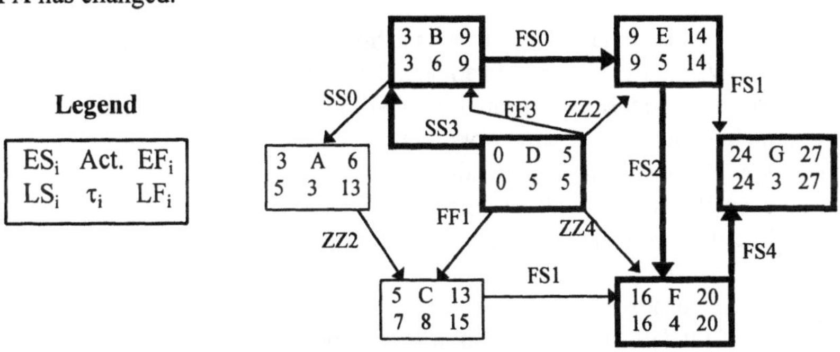

7. Activities A and C are not critical, but the critical path touches them.

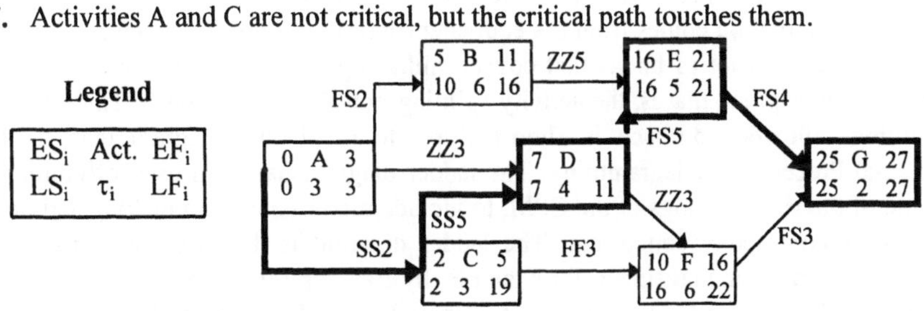

4.14 HISTORICAL REVIEW

The history of Precedence Diagramming Method is dated back to 1958, when a French scientist, B. Roy [58],[59] developed his 'Method of Potentials'. He used the activity-on-node representation, activities were non-interruptable, and he introduced the SS precedence relationship with lead/lag factors. His method was so sophisticated that it allowed the use of maxSS precedence relationship he called the backward acting relationship. Maximal type of relationships always result in loops in mathematical sense, but not in logical sense. We are going to discuss this in Chapter 5: Advanced Precedence Diagramming. A response to his work in Germany was the development of the HMN technique. This technique allowed the use of FF and maxFF relationships. In the non-English speaking part of Europe, the Metra Potential Method (MPM) name is used instead of the Precedence Diagramming Method, honoring Roy's work. These pieces of work were completely parallel with the development of the CPM and PERT, and in Europe this method has been used since the early sixties together with CPM.

In the United States, these works were much ignored as they were mainly published in French and German. In the United States, John Fondahl's work [18] is considered as a forerunner of the PDM method. In his work, he introduced the concept of activity-on-node representation. The first complete description with the four precedence relationships appeared in 1964 in an IBM users manual [39]. One of the principal authors of this work was J. Craig. The term of 'precedence diagramming' also originates form Craig.

The characteristics of a PDM critical path, and the paradoxical behavior of it has been described by Weist [65]. Another remarkable work in this field is the work of Keith C. Crandall [11] who has done a lot for the popularization of the PDM technique in the construction industry. The activity splitting approach has also been developed by Crandall, and it is appeared in the book of Moder, Phillips and Davis [50]. In his approach, he developed a computational procedure which tries to maintain the logical relationship between the activities, not between their extreme points. This method comes from the equivalent activity-on-arrow scheme.

There are a lot of books that cover similar topics, but a lot of them forget to discuss the original, that is, the activity-splitting-not-allowed approach. The activity-splitting- allowed approach is almost closer to the ladder diagram than to the original Precedence Diagramming. The ladder diagram has been used of eliminate some of the shortcomings of the CPM, to include the possibility to making activities partly or completely concurrent. The ladder diagram is discussed in many early books on project management, for example, by Battersby [6]. Lockyer and Gordon [44] also discuss the ladder diagram, and they discuss the Method of Potentials and Precedence Diagramming Method as two different techniques. This approach is also acceptable for us. The Precedence Diagramming Method is on the edge of being a completely new technique or 'just' an improvement over the Method of Potentials. In some countries of the world, mainly in Europe, the PDM is called the MPM and it is said that the three new precedence relationships in the PDM do not make a new technique. In the U.S. where the PDM technique has been developed, it is considered a completely new technique. As we said earlier, both allegations are acceptable for us.

A mathematically correct discussion of the PDM methodology can be found, for example, in the above cited Moder text, and in the book by Ahuja, Dozzi and Abourizk [2].

Chapter 5

ADVANCED PRECEDENCE DIAGRAMMING

In this chapter we introduce the so called maximal type of relationships. Maximal types of relationships were, for example, the parts of the original Method of Potential and the HMN technique, but they were not nicely treated in these early works. It was relatively hard to understand the concept, and the computational procedure was difficult. Here we present the original concept, but we slightly modify the methodology. This modification makes this section easy to understand. Constraints are also discussed in this chapter. Constraints can be used for modifying the activity early and late times without precedence relationships. This is in fact a non-network tool, which is used to modify the results gained from the calculation based on the network logic.

In the first part, Introducing the Sample Project, we present a sample project, which is going to be used to illustrate the use of maximal type of precedence relationships.

The new precedence relationships are the topic of the next section, Maximal Type of Relationships.

In the following chapter, Calculations with Maximal Relationships, we discuss the computational procedure which is able to handle both minimal and maximal type of precedence relationships.

In the fourth chapter, Information Obtained from Calculations, we discuss the changes that have been effected by the application of maximal precedence relationships.

Constraints are discussed in the fifth section, Constraints in PDM.

In the next section, PDM Versus CPM, we compare the CPM and MPM techniques. We are going to show that no equivalent CPM representation can be produced from a PDM network.

In the last two sections we give exercises and a historical review.

5.1 INTRODUCING THE SAMPLE PROJECT

Let us consider the sample project that has been described in the first part of Chapter 4: Introducing the Sample Project. A covered depot has to be built from prefabricated standardized steel elements, in order to protect large industrial equipment, which is used for engine production. The project has been divided into fifteen processes from A to N. We have kept both the original duration estimations, and the precedence relationships of Table 4-1.

Now, we are going to establish some new activity interdependencies.

- The delivery of prefabricated steel structure (I) and its assembling (N) will be carried out by the same subcontractor. Due to its internal schedule the subcontractor can spend only 100 days on this project, but these two activities cannot be shortened. (Observe that now we have 115 days between the early start of I and early/late finish of N.)
- A deadline is given for the start of the foundation work. It has to be started before the 70th day of the project. (At this moment, its late start is on day 85)
- Gravel and cement can be deposited on the site for only 40 days due to lack of enough free space. This means that the expired time between the start of G and finish of L has to be less than 40 days.

These verbally stated activity relationships are difficult to be expressed by any of the well-known minimal relationships. The reason for this is that a minimal type of relationship stands for the minimum necessary time that has to pass between two activities. The above stated conditions require an another type of precedence relationships. To model these kinds of activity interdependencies, we introduce the so called maximal type of precedence relationships. We can use four different maximal relationships between activities.

5.2 MAXIMAL TYPE OF RELATIONSHIPS

We give the definitions for the four maximal precedence relationships and the definition for the most frequently used combination.

Definition 5-1 *$maxSSz_{(ij)k}$ (maxStart-to-Start-$z_{(ij)k}$)*
A $maxSSz_{(ij)k}$ precedence relationship means that activity j can start as maximum $z_{(ij)k}$ lag time elapsed after the start of activity i, or it can start before that.

In Figure 5-1 the graphical portrayal of the precedence relationship and its interpretation by Linear Scheduling Method can be seen. The half-line, which originates from point O, shows the set of the possible starts for activity B. Three possible accomplishments are portrayed on the LSM chart, the latest shown by a continuos, the earlier ones by a dotted line.

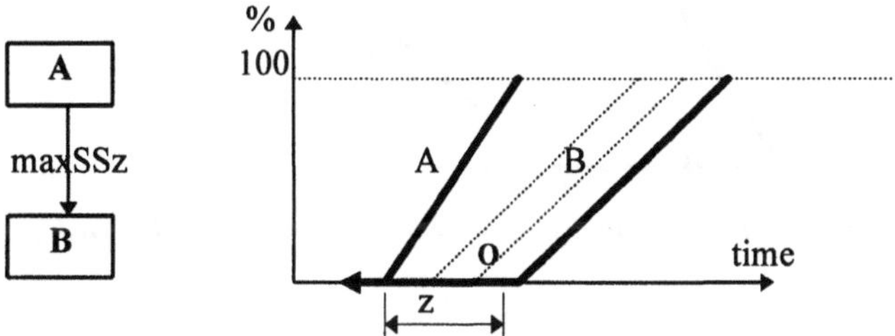

Figure 5-1 *LSM interpretation of maxStart-to-Start precedence relationship.*

Definition 5-2 *$maxFFz_{(ij)k}$ (maxFinish-to-Finish-$z_{(ij)k}$)*
A $maxFFz_{(ij)k}$ precedence relationship means that activity j can finish as maximum $z_{(ij)k}$ lag time elapsed after the finish of activity i, or it can finish before that.

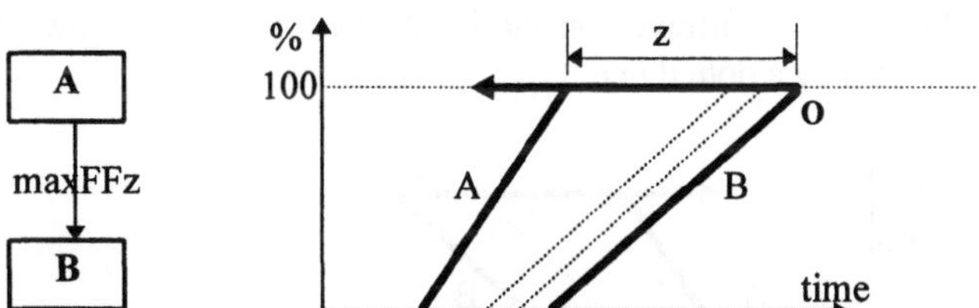

Figure 5-2 *LSM interpretation of maxFinish-to-Finish precedence relationship.*

In Figure 5-2 the graphical portrayal of the precedence relationship and its interpretation by Linear Scheduling Method can be seen. The half-line, which originates from point O, shows the set of the possible finishes of activity B. There are three possible accomplishments are portrayed on the LSM chart, the latest shown by a continuos, the earlier ones by a dotted line.

Definition 5-3 *maxFSz*$_{(ij)k}$ *(maxFinish-to-Start-z*$_{(ij)k}$*)*
A maxFSz$_{(ij)k}$ *precedence relationship means that activity j can start as maximum* $z_{(ij)k}$ *lag time elapsed after the finish of activity i, or it can start before that.*

In Figure 5-3 the graphical portrayal of the precedence relationship and its interpretation by Linear Scheduling Method can be seen. The half-line, which originates from point O, shows the set of the possible finishes of activity B. Three possible accomplishments are portrayed on the LSM chart, the latest shown by a continuos, the earlier ones by a dotted line.

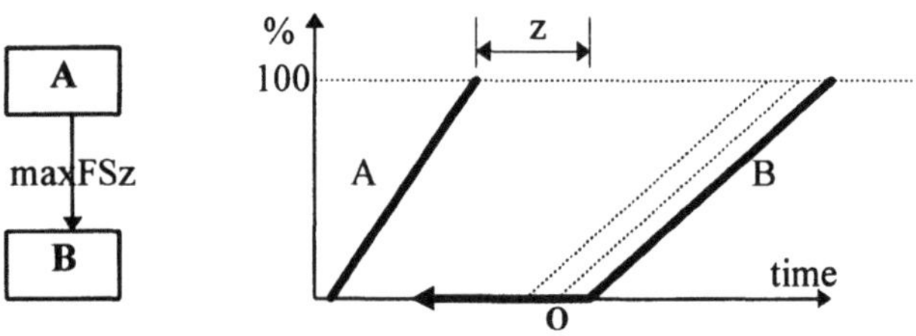

Figure 5-3 *LSM interpretation of maxFinish-to-Start precedence relationship.*

Definition 5-4 *maxSFz*$_{(ij)k}$ *(maxStart-to-Finish-z*$_{(ij)k}$*)*
A maxSFz$_{(ij)k}$ *precedence relationship means that activity j can finish as maximum* $z_{(ij)k}$ *lag time elapsed after the start of activity i, or it can finish before that.*

In Figure 5-4 the graphical portrayal of the precedence relationship and its interpretation by Linear Scheduling Method can be seen. The half-line, which originates from point O, shows the set of the possible finishes of activity B. Three possible accomplishments are portrayed on the LSM chart, the latest shown by a continuos, the earlier ones by a dotted line.

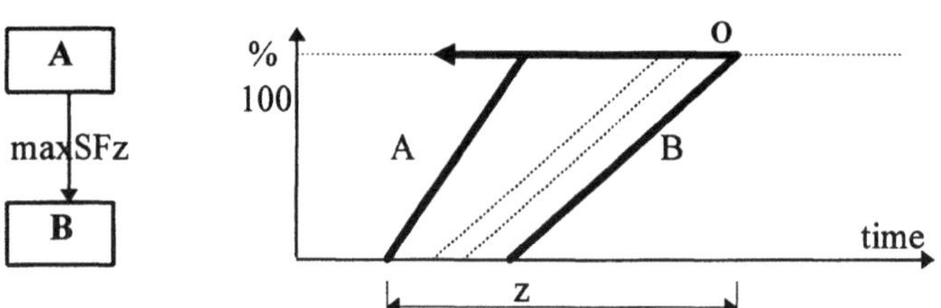

Figure 5-4 *LSM interpretation of maxStart-to-Finish precedence relationship.*

Definition 5-5 *maxZZz*$_{(ij)k}$ *(A frequently used combination)*
A maxZZz$_{(ij)k}$ *precedence relationship means that there should be maximum z or less lag time between activity j and activity i at any stages of i and j. This is nothing else than the application of maxSSz and maxFFz precedence relationships together.*

In Figure 5-5 the graphical portrayal of the precedence relationship and its interpretation by Linear Scheduling Method can be seen. The half-line, which originates from point O, shows the set of the possible finishes of activity B. If $\tau_i>\tau_j$, then the application of maxZZz gives the same result as the application of only a maxSSz precedence relationship. If $\tau_i<\tau_j$, then the use of maxZZz gives the same result as we get by using a maxFFz precedence relationship only. One can argue that we should simply examine the activity durations and then to add maxSSz or maxFFz relationships to the plan instead of using maxZZ relationship. We disagree with this argument. Precedence relationships always have to reflect the real logical interdependencies, and the use of mathematical equivalents are not advisable. We have discussed this issue when we introduced the minimal ZZ relationship. The shaded area indicates the forbidden zone for B.

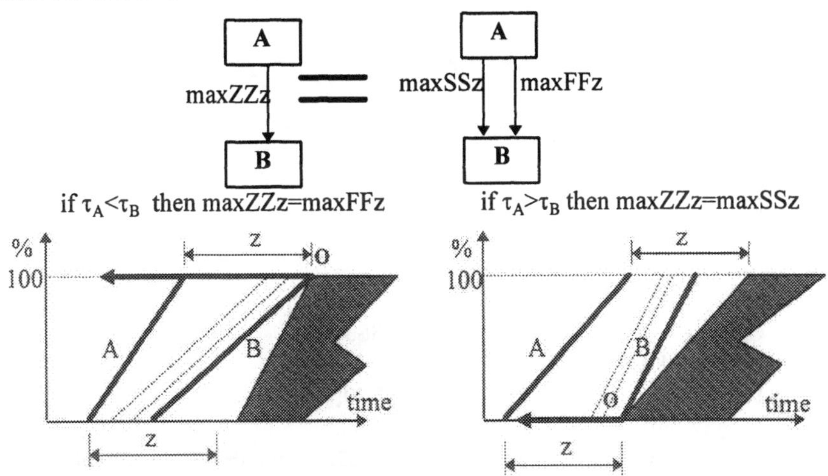

Figure 5-5 *LSM representation of maxZZ combination.*

These precedence relationships are called the *maximal precedence relationships* because the *z* value stands for the maximal allowable distance between the two distinguished points -- start or finish -- of activities. This means that the lag time between these points can be less than *z*, but never more.

A maximal precedence relationship from A to B means that activity B logically depends on activity A.

Let us see, how the verbally drafted activity relationships can be turned into PDM precedence relationships.

- The delivery of prefabricated steel structure (I) and its assembling (N) will be done by the same subcontractor. Due to its own schedule the subcontractor can spend only 100 days in this project. *This is a maxSF100 precedence relationship originating from activity I and terminating in activity N.*

- A deadline is given for the start of the foundation work. It has to be started before the 70th day of the project. *This is a maxSS70 precedence relationship from A to K.*

- Gravel and cement can be deposited on the site for only 40 days due to the lack of enough free space. This means that the time span between the start of G and finish of L has to be less than 40 days. *This is a maxSF40 precedence relationship from H to L.*

The complete preceding activity list can be seen in Table 5-1. The newly added precedence relationships are printed in boldface format.

Activity	Duration in days	Preceding activities with precedence relations
A	30	-
B	20	A FS0;
C	30	B FS0; E FS0;
D	10	A FS0;
E	14	A FS0;
F	30	E SS0;
G	20	D FS0; E SS3;
H	80	D FS0; E SS0;
I	90	D FS0; E SS0;
J	10	F FS0; E FS0;
K	30	J ZZ5; G ZZ3; **G maxSF40; A maxSS70**
L	20	K ZZ10; H ZZ10;
M	5	L FS0; C FS0; K ZZ20;
N	20	M FS0; I SS2; I FF0; **I maxSF100**
O	15	N FF0; SS0

Table 5-1 *Input data for the sample project using the PDM technique.*

The network representation of the enlarged preceding activity list can be seen in Figure 5-6. The newly added precedence relationships are depicted in gray.

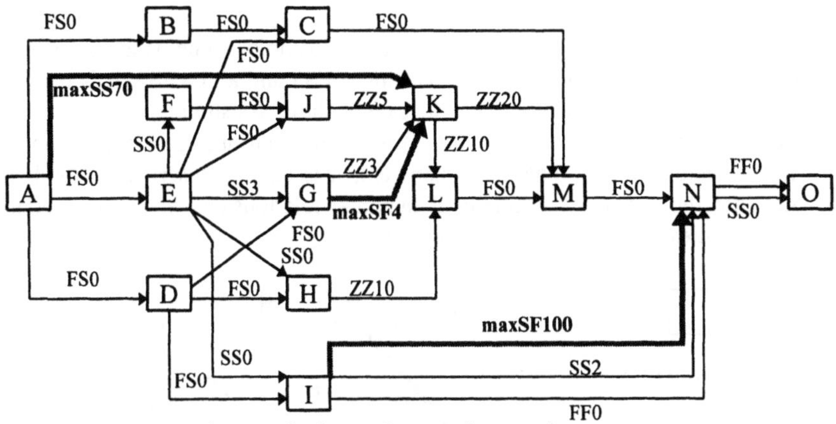

Figure 5-6 *Maximal type of relationships in the sample project.*

Using the maximal type of precedence relationships the network remains easy-to-survey because neither loops nor negative lag factors have been added to the network.

How can the maximal type of precedence relationships be used for modeling a project, and what kinds of problems can arise during the application? We are going to examine this question in the following.

The most frequent use of maximal relationships is when an activity, for example, activity B has to start exactly when an another activity, for example, A has finished. This situation can be seen in Figure 5-7. In this case one has to apply a FS0 and a maxFS0 precedence relationship from A to B. The FS0 gives a possible set for the start of B, which is a half line originating at the time of finish of B and tends to right. The maxFS0 gives a possible set for the start of B, which is a half line originating at the time of finish of B and directs towards the left. The common part of these two half lines determines the possible set of starts for activity B. As they have only one common point, activity B has to start at the exact time when activity A is finished.

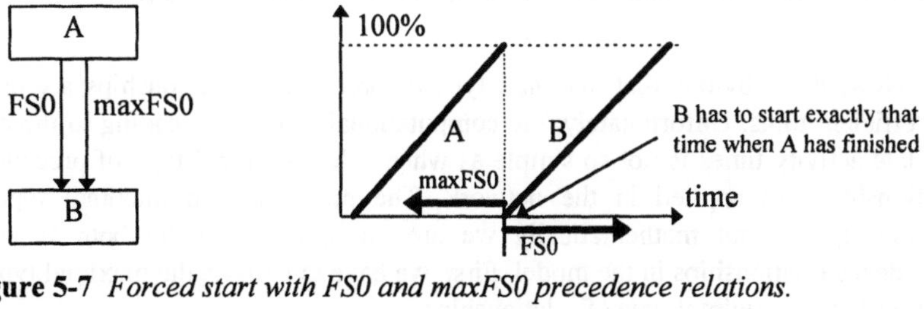

Figure 5-7 *Forced start with FS0 and maxFS0 precedence relations.*

The application of maximal type of relationships holds some dangers. Imagine that activity B is preceded by two relationships, a FS5 and a maxFS3. It is obvious that there is no schedule that is able to satisfy these constraints, as the sets of feasible solutions have no common part. This can be seen in Figure 5-8. Although the illogical nature of this diagram is apparent, it is easy to insert these kinds of mistakes into the network, especially if maximal relationships make connections between activities that are far away from each other.

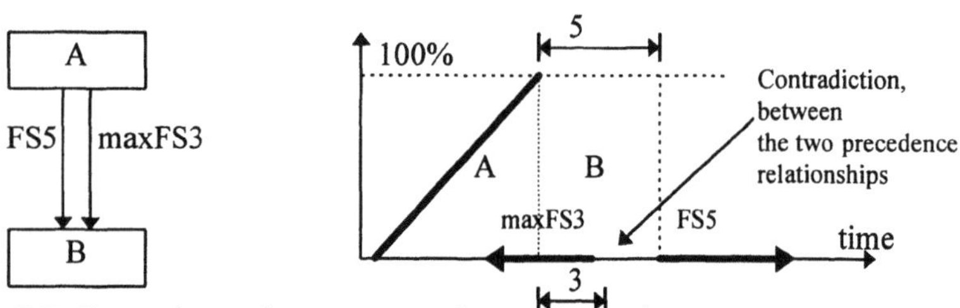

Figure 5-8 *Contradiction between precedence relationships*

Maximal type of precedence relationships can be used to prescribe a time interval for any activity start or finish. For example, if activity B can start after al least 2, but less than 5 days elapsed since the finish of A, then we have to use a FS2 and maxFS5 combination between A and B. This combination will force the start of B into the desired interval, as it can be seen in Fig. 5-9. The earliest and latest possible accomplishments of B are depicted in gray.

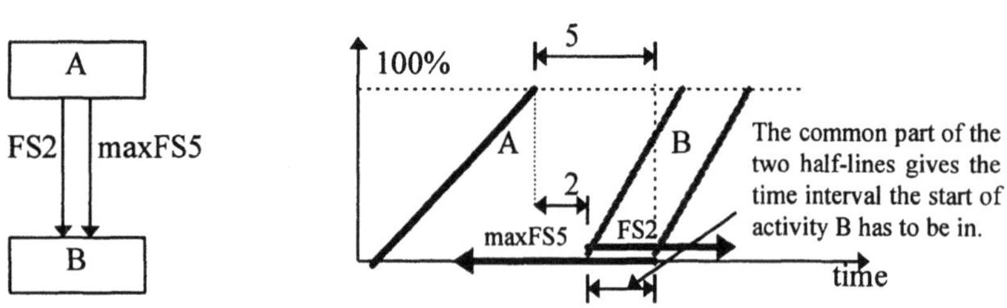

Figure 5-9 *Describing time interval for the possible start of an activity.*

Now, it is obvious that maximal type of precedence relationships are useful and efficient tools. Unfortunately, the computational procedure leading to the early and late activity times is not so simple as when solely minimal type of precedence relationships are applied in the network. The problem with maximal type of relationships is that mathematically we are not able to handle both types of precedence relationships in the model. First, we have to convert the maximal type of relationships to minimal type of relationships.

A minimal type of FSz relationship between an arbitrary activity *i* and *j*, where *j* depends on *i*, can be described as $\pi_{jS}-\pi_{iF}\geq z_{ij}$. A maxFSz relationship between *i* and *j*, where *j* depends logically on *i*, can be described as $\pi_{jS}-\pi_{iF}\leq z_{ij}$. To make a maximal type of relationship a minimal type of relationship, we simply have to multiply the condition by '-1'. Thus we get the following: $\pi_{iF}-\pi_{jS}\geq -z_{ij}$. This is nothing else but a minimal type of SF relation leading from *j* to *i* with negative lag (-z). The result of this argument can be seen in Figure 5-10.

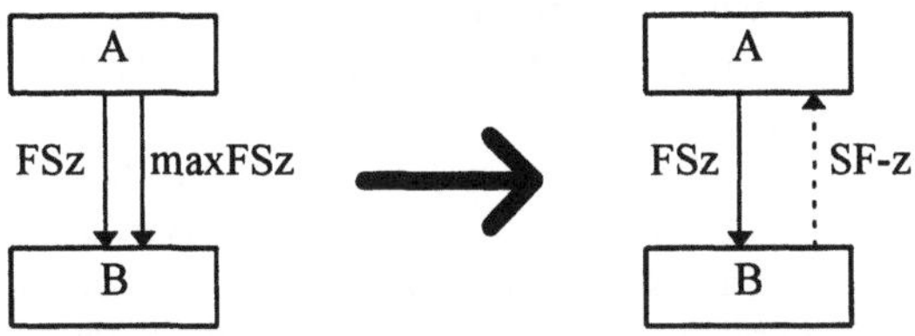

Figure 5-10 *The mathematical equivalent of maxFSz precedence relationship.*

Converting the maximal type relationships to minimal type relationships, loops arise in the network. However, these are loops only in a mathematical sense, but not in logical sense. Logically, B depends on A by a FSz and maxFSz relationship, and A does not depend on B. To distinguish these 'mathematical' loops from the logical loops, which have been described in the previous chapter, we depict the converted precedence relationships by dotted lines.

The mathematically equivalent minimal type of precedence relationships are summarized in Table 5-2.

Maximal type of precedence relationship	Mathematically equivalent minimal type of precedence relationship
$maxFSz_{ij}$	$SF\text{-}z_{ij}$
$maxSSz_{ij}$	$SS\text{-}z_{ij}$
$maxSFz_{ij}$	$FS\text{-}z_{ij}$
$maxFFz_{ij}$	$FF\text{-}z_{ij}$
$maxZZz_{ij}$	$ZZ\ \text{-}z_{ij}$

Table 5-2 *Maximal type of precedence relationships and their equivalents*

The use of maximal type of precedence relationship is not new in PDM. It was applied in the original PDM methodology. Two types of precedence relationships were applied in the network, the so called forward acting precedence relationships, and the so called backward acting precedence relationships. The forward acting precedence relationship was equivalent to the minimal type of precedence relationships. The backward acting precedence relationships were minimal relationships with negative lag, which always were in a loop. Backward acting relationships represented a reverse logic. In fact, they were the mathematical equivalent of the maximal type of precedence relationships. The equivalents have been displayed on the right side of Table 5-2. Of course, in case of backward acting precedence relationships, the computational procedure given in Chapter 4 does not work.

Using the backward acting relationships it was hard to distinguish the real logical loops from the loops caused by the application of backward acting relationships. It was burdensome to work with negative lag, and switch to reversed logic. Becauseof the reverswe logic and the different computational procedure their application was especially troublesome.

Using the maximal type of precedence relationships, we do not have to use negative lag; the real logical loops are easy to filter from the network, and the network logic always remains clear. That is why we emphasize that maximal type of precedence relationships are advisable to use instead of their mathematical equivalents.

During the computations the maximal type of precedence relationships have to be converted, which will cause loops in the network, but during the planning phase the network logic remains clear and easy-to-survey.

The sample network prepared for calculation can be seen in Figure 5-11.

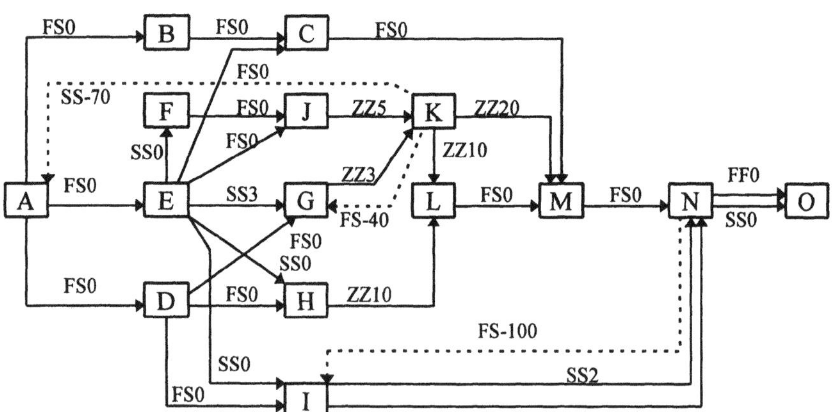

Figure 5-11 *The sample project using the mathematical equivalent relationships.*

5.3 CALCULATIONS WITH MAXIMAL RELATIONSHIPS

The application of the maximal type of relationships causes problems in the computational procedure. According to the rules, which have been described in Chapter 4, an activity can be chosen for forward calculation if the early start and finish times are known for all its predecessor activities. In case of using maximal type of relationships, as we cannot handle two different types in one model, these relationships have to be transformed into minimal type of relationships. A small sample can be seen in Figure 5-12. Following the rules given in Chapter 4: PDM Calculations, the early start and finish of activity A can be calculated, but after this no activity can be chosen for which all the predecessors are calculated. The procedure will stop with an error message: There is a loop in the network.

In case of maximal precedence relationship another algorithm is necessary, which is going to be presented below. The idea behind the algorithm is that we will satisfy the conditions prescribed by the precedence relationships in an iterative process. The activities can be chosen for calculation in an arbitrary order. It does not matter if one of its preceding activities has no final early start and finish times. The early start and finish times have to be determined on the basis of existing data. This calculation has to be done for each activity. When all activities have been included in the calculations, then the iteration is finished. If there were any changes during the iteration, it has to be repeated. The iterative process has to be repeatedly done until the result remains the same as in the previous iteration.

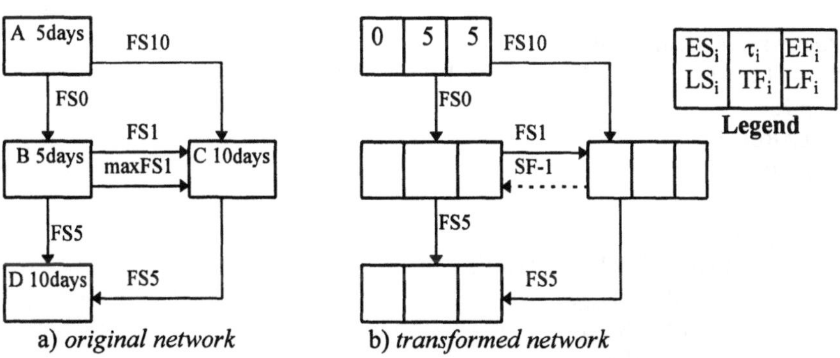

Figure 5-12 *Computational procedure stops after A has been calculated.*

It is simpler than the originally given PDM algorithm because one does not have to care about choosing the next activity for computation. However, this relative simplicity requires with more computational steps.

The algorithm can be summarized in the following manner:

PDM Calculation: Maximal Relationships Allowed: No Splitting Allowed
Forward Calculation

Step1. Let $ES_s=0$; $EF_s=ES_s+\tau_s$; $ES_i= -\infty=$ INITIAL TIME $\forall\ i\in\mathbf{N}\text{-}s$, No_of_Iter:=0

Step 2.

LABEL 1:

$f:=0$

REPEAT

$f:=f+1$

$ES_j=\max\{$INITIAL TIME;

$ES_i+SSz_{(ij)k}$;

$EF_i+FSz_{(ij)k}$;

$ES_i+SFz_{(ij)k}-\tau_j$;

$EF_i+FFz_{(ij)k}-\tau_j\}\quad \forall\ \{(ij)k\}\in\mathbf{A}$

$EF_j=ES_j+\tau_j$

UNTIL $f=n$

No_of_Iter:=No_of_Iter+1;

Step 3.

IF No_of_Iter>n *THEN* HALT (There is no feasible activity time system)

IF there was a change in any ES or EF values *THEN* go to 1

IF there were no changes THEN the forward pass has been finished.

The backward pass is completed by working from the terminal activity to the initial activity, in the reverse direction of arrows. It is summarized as follows:

Backward Calculations

Step1. Let $LF_t=p=ES_t$; $LS_t=LS_t-\tau_t$; $LF_i=\infty=$TERMINAL TIME $\forall\ i\in\mathbf{N}\text{-}t$

Step 2.

LABEL 2:

$f:=0$

REPEAT

$f:=f+1$

$LF_i=\min\{$TERMINAL TIME;

$LF_j-FFz_{(ij)k}$,

$LS_j-FSz_{(ij)k}$;

$LF_j-SFz_{(ij)k}+\tau_i$;

$LS_j-SSz_{(ij)k}+\tau_i\}\quad \forall\ \{(ij)k\}\in\mathbf{A}$

$LS_i=LF_i-\tau_i$

UNTIL $f=n$

Step 3.

IF there was a change in any LS, LF values *THEN* go to 2

IF there were no changes THEN the backward pass has been finished.

Remarks to the algorithm:

- *The number of necessary iterations is greatly influenced by the order of activities that have been taken into consideration. In* Figure 5-13 *we performed the forward calculation process for the small sample, given in* Figure 5-12. *We needed 3 iterations, and got the correct results in the second iteration, which has been confirmed in the third iteration. The A,C,B,D order would have been resulted in right ES, EF values only in two iterations.*
- *The algorithm can be made faster by a properly chosen order, and properly chosen initial and terminal values. For this, one has to make a calculation without the maximal type of relationships and keep the order of activities of this calculation. (This also helps to filter the real logical loops) This order has to be followed in the iteration process. Set an initial value for all the activities equal to their* ES_i *value, which has been calculated without the maximal relationships. Start the computation with this order and these values. For the backward calculation, the reverse order of the forward calculation is suggested to follow. The TERMINAL TIMES for each activity can be set equal to the* LF_i *values, which are calculated after finishing the backward calculation without the maximal type of relationships. It is worth to use this method in case of hand calculations or in case of a large schedule, i.e., more than 1000 activities.*
- *Sometimes there is no solution to the network. Consider the two-activities network in* Figure 5-8. *The feasible sets of possible starts determined by each precedence relationship have no common part, that is, no possible solution exists. In this case the iteration process will never stop as the ES, and EF values will increase in each iteration,* ***because the sum of the values on the loop is positive****. During the calculation process maximum n iteration steps are necessary to obtain the right solution if there is one, even in the case of a bad order of activities. Therefore, we count the number of iterations, and if it is greater than the number of nodes, then the algorithm will stop because there is no solution to the network.*
- *The exact maximal number of necessary iterations is the number of the levels in the network plus one. To determine this number, some new calculations are necessary but it makes the error detecting much faster. (Imagine that there are thousands of activities in the network but the longest chain consists of twenty activities, that is, the number of levels is twenty. In case of wrong network, the algorithm would stop after 20 iterations instead of 1000.)*
- *We do not provide here the algorithm for the splitting-allowed case. The reader should be able to do it alone.*

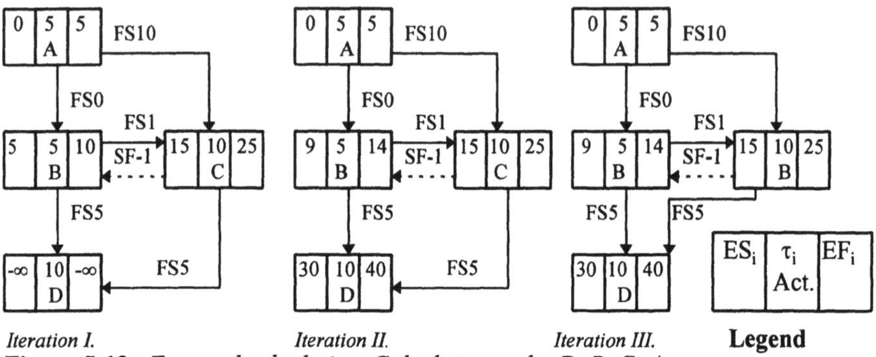

Figure 5-13 *Forward calculation. Calculation order D, B, C, A*

The result of the time analysis for the depot construction project can be seen in Figure 5-14. We have also copied Figure 4-10 to Figure 5-15 for comparing the results of the two cases. The computational procedure can be followed below. We followed the same order as we would follow without the maximal type of relationships in order to decrease the number of iterations.

Forward pass calculations: (ES_i, EF_i values)

Iteration I.

$ES_A = \max\{0, ES_K - 70\} = \max\{0, -\infty\} = 0$ $EF_A = 30;$

$ES_B = \max\{EF_A + 0\} = 30$ $EF_B = 50;$

$ES_D = \max\{EF_A + 0\} = 30$ $EF_D = 40;$

$ES_E = \max\{EF_A + 0\} = 30$ $EF_E = 44;$

$ES_F = \max\{ES_E + 0\} = 30$ $EF_F = 60;$

$ES_C = \max\{EF_E + 0;\ EF_B + 0\} = \max\{44, 50\} = 50$ $EF_C = 80;$

$ES_J = \max\{EF_E + 0;\ ES_F + 0\} = \max\{44, 60\} = 60$ $EF_J = 70;$

$ES_G = \max\{ES_E + 3;\ EF_D + 0; EF_K - 40\} = \max\{33, 40, -\infty\} = 40$ $EF_G = 60;$

$ES_H = \max\{ES_E + 0;\ EF_D + 0\} = \max\{30, 40\} = 40$ $EF_H = 120;$

$ES_I = \max\{ES_E + 0;\ EF_D + 0; EF_N - 100\} = \max\{30, 40, -\infty\} = 40$ $EF_I = 130;$

$ES_K = \max\{ES_J + 5; EF_J + 5 - \tau_K; ES_G + 3; EF_G + 3 - \tau_K\} = 65$ $EF_K = 95;$

$ES_L = \max\{ES_K + 10; EF_K + 10 - \tau_L; ES_H + 10;\ EF_H + 10 - \tau_L\} = 110$ $EF_L = 130;$

$ES_M = \max\{EF_C + 0; EF_L + 0; ES_K + 20; EF_K + 20 - \tau_M\} = 130$ $EF_M = 135;$

$ES_N = \max\{EF_M + 0;\ ES_I + 2;\ EF_I + 0 - \tau_N\} = \max\{135, 42, 110\} = 135$ $EF_N = 155;$

$ES_O = \max\{EF_N + 0 - \tau_0;\ ES_N + 0) = 140;$ $EF_O = 155;$

All the ES and EF values have been changed, except ES_A and EF_A, so new iteration is necessary.

Iteration II.

$ES_A=\max\{0, ES_K-70\}=\max\{0,-5\}=0$	$EF_A=30$;
$ES_B=\max\{EF_A+0\}=30$	$EF_B=50$;
$ES_D=\max\{EF_A+0\}=30$	$EF_D=40$;
$ES_E=\max\{EF_A+0\}=30$	$EF_E=44$;
$ES_F=\max\{ES_E+0\}=30$	$EF_F=60$;
$ES_C=\max\{EF_E+0; EF_B+0\}=\max\{44,50\}=50$	$EF_C=80$;
$ES_J=\max\{EF_E+0; ES_F+0\}=\max\{44,60\}=60$	$EF_J=70$;
$\mathbf{ES_G=\max\{ES_E+3; EF_D+0;EF_K-40\}=\max\{33,40, 55\}=55}$	$\mathbf{EF_G=75}$**;**
$ES_H=\max\{ES_E+0; EF_D+0\}=\max\{30,40\}=40$	$EF_H=120$;
$\mathbf{ES_I=\max\{ES_E+0; EF_D+0;EF_N-100\}=\max\{30,40, 55\}=55}$	$\mathbf{EF_I=145}$**;**
$ES_K=\max\{ES_J+5;EF_J+5-\tau_K;ES_G+3;EF_G+3-\tau_K\}=65$	$EF_K=95$;
$ES_L=\max\{ES_K+10;EF_K+10-\tau_L;ES_H+10; EF_H+10-\tau_L\}=110$	$EF_L=130$;
$ES_M=\max\{EF_C+0;EF_L+0;ES_K+20;EF_K+20-\tau_M\}=130$	$EF_M=135$;
$ES_N=\max\{EF_M+0; ES_I+2; EF_I+0-\tau_N\}=\max\{135,42,110\}=135$	$EF_N=155$;
$ES_O=\max\{EF_N+0-\tau_0; ES_N+0)=140$;	$EF_O=155$;

Some values have been changed, therefore new iteration is necessary.

Iteration III.

$ES_A=\max\{0, ES_K-70\}=\max\{0,-5\}=0$	$EF_A=30$;
$ES_B=\max\{EF_A+0\}=30$	$EF_B=50$;
$ES_D=\max\{EF_A+0\}=30$	$EF_D=40$;
$ES_E=\max\{EF_A+0\}=30$	$EF_E=44$;
$ES_F=\max\{ES_E+0\}=30$	$EF_F=60$;
$ES_C=\max\{EF_E+0; EF_B+0\}=\max\{44,50\}=50$	$EF_C=80$;
$ES_J=\max\{EF_E+0; ES_F+0\}=\max\{44,60\}=60$	$EF_J=70$;
$ES_G=\max\{ES_E+3; EF_D+0;EF_K-40\}=\max\{33,40, 55\}=55$	$EF_G=75$;
$ES_H=\max\{ES_E+0; EF_D+0\}=\max\{30,40\}=40$	$EF_H=120$;
$ES_I=\max\{ES_E+0; EF_D+0;EF_N-100\}=\max\{30,40, 55\}=55$	$EF_I=145$;
$ES_K=\max\{ES_J+5;EF_J+5-\tau_K;ES_G+3;EF_G+3-\tau_K\}=65$	$EF_K=95$;
$ES_L=\max\{ES_K+10;EF_K+10-\tau_L;ES_H+10; EF_H+10-\tau_L\}=110$	$EF_L=130$;
$ES_M=\max\{EF_C+0;EF_L+0;ES_K+20;EF_K+20-\tau_M\}=130$	$EF_M=135$;
$ES_N=\max\{EF_M+0; ES_I+2; EF_I+0-\tau_N\}=\max\{135,42,110\}=135$	$EF_N=155$;
$ES_O=\max\{EF_N+0-\tau_0; ES_N+0)=140$;	$EF_O=155$;

There were no changes in this iteration, so the forward pass has been finished.

The project duration is equal to the early finish time of the finishing activity, that is, $p=EF_{t=O}=155$ days. The backward calculation has to be done in the same way. The reverse order of the forward calculation can be used. We do not detail the backward calculations, but recommend the reader to do it alone. The results can be seen in Figure 5-14.

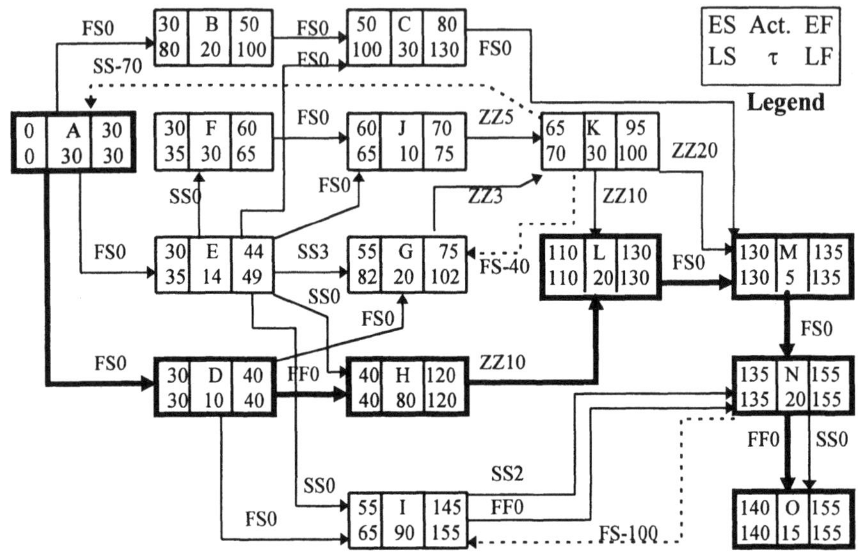

Figure 5-14 *PDM calculation for the sample project with maximal relationships.*

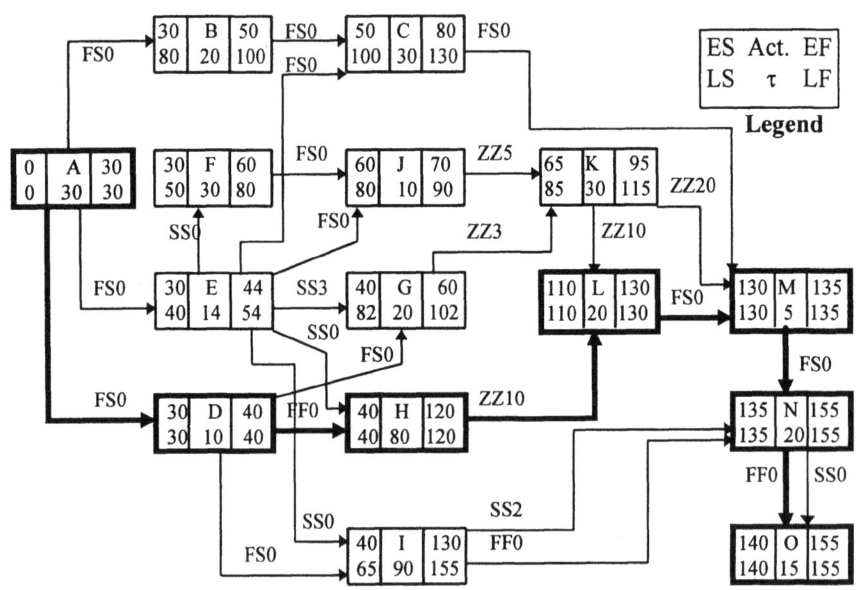

Figure 5-15 *PDM calculation without maximal relationships.*

5.4 INFORMATION OBTAINED FROM CALCULATIONS

When maximal type of relationships are used, one can define the same floats for activities, when only minimal types of precedence relationships were applied. The definition of floats, critical activities, and characteristics of critical activities are the same. The only difference is the form of the critical path. So far, a critical path has always been a chain from the initial to the terminal activity. When maximal types of precedence relationships are used, a critical path can form a tree. There should be a chain from the start to the end, the trunk of the tree, but there can be activities which are critical and connected to this chain like branches. We have illustrated this in Figure 5-16.

For better understanding the nature of the critical path one should recall that minimum two determinant precedence relationships are necessary to make an activity critical. (Except for the start and finish activities.)

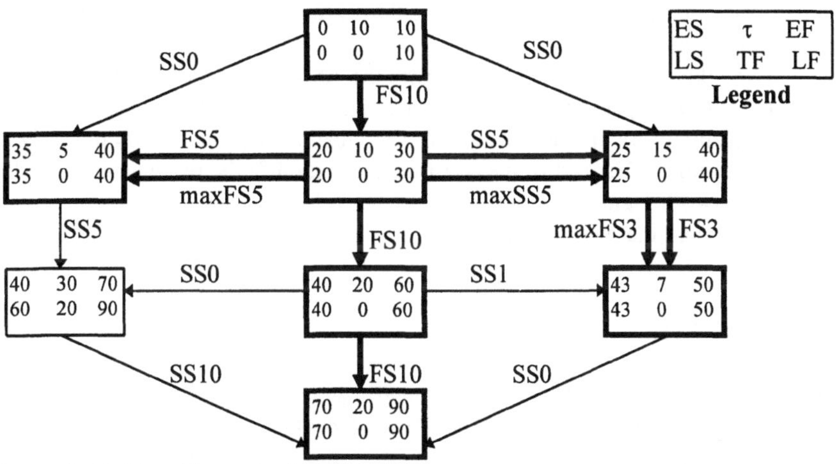

Figure 5-16 *Tree-like critical path*

One, which is a preceding relationship, defines the early activity times, and the other, a successor relationship, defines the late activity times. If these times are identical, then the activity is critical, and a critical path goes through the activity. It enters with the preceding determinant relationship and leaves with the succeeding determinant relationship. The situation is the same with the tree-like critical paths. The critical path does not end at the branches, but turns back and leaves these activities through the maximal type of precedence relationships. The critical path always goes through in the direction of the minimal type of precedence relationship, and the reverse direction of the maximal type of precedence relationship.

5.5 CONSTRAINTS IN PDM

In Chapter 4: PDM Calculations, we mentioned that the application of constraints is an efficient but dangerous tool. The reason is that constraints usually modify the results, but they do not appear in the network in a visible form. The user can easily forget where these constraints were applied, and the results without the knowledge of the proper place and type of the constraints can seem strange. Second, it is easy to apply constraints that collide with the network logic or each other, and can result in non-interpretable results. We also showed that the INITIAL TIME and TERMINAL TIME used in the calculation is nothing else than a 'Must start on or later' and a 'Must finish on or before' constraint, respectively.

We can classify these constraints into three groups. The first group contains the constraints, which force the activity to start or finish on a given date or later, and we call them minimal type of constraints. The second group contains the constraints that force the activity time and finish to a given date or earlier. These constraints are called maximal type of constraints. The third group of constraints is a mixture of the above mentioned two.

Two additional constraints are known, the ASAP, when activities have to start as soon as possible, and the ALAP, when activities have to start as late as possible. These constraints are unique because their application does not modify the early and late activity times. The rest of the constraints can modify the results and can collide with the network logic.

We discuss these constraints below.

Minimal Type Of Constraints

Definition 5-6 *(Must Start On Or Later constraint)*
The activity provided with this constraint can start at the date given in the constraint or later.

This constraint is also known as 'Start No Earlier Than' constraint. If such a constraint is used, then an INITIAL TIME has to be assigned to the early start of the activity in question, which value is equal to the value of the constraint. It can be replaced by an SS relationship from the initial activity to the activity in question.

Definition 5-7 *(Must Finish On Or Later constraint)*
The activity provided with this constraint can finish at the date given in the constraint or later.

This constraint is also known as 'Finish No Earlier Than' constraint. If such a constraint is used, then an INITIAL TIME has to be set to the early start of the activity in question, which value is equal to the value of the constraint minus the activity duration. It can be replaced by an FF relationship from the initial activity to the activity in question with the same lag value, which has been set in the constraint.

Maximal Type Of Constraints

Definition 5-8 *(Must Start On Or Before constraint)*
The activity provided with this constraint can start at the date given in the constraint or before.

This constraint is also known as 'Start No Later Than' constraint. If such a constraint is used, then a TERMINAL TIME has to be set to the late finish of the activity in question, which value is equal to the value of the constraint plus the duration of the activity. It can be replaced by a maxSS relationship from the initial activity to the activity in question.

Definition 5-9 *(Must Finish On Or Before constraint)*
The activity provided with this constraint can finish at the date given in the constraint or before.

This constraint is also known as 'Finish No Later Than' constraint. If such a constraint is used, then a TERMINAL TIME has to be set to the late finish of the activity in question, which value is equal to the value of the constraint. This constraint can be replaced by a maxFF relationship from the initial activity to the activity in question with the same lag value, which has been set in the constraint.

Mixed Type Of Constraints

These constraints can be replaced by a minimal and a maximal type of constraints, that is, by a minimal and a maximal type of precedence relationship.

Definition 5-10 *(Must Start On)*
The activity provided with this constraint must start at the date given in the constraint.

If such a constraint is used, then an INITIAL TIME and a TERMINAL TIME have to be set to the early start and to the late finish of the activity in question. The value of the INITIAL TIME has to be equal to the value of the constraint. The value of the TERMINAL TIME has to be equal to the value of the constraint plus the activity duration. This constraint can be replaced by an SS and a maxSS relationship from the initial activity to the activity in question, with the same lag values.

Definition 5-11 *(Must Finish On)*
The activity provided with this constraint has to finish on the date given in the constraint.

If such a constraint is used, then an INITIAL TIME and a TERMINAL TIME has to be set to the early start and to the late finish of the activity in question. The value of the INITIAL TIME is equal to the value of the constraint minus the activity duration. The value of the TERMINAL TIME will be equal to the value of the constraint. This constraint can be replaced by the use of an FF relationship and a maxFF relationship from the initial activity to the activity in question with the same lag value which has been set in the constraint.

Definition 5-12 *(Must Start Between)*
The activity provided with this constraint has to start within the given date interval given in the constraint.

If such a constraint is used, then an INITIAL TIME and a TERMINAL TIME have to be set to the early start and to the late finish of the activity in question.The value of the INITIAL TIME has to be equal to the smaller value of the constraint. The value of the TERMINAL TIME has to be equal to the greater value of the constraint plus the activity duration. This constraint can be replaced by an SSx and a maxSSy relationships from the initial activity to the activity in question, where x is the smaller date and y is the greater date.

Definition 5-13 *(Must Finish Between)*
The activity provided with this constraint has to finish within the dates given in the constraint.

If such a constraint is used, then an INITIAL TIME and a TERMINAL TIME have to be set to the early start and to the late finish of the activity in question. The value of the INITIAL TIME is equal to the smaller value of the constraint minus the activity duration. The value of the TERMINAL TIME will be equal to the value of the constraint. This constraint can be replaced by the use of an FFx relationship and a maxFFy relationship from the initial activity to the activity in question. The value of x stands for the smaller date in the constraint, and the value of y stands for the greater value in the constraints.

There are other types of constraints too, but the above mentioned are the most frequently used. The application of constraints holds a couple of dangers which are better to avoid.

Our main concern is about the original concept of network-based project methodology. The basic concept of this methodology is to set apart the planning and scheduling phases which have been taken into consideration simultaneously in non-network techniques. Constraints give the possibility to force the start and finish of activities on the desired date. If during the calculation it becomes evident that some of the constraints collide with the network logic, the bad practice is that the scheduler rather deletes or changes the relationships instead of examining the reason of the illogical results. This is a bad but convenient practice. In extreme cases one can produce a schedule solely using the constraints and not precedence logic. The less the number of constraints in the planning phase, the better the plan will be.

The other problem with using constraints that very few schedulers are aware of the fact that minimal type of constraints can be replaced by minimal type of precedence relationships, and maximal type of constraints can be replaced by maximal type of precedence relationships. In case of using constraints, one must always interpret these constraints like relationships and check if these relationships would really describe an existing logic. We have mentioned in Chapter 4: Notes on Calculation Rules that some authors suggest to set an INITIAL TIME equal to zero and a TERMINAL TIME equal to the project duration to each activity. This is nothing else but an SS0 and a maxFF*p* precedence relationship to all activities from the initial node. We are not sure that all the schedulers want this kind of logic automatically included in the network. One must examine very carefully the real logical dependencies of the project and apply precedence relationships that are closest to this logic.

The application of minimal type of constraints solely will never result in paradoxical results, e.g., a smaller late than early start time, if there are only minimal types of precedence relationships applied in the network. However, the application of the other two types of constraints can result in strange solutions.

If splitting is allowed in the network, then constraints have to be handled in different ways. In this cases two INITIAL TIMES and two TERMINAL TIMES have to be applied in the network.

5.6 PDM VERSUS CPM

In this chapter we discuss the differences between the CPM and PDM techniques. For this, we use the four-activities sewer construction sample we used in Chapter 2:Pros and Cons of CPM Technique.

A sewer has to be built in 1 km length, with a pumping plant. Activities have to be carried on continuously, without interruption, due to the resource limits. The project can be divided into four activities, and **splitting of activities is not allowed**.

A- groundwork (10 days)
B- pipe lay down (8 days)
C- refilling (10 days)
D - construction of pumping plant (20 days)

The safety distances between activities A, B, and C have to be two days. Activity A precedes B by ZZ2, B precedes C by ZZ2. Activity D does not depend on any other activities. The PDM network, and the result of the time analysis can be seen in Figure 5-17.

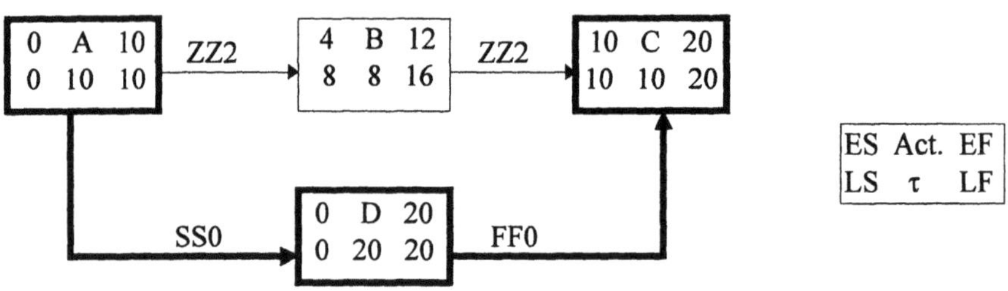

Figure 5-17 *PDM network for the sewer construction.*

Now, we are trying to model this project in CPM network. Remember, the problem is that due to the original CPM definitions (Definition 2-1 and Definition 2-2) an activity can start if all the preceding activities have finished.

A possible solution can be seen in Figure 5-18. We prescribed the two days overlapping between A and B by dividing activity A into two parts. Activity B can start when the first part of A, which takes two days, finished. After finishing of A can start the last two-day long part of B. This logic has to be used to ensure the necessary two days overlapping between activity B and C. As a result of this argument, a network consistsing of eight activities and six events has emerged.

Observing the CPM diagram (Figure 5-18) it can be seen that the early finish and start of activities B and C are not the same as in Figure 5-17. They are identical with the results of PDM calculation when activity splitting is allowed. No wonder, in the CPM representation we have broke the activities into pieces of different length, therefore we allowed the splitting of activities.

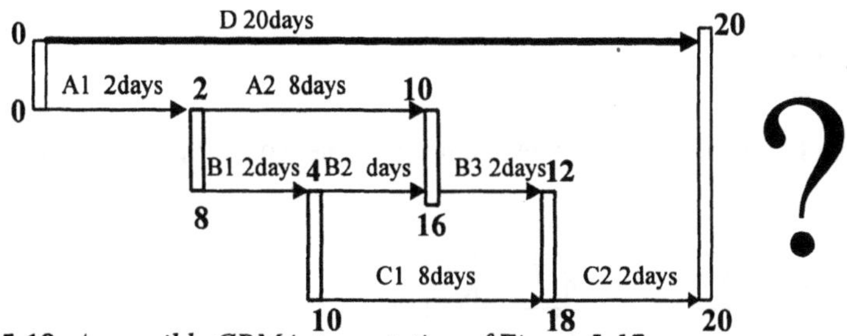

Figure 5-18 *A possible CPM interpretation of Figure 5-17*

The LSM interpretation of a possible accomplishment of A and B, which conforms to the CPM results can be seen in Figure 5-19. According to the CPM schedule, part of activity B can be accomplished before A. It means that pipe laying can be done where there is no ditch. This is obviously a wrong schedule. The reason for this is, than with activity splitting the possibility for making connections between whole activities and not just between their certain points, is lost. The other reason is that according to the CPM definitions there is no logical connection between activities A2 and B1 and between A2 and B2.

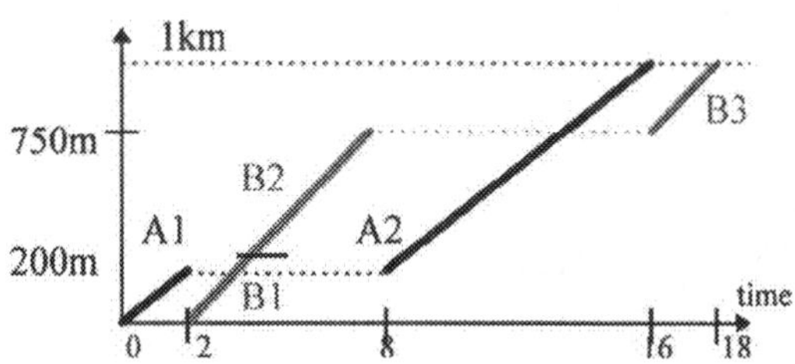

Figure 5-19 *A possible LSM interpretation of activities A and B*

Now, it is clear that, concurrent activities cannot be modeled in CPM if we use the original, that is, the splitting-not-allowed concept. If it is still not enough to prove that a PDM network cannot be a hundred percent translated into CPM, then we suggest to the skeptics to model the anomalous behavior of PDM by using CPM. Recall, in PDM, shortening an activity in the critical path will result in the increasing of the project duration if the activity is reverse critical.

Summarizing the above mentioned, PDM is a more efficient and flexible tool for modeling life-like situations, therefore it is recommended to be used instead of CPM for any type of projects.

5.7 Practical Problems and Solutions

Problems

1. Perform PDM calculations on the below given network! (Splitting is not allowed.)

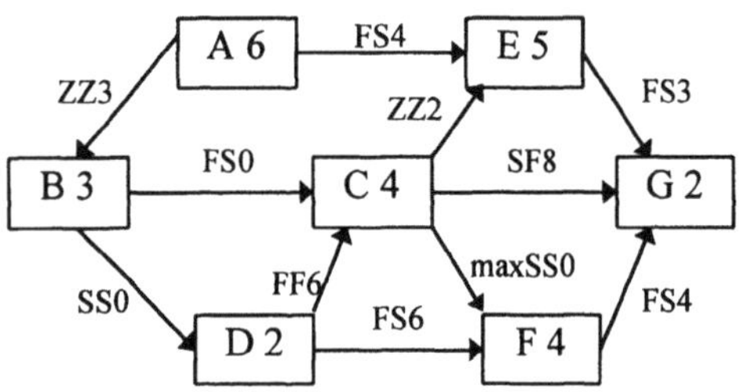

2. Perform PDM calculations on the below given network! (Splitting is not allowed.)

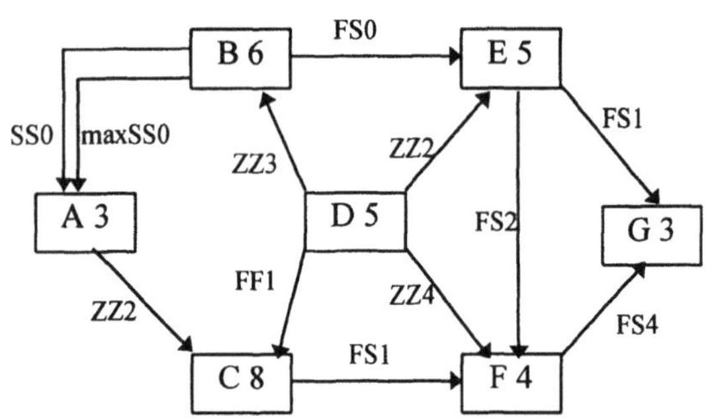

3. Perform PDM calculations on the below given network! (Splitting is not allowed.)

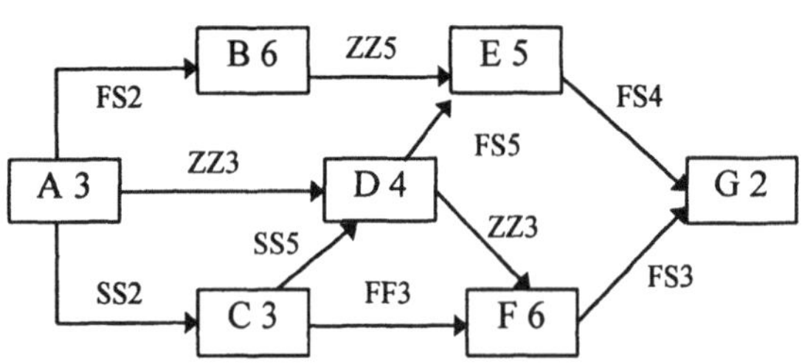

4. Perform PDM calculations on the below given network! (Splitting is not allowed.)

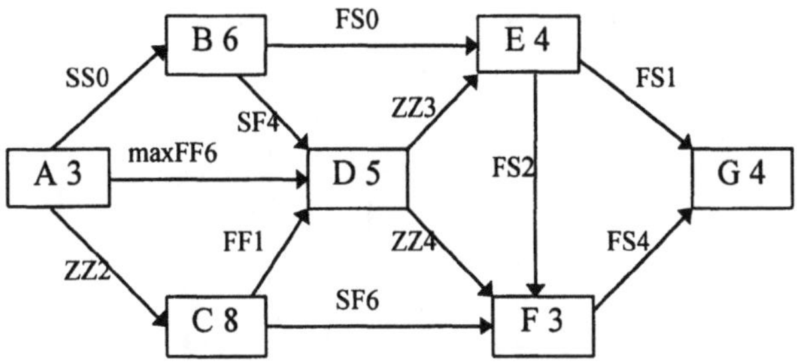

5. Perform PDM calculations on the below given network! (Splitting is not allowed.)

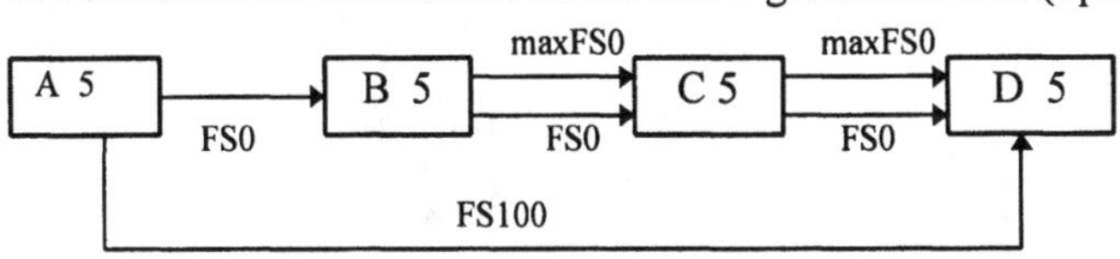

6. Give an algorithm for PDM time analysis which can handle both types of precedence relationships and both splittable and non-splittable activities.

7. Solve the network given in 1. with the activity splitting allowed assumption!

Solutions

1. The results can be seen below. Observe that all the activities are critical, and the critical path goes backward on the maxSS0 relationship.

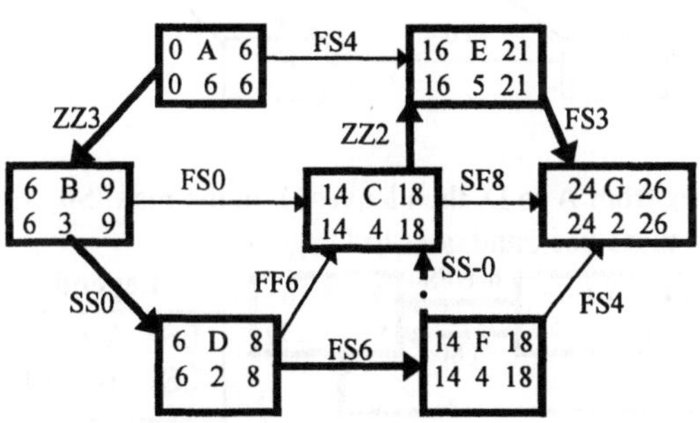

Legend

ES_i	Act.	EF_i
LS_i	τ_i	LF_i

2. The results can be seen below. Activity D is start critical, activities B, E, F and G are normal critical. The critical path is tree-like.

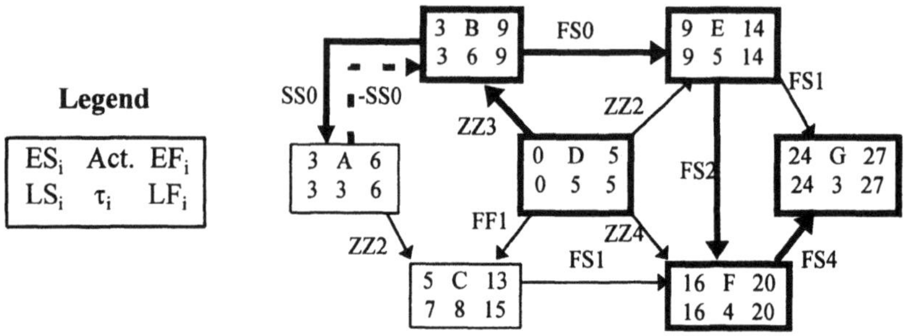

3. The results can be seen below. Activities A and C are start critical, D, E and G are normal critical.

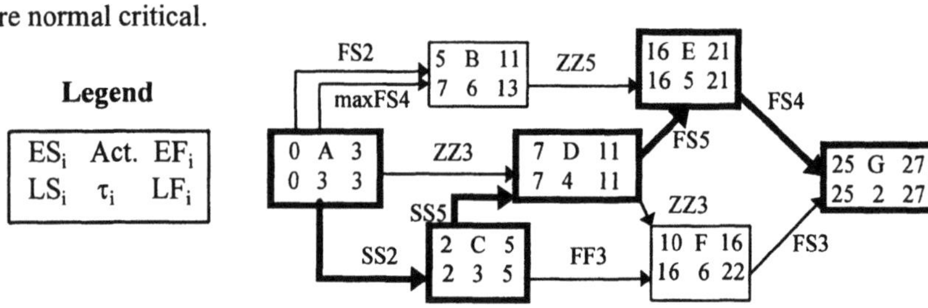

4. There is no feasible time policy for this network. The value of the loop (A;SS2;C;FF1;D;FF-6;A) is positive, that is, the iteration would never stop. (2+8+1-6-3=2)

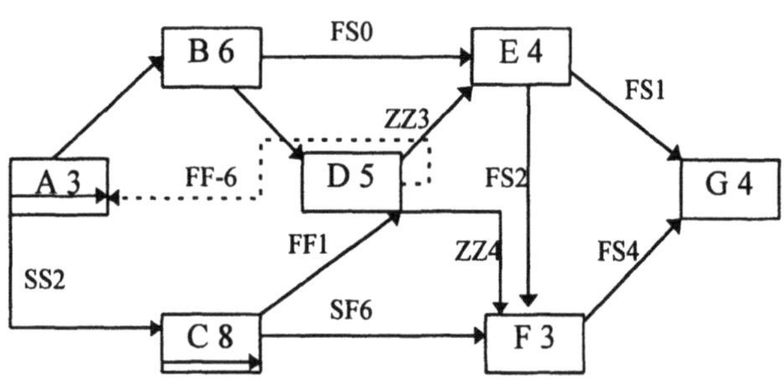

5. Observe the critical path. It goes from A to D, then leaves D on the maxFS0, then leaves C on the maxFS0. At B, it turns back and goes to D again.

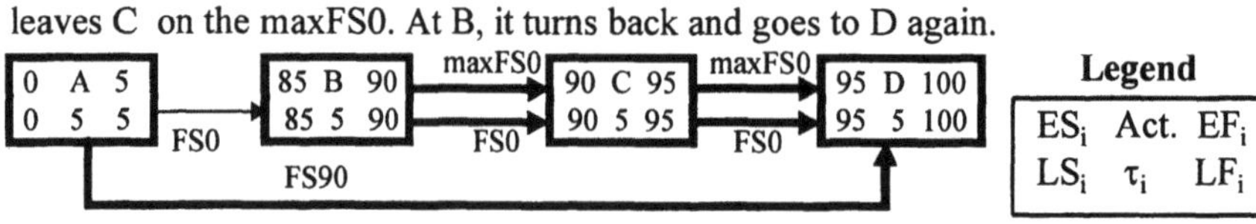

6. PDM Calculation: Maximal Relationships Allowed: Splitting Allowed

Forward Calculation

Step1. Let ES_s=0; $EF_s=ES_s+\tau_s$; $ES_i=-\infty$= INITIAL TIME for all the remaining activities, No_of_Iter:=0

Step 2.

LABEL 1;

f:=0

REPEAT

f:=*f*+1

IF there are neither SS nor FS relations run into *j AND j* is splittable

THEN ES_j=INITIAL TIME

ELSE $ES_j=\max\{ES_i+SSz_{ijk}, EF_i+FSz_{ijk}\}$ $\forall\{(ij)k\}\in\mathbf{A}$ and SS or FS.

$EF_j=\max\{ES_j+\tau_j, ES_i+SFz_{ijk}, EF_i+FFz_{ijk}\}$ $\forall\{(ij)k\}\in\mathbf{A}$ and SF or FF

IF j is non-splittable then $ES_j=EF_j-\tau_j$

IF j is splittable but splitting is limited then $ES_j:=\max\{ES_j; EF_j-\tau_j-d_j\}$

UNTIL f=n

No_of_Iter:=No_of_Iter+1;

Step 3.

IF No_of_Iter>*n THEN* HALT (There is no feasible activity time system)

IF there was a change in any ES or EF values *THEN* go to 1

IF there was no change in any value *THEN* the forward pass has been finished

Backward Calculations

Step1. Let $LF_t=p=ES_t$; $LS_t=LS_t-\tau_t$; $LF_i=\infty$=TERMINAL TIME $\forall\ i\in\mathbf{N}$-*t*

Step 2.

LABEL 2; f:=0

REPEAT

f:=*f*+1

IF there are neither FF nor FS relations run out from *i AND i* is splittable

THEN LF_i =TERMINAL TIME)

ELSE $LF_i=\min\{LF_j-FFz_{ijk}, LS_j-FSz_{ijk}\}$ for all FF, FS immediate succeed *i*.

$LS_i=\min\{LF_i-\tau_i, LF_j-SFz_{ijk}, LS_j-SSz_{ijk}\}$ for all SF, SS immediate succeed *j*.

IF j is non-splittable *THEN* $LF_i=LS_i+\tau_i$

IF j is splittable but splitting is limited *THEN* $LF_i:=\min\{LF_i; LS_i+\tau_i+d_i\}$

UNTIL f=n

Step 3.

IF there was a change in any LS, LF values *THEN* go to 2

IF there was no change in any value *THEN* the backward pass has been finished

7.

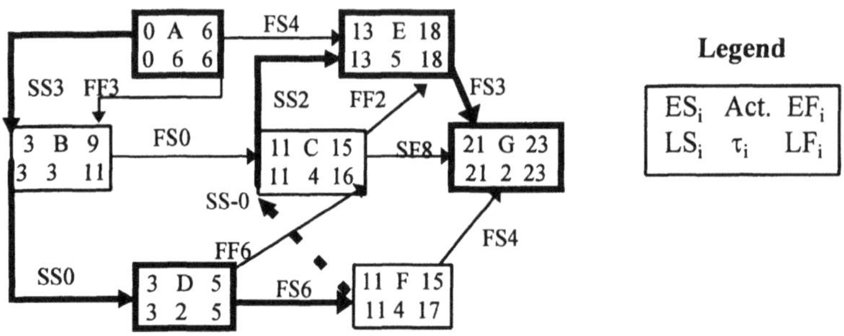

5.8 HISTORICAL REVIEW

As at the beginning of this chapter it has been mentioned that the application of the maximal type of relationships is not new. The original Method of Potentials, which was developed by Roy [58],[59] allowed the use of SS precedence relationship and maxSS relationship as well. He called it backward acting precedence relationships, and these were nothing else but the equivalent minimal relationships of the maximal type of precedence relationships. The approach followed in this book, that is, the application of maximal type of precedence relationships instead of their equivalent minimal relationship, differs from the original approach. We think this makes the methodology and the network logic easier to understand.

The lack of computers in the sixties and seventies and the time consuming hand computational procedure have worked against the application of maximal type of precedence relationships, therefore these types of precedence relationships have been ignored in practice. The algorithm applied in this text is a modification of the algorithm developed for finding the shortest paths to all nodes from the source. This algorithm was developed by Ford [19] and Bellman [7] and Roy integrated it into his method. For this, he had to modify the algorithm to search for longest paths in a network. The Ford-Bellman algorithm has been improved by Yen [67] in many aspects. One of his improvements has been adopted in our algorithm. In the original Ford-Bellman algorithm, during the kth iteration the ES/EF values of the k-1th have been used. Yen showed that using of the values of the kth iteration, if they have already been calculated, will result in a faster algorithm. Another significant improvement in speed that during the computational process each new iteration will result in values, which will not change, therefore one does not have to deal with them in the following iterations. We have not implemented this modification in our algorithm because it makes the algorithm longer and the understanding of it demands more mathematical considerations.

During the transformation of maximal type of precedence relationships to equivalent minimal type of precedence relationships, loops will be formed. If the value of a loop is positive, there is no feasible solution for the network. A nice proof of this feasibility theorem was developed by Kas [40].

Chapter 6

PRECEDENCE DIAGRAMMING WITH BOUNDED ACTIVITY DURATION

Making the first version of a schedule, planners usually count with normal crew sizes on normal work days, that is, they count with normal activity duration times. If the calculated project duration is beyond the deadline, then it has to be decreased. The planner has two possibilities to compress the project time span within the given boundaries. First, he or she can change the precedence relationships, make more activities partially or completely concurrent, or organize the resources more efficiently, which can result in deleting some precedence relationships. The second possibility is to shorten the activity durations, as much as needed for the schedule to meet the deadline. It would be a great help to know how much can the project duration be shortened, i.e., what the minimum project duration is. If the schedule according to the minimum project duration is still behind the desired schedule, then the only solution is to change the network logic.

The problem of the minimal and the maximal project duration is not so simple as it was in CPM technique. In PDM the minimal project duration can be less than the project duration calculated from the crash duration times, and the maximal project duration can be greater than the project duration calculated from normal duration times.

In the first section, we establish a sample project, which is going to be used through the chapter, for illustration.

In the second part, we give the calculation rules which govern the algorithm that can be applied when solely minimal type of precedence relations are used in the network.

In the third part, we discuss the algorithm when both minimal and maximal types of relationships are allowed.

In the fourth section, practical problems and their solutions can be found. In the last part we give a historical review.

6.1 INTRODUCING THE SAMPLE PROJECT

A covered depot has to be built to protect a large machine. The activities and the logical relationships between the activities are the same as they were in Chapter 4: Introducing the Sample Project. The activity durations of Chapter 4 will serve as normal durations. The crash durations are the same that were used in Chapter 3: CPM Least Cost Scheduling. The input data of this project are summarized in Table 6-1.

Activity	a_i Crash duration	b_i (Normal duration)	Preceding activities and precedence relations
A	20	30	-
B	15	20	A FS0;
C	20	30	B FS0; E FS0;
D	6	10	A FS0;
E	10	14	A FS0;
F	25	30	E SS0;
G	12	20	D FS0; E SS3;
H	50	80	D FS0; E SS0;
I	50	90	D FS0; E SS0;
J	5	10	F FS0; E FS0;
K	25	30	J ZZ5; G ZZ3;
L	15	20	K ZZ10; H ZZ10;
M	5	5	L FS0; C FS0; K ZZ20;
N	12	20	M FS0; I SS2; I FF0;
O	10	15	N FF0; SS0

Table 6-1 *Input data for the sample project.*

What is the minimum available project duration (p_{min}) in the PDM network which contains these input data? We answer the question in the following parts of this chapter. We premise that p_{min} is not necessarily equal to the project duration calculated from the crash durations (p_a). To prove this, let us examine Figure 6-1. The project duration calculated by crash durations is 20 days. If we change the activity duration of B to its normal duration, the project duration will change to 15 days, that is, increasing an activity duration can result in a decreasing of the project duration. The maximal project duration is also not necessarily equal to the project duration calculated by the normal durations. This can be seen in Figure 6-1.

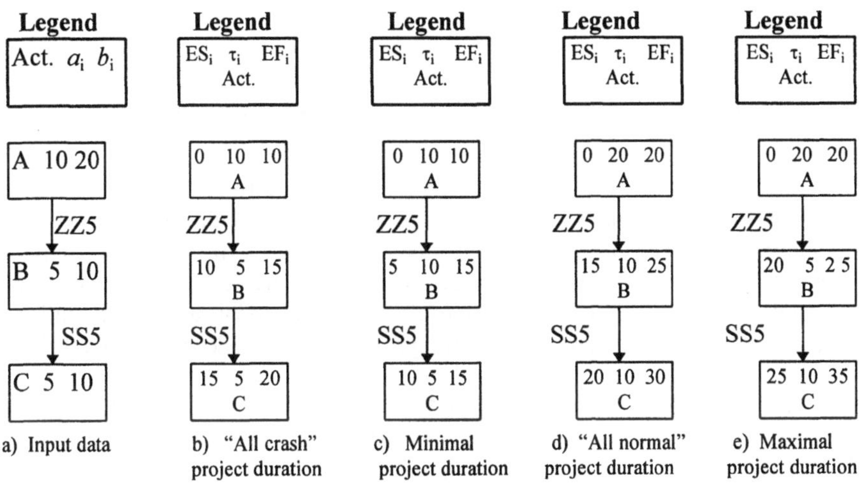

Figure 6-1 *p_a is not necessarily equal to p_{min}, and p_b is not necessarily equal to p_{max}.*

We use the following notations during the chapter:

n - number of activities
i - activity (i=1..n)
$\{(ij)k\}$ - the kth precedence relationship between i and j
N,**A** - set of activities and set of precedence relationships
s - start activity, no precedence relation runs into this activity
t - finish activity, no precedence relationship runs out of this activity
τ_i - duration time of activity i
a_i/b_i - crash duration and normal duration for an activity
ES/EF_i - early start/finish of activity i
LS/LF_i - late start/finish of activity i
π_{iS}, π_{iF} - arbitrary start and finish of activity i
p - project duration

p_{min} - minimal available project duration in the network
p_{max} - maximal available project duration in the network
p_a, p_b - project duration calculated from the crash/ normal durations
S_i, F_i - start/finish of an activity if splitting is allowed for the activity

The minimal project duration (p_{min}) is never equal to p_a if reverse critical activities can be found in the network. At the same time it indicates that $p_a=p_{min}$ and $p_b=p_{max}$ if the activity splitting allowed concept is used. In this case, activities cannot be reverse critical, therefore $p_{min}=p_a$ and $p_{max}=p_b$.

6.2 CALCULATIONS WITH MINIMAL RELATIONSHIPS

The mathematical model of the problem can be formulated in the following way. A PDM network is given. No loops or activity splitting are allowed. Any number of precedence relationships can be given between two activities. Activities have to be accomplished within their time boundaries, that is, between the crash and normal durations. Only minimal type of precedence relationships can be applied in the network. Knowing these, the following mathematical model can be established:

$a_i \geq \tau_i \geq b_i$	$\forall i \in \mathbf{N}$	1.
$\pi_{jS}-\pi_{iS} \geq z_{ijk}$	$\forall \{(i,j)k\} \in \mathbf{A}$ and SSz_{ijk}	2.a.
$\pi_{jF}-\pi_{iF} \geq z_{ijk}$	$\forall \{(i,j)k\} \in \mathbf{A}$ and FFz_{ijk}	2.b.
$\pi_{jF}-\pi_{iS} \geq z_{ijk}$	$\forall \{(i,j)k\} \in \mathbf{A}$ and SFz_{ijk}	2.c.
$\pi_{jS}-\pi_{iF} \geq z_{ijk}$	$\forall \{(i,j)k\} \in \mathbf{A}$ and FSz_{ijk}	2.d.
There should be at least one s→t path with $\pi_{tF}-\pi_{sS}$ length.		3.

In case of searching for minimal project duration:

$\pi_{tF}-\pi_{sS} \rightarrow \min$ 4.

In case of searching for maximal project duration:

$\pi_{tF}-\pi_{sS} \rightarrow \max$ 5.

The third condition ensures that at least one critical path will be in the network. In case of minimizing the objective function, this condition is automatically satisfied so it can be omitted from the model. In case of maximizing the objective function, it cannot be deleted because it would lead to an infinite value of π_{tF}.

Determining the Minimal Project Duration

In the solution of these problems we use Bellmann's dynamic programming theorem, that is, the project duration will be minimal if the values π of the immediate predecessors are minimal, and they will be minimal if the values π of their predecessors are minimal, etc.

During the algorithm we divide the set of activities into two sets, **S** and **T**. Each of the activities has to be either **S** or **T**, and none of them can be in both sets at the same time ($\mathbf{S}\cap\mathbf{T}=0$, $\mathbf{S}\cup\mathbf{T}=\mathbf{N}$). The activities in **S** have already been taken into consideration and they start and finish as soon as possible. At the beginning, activity s is the only element of S, $\pi_{sS}:=0$ and $\pi_{sF}:=a_s$. In the algorithm set **S** will be enlarged, set **T** will be shrunk , and the algorithm will finish if $\mathbf{S}\equiv\mathbf{N}$.

The steps of the algorithm can be summarized in the following manner:

Step 1. Let $S_i:=F_i:=-\infty$ $\forall i\in\mathbf{N}$; {place for INITIAL TIME}
$s\in\mathbf{S}$, $\mathbf{N}\text{-}s\in\mathbf{T}$; $\pi_{sS}=0$ and $\pi_{sF}=a_i$;

Step 2.
REPEAT
Take an activity j, which is not calculated yet,
but all its predecessors are in **S**.
IF there is no such an activity then *GO TO Step 3*.
$S_j:=\max\{\pi_{iS}+z_{ijk} : i\in\mathbf{S}, (ijk)\in\mathbf{A} \text{ and } SSz_{ijk};\ \pi_{iF}+z_{ijk}: i\in\mathbf{S}, (ijk)\in\mathbf{A} \text{ and } FSz_{ijk}\}$
$F_j:=\max\{\pi_{iS}+z_{ijk} : i\in\mathbf{S}, (ijk)\in\mathbf{A} \text{ and } SFz_{ijk};\ \pi_{iF}+z_{ijk}: i\in\mathbf{S}, (ijk)\in\mathbf{A} \text{ and } FFz_{ijk}\}$
IF $F_i-S_i\geq b_i$ *THEN* $(\pi_{iS}:=F_i-b_i,\ \pi_{iF}:=F_i,\ \tau_i:=b_i)$
IF $F_i-S_i\leq a_i$ *THEN* $(\pi_{iS}:=S_i,\ \pi_{iF}:=S_i+a_i,\ \tau_i:=a_i)$
IF $b_i>F_i-S_i>a_i$ *THEN* $(\pi_{iS}:=S_i,\ \pi_{iF}:=F_i,\ \tau_i:=F_i-S_i)$
S has to be enlarged by i, **T** has to be decreased by i.
UNTIL $\mathbf{S}\equiv\mathbf{N}$

Step 3. IF $\mathbf{S}\neq\mathbf{N}$ THEN HALT (There is a loop in the network)
ELSE $p_{\min}=\pi_{tF}$

Remarks to the algorithm:

- *Activities have been taken into account in the same order as in the pure PDM time analysis.*
- *The values of π_{iS} are equal to the values of ES_i and $\pi_{iF}=EF_i$.*

We illustrate the algorithm by solving the sample project. The calculation can be followed below, the results can be seen in Figure 6-2. For better understanding, we use the ES_i and EF_i notations instead of π_{iS} and π_{iF}. The minimal project duration is $p_{\min}=107$ days.

$ES_A = 0 \quad EF_A = 20;$

$S_B = \max\{EF_A+0\}=20$	$F_B = -\infty;$	$\Rightarrow ES_B=20$	$EF_B=35;$
$S_D = \max\{EF_A+0\}=20$	$F_D = -\infty;$	$\Rightarrow ES_D=20$	$EF_D=26;$
$S_E = \max\{EF_A+0\}=20$	$F_E = -\infty;$	$\Rightarrow ES_E=20$	$EF_E=30;$
$S_F = \max\{ES_E+0\}=20$	$F_F = -\infty;$	$\Rightarrow ES_F=20$	$EF_F=45;$
$S_C = \max\{EF_B+0;\ EF_E+0\}=35$	$F_C = -\infty;$	$\Rightarrow ES_C=35$	$EF_C=55;$
$S_G = \max\{ES_E+3;\ EF_D+0\}=26$	$F_G = -\infty;$	$\Rightarrow ES_G=26$	$EF_G=38;$
$S_H = \max\{EF_D+0;\ ES_E+0\}=26$	$F_H = -\infty;$	$\Rightarrow ES_H=26$	$EF_H=76;$
$S_I = \max\{EF_D+0;\ ES_E+0\}=26$	$F_E = -\infty;$	$\Rightarrow ES_I=26$	$EF_I=76;$
$S_J = \max\{EF_E+0;\ EF_F+0\}=45$	$F_J = -\infty;$	$\Rightarrow ES_J=45$	$EF_J=50;$
$S_K = \max\{ES_J+5;\ ES_G+3\}=50$			
	$F_K=\max\{EF_J+5; EF_G+5\}=55;$	$\Rightarrow ES_K=50$	$EF_K=75;$
$S_L = \max\{ES_K+10;\ ES_H+10\}=60$			
	$F_L=\max\{EF_K+10; EF_H+10\}=86;$	$\Rightarrow ES_L=66$	$EF_L=86;$
$S_M = \max\{EF_C+0;\ ES_K+20; EF_L+0\}=85$			
	$F_M=\max\{EF_K+20\}=95;$	$\Rightarrow ES_M=90$	$EF_M=95;$
$S_N = \max\{EF_M+0;\ ES_I+2\}=95$	$F_N=\{EF_I+0\}=76;$	$\Rightarrow ES_N=95$	$EF_N=107;$
$S_O = \max\{ESN+0\}=95$	$F_O=\{EF_N+0\}=107;$	$\Rightarrow ES_O=95$	$EF_O=107;$

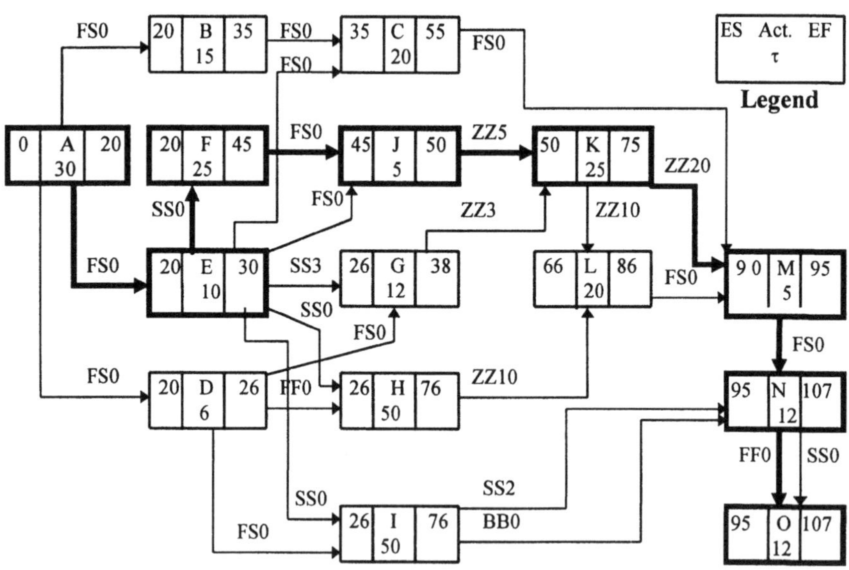

Figure 6-2 *Minimal project duration calculations for the sample project.*

Determining the Maximal Project Duration

The algorithm for determining the maximal project duration is based on the same idea as the algorithm determining the minimal project duration, but slightly differs it. In the algorithm we divide the set of activities into two sets, **S** and **T**. Each of the activities has to be either **S** or **T**, and none of them can be in both sets at the same time ($S \cap T=0$, $S \cup T=N$). Activities of **S** have already been taken into consideration and they start and finish as soon as possible. At the beginning, activity s is the only element of **S**, $\pi_{sS}:=0$ and $\pi_{sF}:=b_s$. In the algorithm, set **S** will be enlarged, set **T** will be shrunk, and the algorithm will finish if $S \equiv N$.

The steps of the algorithm can be summarized in the following manner:

Step 1. Let be $S_i:=F_i:=-\infty \ \forall i \in \mathbf{N}$; {place for INITIAL TIME}
$s \in \mathbf{S}$, $\mathbf{N}\text{-}s \in \mathbf{T}$; $\pi_{sS}=0$ and $\pi_{sF}=b_i$;

Step 2.
REPEAT
Take an activity j, which is not calculated yet,
but all its predecessors are in **S**.
IF there is no such an activity then *GO TO Step 3.*
$S_j:=\max\{\pi_{iS}+z_{ijk} : i \in \mathbf{S}, (ijk) \in \mathbf{A} \text{ and } SSz_{ijk};\ \pi_{iF}+z_{ijk}: i \in \mathbf{S}, (ijk) \in \mathbf{A} \text{ and } FSz_{ijk}\}$
$F_j:=\max\{\pi_{iS}+z_{ijk} : i \in \mathbf{S}, (ijk) \in \mathbf{A} \text{ and } SFz_{ijk};\ \pi_{iF}+z_{ijk}: i \in \mathbf{S}, (ijk) \in \mathbf{A} \text{ and } FFz_{ijk}\}$
IF $F_i-S_i \geq b_i$ *THEN* ($\pi_{iS}:=F_i-a_i$, $\pi_{iF}:=F_i$, $\tau_i:=a_i$)
IF $F_i-S_i \leq a_i$ *THEN* ($\pi_{iS}:=S_i$, $\pi_{iF}:=S_i+b_i$, $\tau_i:=b_i$)
IF $b_i>F_i-S_i>a_i$ *THEN* we double the activity (see Figure 6-3)
(for the first activity $\pi_{iS}:=F_i-a_i$, $\pi_{iF}:=F_i$, $\tau_i:=a_i$
for the second activity ($\pi_{iS}:=S_i$, $\pi_{iF}:=S_i+b_i$, $\tau_i:=b_i$)
S has to be enlarged by i, **T** has to be decreased by i.
(This means two nodes in case of $b_i>F_i-S_i>a_i$)
UNTIL $S \equiv N$

Step 3. IF $S \neq N$ THEN STOP (There is a loop in the network)
ELSE $p_{max}=\pi_{tF}$

Remarks to the algorithm:

- *We have taken into accoun the activitiest in the same order as we used in the PDM time analysis.*
- *The π_{iS}, π_{iF} values are equal to the ES_i, EF_i values, which can be calculated by the τ_i system.*
- *In case of doubled activities, it is not clear which activity has to be chosen in order to achieve the maximal project duration. To decide this, a simple time analysis has to be made on the network, which contains the doubled activities, and also the critical path has to be identified. If none of the doubled activities*

are laying on the critical path, then any of the durations can be chosen. If one of them is laying on the critical path, then this activity determines the project duration, so this one has to be chosen. If both of them are laying on the critical path, then either can be chosen.

We illustrate the algorithm by solving the sample project. The calculation can be followed below and the results can be seen in Figure 6-4. For better understanding, we have used the ES_i and EF_i notations instead of π_{iS} and π_{iF}. The maximal project duration is p_{max}=155 days.

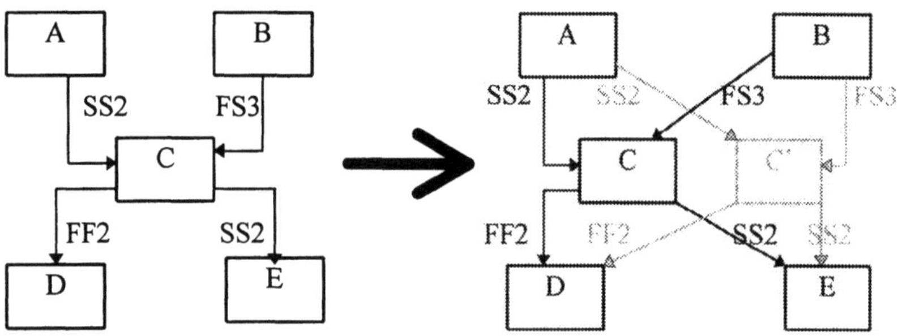

Figure 6-3 *Doubled activities in the calculation of the maximum project duration*

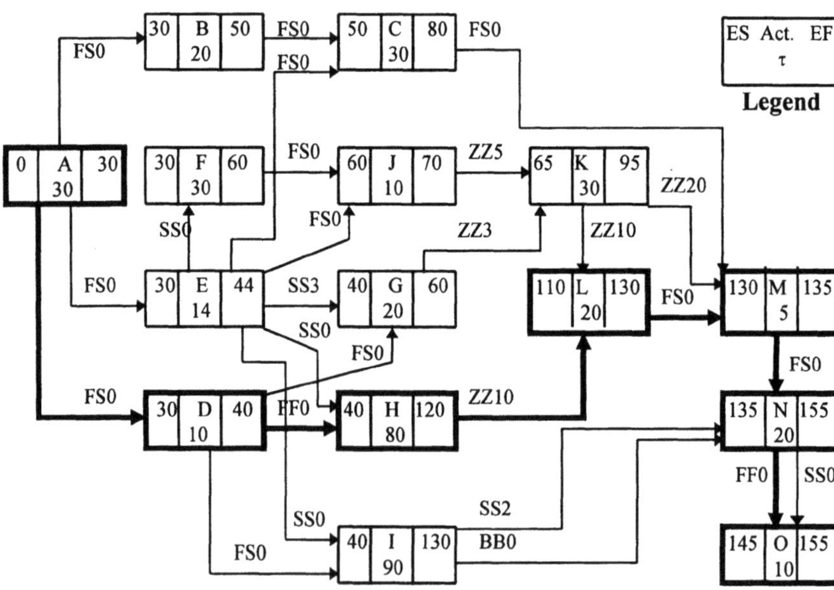

Figure 6-4 *Maximal project duration calculation of the sample project*

$ES_A = 0 \quad EF_A = 30;$
$S_B = \max\{EF_A+0\} = 30 \quad F_B = -\infty; \quad \Rightarrow ES_B = 30 \quad EF_B = 50;$
$S_D = \max\{EF_A+0\} = 30 \quad F_D = -\infty; \quad \Rightarrow ES_D = 30 \quad EF_D = 40;$
$S_E = \max\{EF_A+0\} = 30 \quad F_E = -\infty; \quad \Rightarrow ES_E = 30 \quad EF_E = 44;$
$S_F = \max\{ES_E+0\} = 30 \quad F_F = -\infty; \quad \Rightarrow ES_F = 30 \quad EF_F = 60;$
$S_C = \max\{EF_B+0;\ EF_E+0\} = 50 \quad F_C = -\infty; \quad \Rightarrow ES_C = 50 \quad EF_C = 80;$
$S_G = \max\{ES_E+3;\ EF_D+0\} = 40 \quad F_G = -\infty; \quad \Rightarrow ES_G = 40 \quad EF_G = 60;$
$S_H = \max\{EF_D+0;\ ES_E+0\} = 40 \quad F_H = -\infty; \quad \Rightarrow ES_H = 40 \quad EF_H = 120;$
$S_I = \max\{EF_D+0;\ ES_E+0\} = 40 \quad F_E = -\infty; \quad \Rightarrow ES_I = 40 \quad EF_I = 130;$
$S_J = \max\{EF_E+0;\ EF_F+0\} = 60 \quad F_J = -\infty; \quad \Rightarrow ES_J = 60 \quad EF_J = 70;$
$S_K = \max\{ES_J+5;\ ES_G+3\} = 65$
$F_K = \max\{EF_J+5; EF_G+5\} = 75; \quad \Rightarrow ES_K = 65 \quad EF_K = 95;$
$S_L = \max\{ES_K+10;\ ES_H+10\} = 75$
$F_L = \max\{EF_K+10; EF_H+10\} = 130; \Rightarrow ES_L = 110 \quad EF_L = 130;$
$S_M = \max\{EF_C+0;\ ES_K+20; EF_L+0\} = 130$
$F_M = \max\{EF_K+20\} = 115; \quad \Rightarrow ES_M = 130 \quad EF_M = 135;$
$S_N = \max\{EF_M+0;\ ES_I+2\} = 135 \quad F_N = \{EF_I+0\} = 130; \Rightarrow ES_N = 135 \quad EF_N = 155;$
$S_O = \max\{ES_N+0\} = 135 \quad F_O = \{EF_N+0\} = 155; \Rightarrow ES_O = 145 \quad EF_O = 155;$

6.3 CALCULATIONS WITH MAXIMAL RELATIONSHIPS

Searching for the minimal and maximal project duration when maximal type of precedence relationships are allowed, we face the same problem we had in the time analysis. In case of maximal type of precedence relationships, the above given algorithms will stop with the message that there is a loop in the network.

The problem is the same and the solution is the same too. We take the activities into account in any order, and perform the calculations for them. If we finished the calculations for all the activities and there were changes in any value then the iteration has to be done again.

The algorithms for the two cases are summarized below.

Determining the Minimal Project Duration

Step 1.

Let $S_i := F_i := -\infty \ \forall i \in \mathbf{N}$; {place for INITIAL TIME}
$\pi_{sS} = 0$ and $\pi_{sF} = a_i$; No_of_Iter:=0

Step 2.
LABEL 1
f:=0
REPEAT
f:=f+1;
Take any j activity, which has not been calculated yet in this iteration.
$S_j:=\max\{\pi_{iS}+z_{ijk} : i\in\mathbf{S}, (ijk)\in\mathbf{A} \text{ and } SSz_{ijk} ;\ \pi_{iF}+z_{ijk}: i\in\mathbf{S}, (ijk)\in\mathbf{A} \text{ and } FSz_{ijk}\}$
$F_j:=\max\{\pi_{iS}+z_{ijk} : i\in\mathbf{S}, (ijk)\in\mathbf{A} \text{ and } SFz_{ijk};\ \pi_{iF}+z_{ijk}: i\in\mathbf{S}, (ijk)\in\mathbf{A} \text{ and } FFz_{ijk}\}$
IF $F_i-S_i \geq b_i$ *THEN* $(\pi_{iS}:=F_i-b_i, \pi_{iF}:=F_i, \tau_i:=b_i)$
IF $F_i-S_i \leq a_i$ *THEN* $(\pi_{iS}:=S_i, \pi_{iF}:=S_i+a_i, \tau_i:=a_i)$
IF $b_i>F_i-S_i>a_i$ *THEN* $(\pi_{iS}:=S_i, \pi_{iF}:=F_i, \tau_i:=F_i-S_i)$
UNTIL f=n
Step 3. No_of_Iter:=No_of_Iter+1;
IF No_of_Iter>n *THEN* HALT (There is no feasible activity time system)
IF there was a change in any π_{iS} or π_{iF} value *THEN* GOTO 1
ELSE $p_{min}=\pi_{tF}$

Determining the Maximal Project Duration

Step1. Let $S_i:=F_i:= -\infty\ \forall i\in\mathbf{N}$; {place for INITIAL TIME}
$\pi_{sS}=0$ and $\pi_{sF}=a_i$; No_of_Iter:=0; f:=0
Step 2.
LABEL 1:
f:=0;
REPEAT
f:=f+1
Take any activity, which has not been calculated yet.
$S_j:=\max\{\pi_{iS}+z_{ijk} : (ijk)\in\mathbf{A} \text{ and } SSz_{ijk} ;\ \pi_{iF}+z_{ijk}: (ijk)\in\mathbf{A} \text{ and } FSz_{ijk}\}$
$F_j:=\max\{\pi_{iS}+z_{ijk} : (ijk)\in\mathbf{A} \text{ and } SFz_{ijk};\ \pi_{iF}+z_{ijk}: (ijk)\in\mathbf{A} \text{ and } FFz_{ijk}\}$
IF $F_i-S_i \geq b_i$ *THEN* $(\pi_{iS}:=F_i-a_i, \pi_{iF}:=F_i, \tau_i:=a_i)$
IF $F_i-S_i \leq a_i$ *THEN* $(\pi_{iS}:=S_i, \pi_{iF}:=S_i+b_i, \tau_i:=b_i)$
IF $b_i>F_i-S_i>a_i$ *THEN* the activity has to be doubled (see Fig. 6-3)
for the firs activity $(\pi_{iS}:=F_i-a_i, \pi_{iF}:=F_i, \tau_i:=a_i)$
for the second activity $(\pi_{iS}:=S_i, \pi_{iF}:=S_i+b_i, \tau_i:=b_i)$
UNTIL f=n
Step 3. No_of_Iter:=No_of_Iter+1;
IF No_of_Iter>n *THEN* HALT (There is no feasible activity time system)
IF there was a change in any π_{iS} or π_{iF} value *THEN* GOTO 1

Although the above given algorithms are easy to understand, it is advisable to implement them using computers. The algorithms giving the maximal project durations are especially troublesome, because in extreme cases the number of activities and precedence relationships can be doubled. The algorithms, which can handle the maximal type of precedence relationships are also not suggested for hand computations.

6.4 PRACTICAL PROBLEMS AND SOLUTIONS

Problems

1. Determine the minimal and maximal project durations in the below given network!

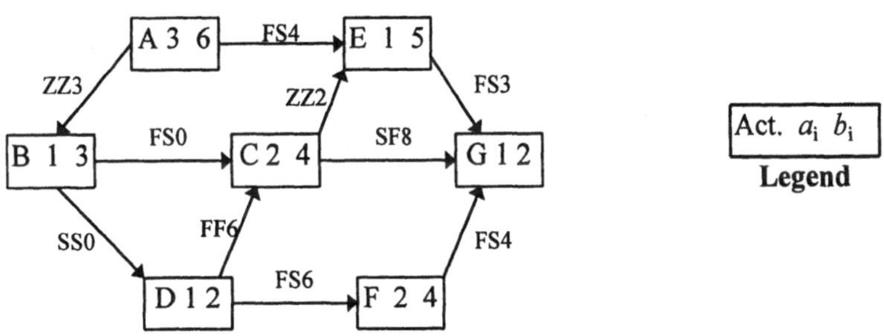

2. Determine the minimal and maximal; project duration in the below given network!

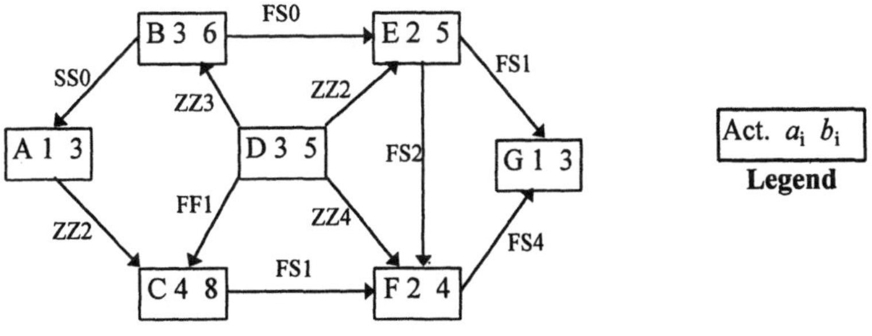

3. Determine the minimal and maximal project durations in the below given network!

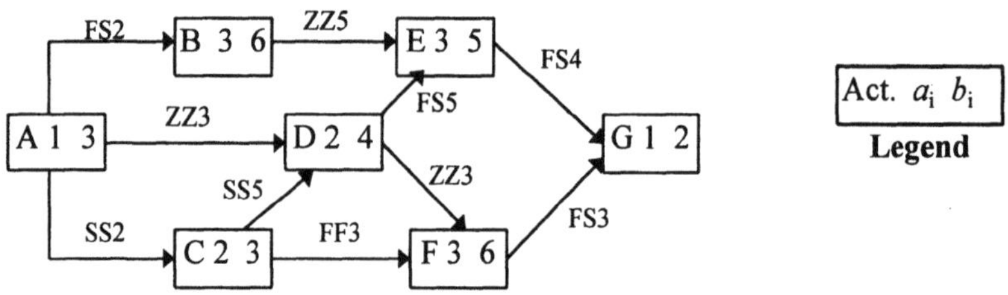

4. Determine the minimal and maximal project durations in the below given network!

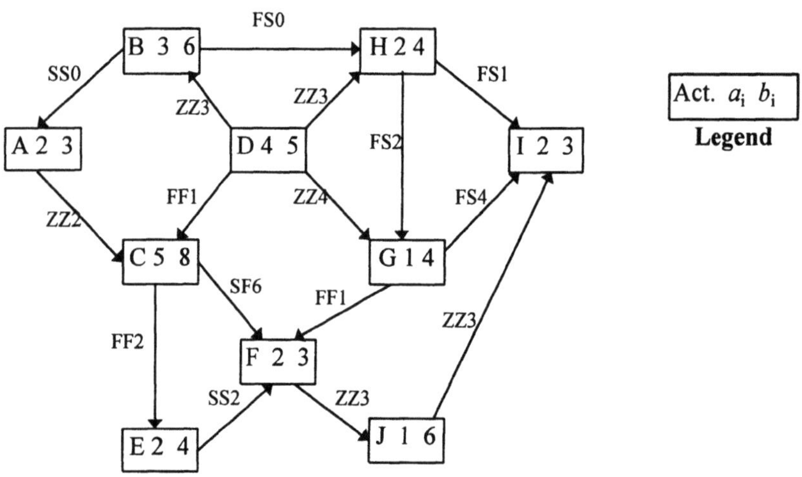

Solutions

1. p_{min}=17 days, p_{max}=26 days.

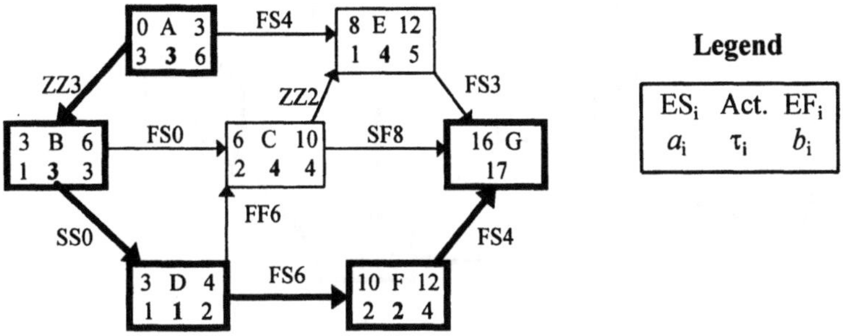

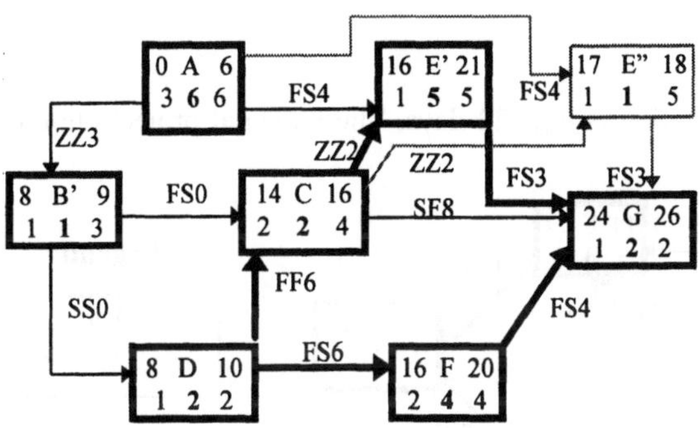

2. The minimal project duration is p_{min}=17 days, p_{max}=27 days. Observe that in the second case both B' and B" are critical, that is, the maximal project duration is available in two different ways, one when $\tau_B=a_B$ and one when $\tau_B=b_B$. Choosing τ_B between its boundaries will result in a smaller project duration.

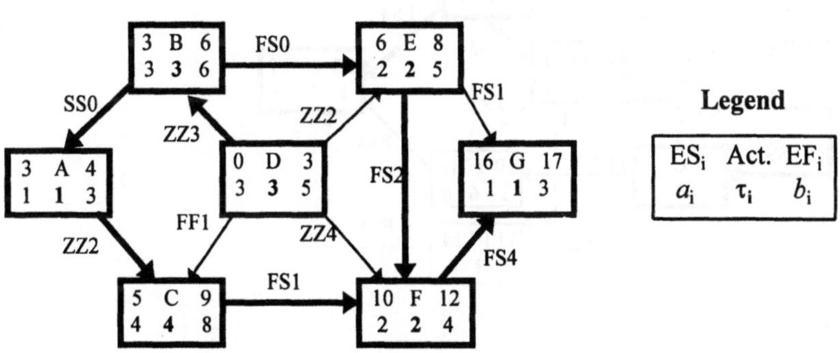

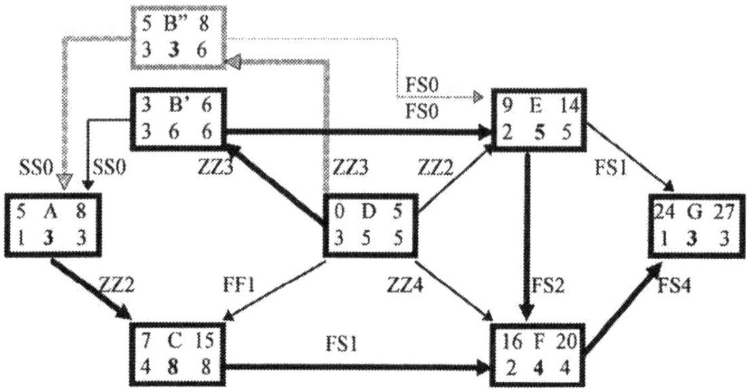

3. The minimal project duration is p_{min}=22 days. The maximal project duration is p_{max}=27 days.

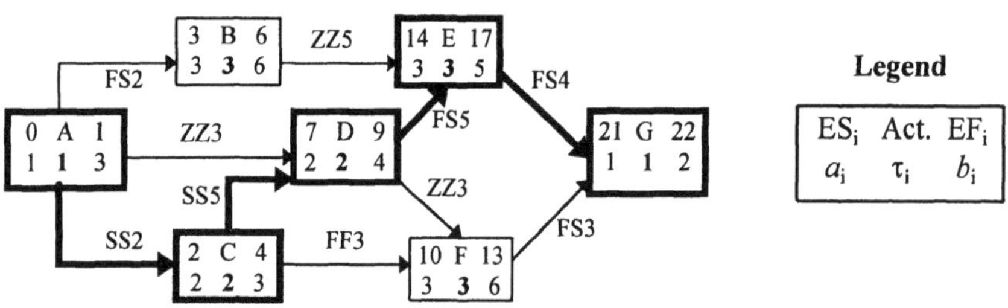

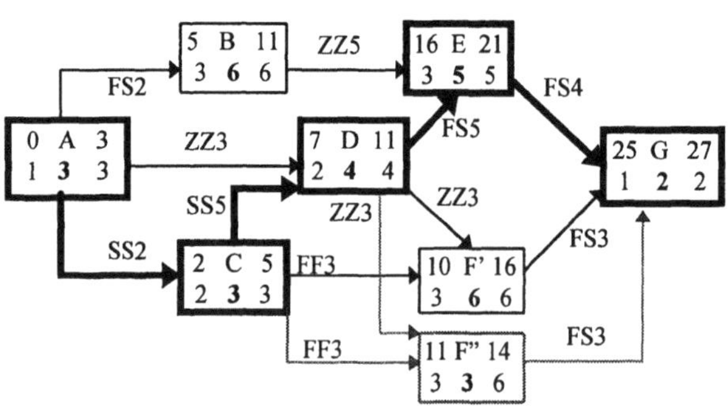

4. The minimal project duration is p_{min}=19 days. The maximal project duration is p_{max}=30 days.

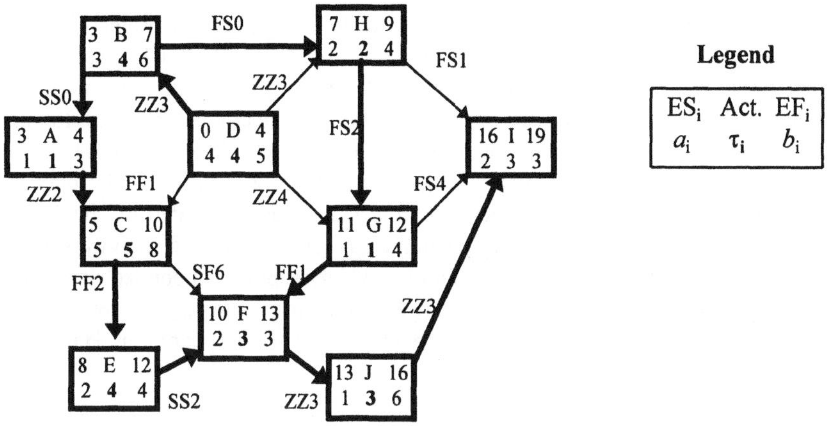

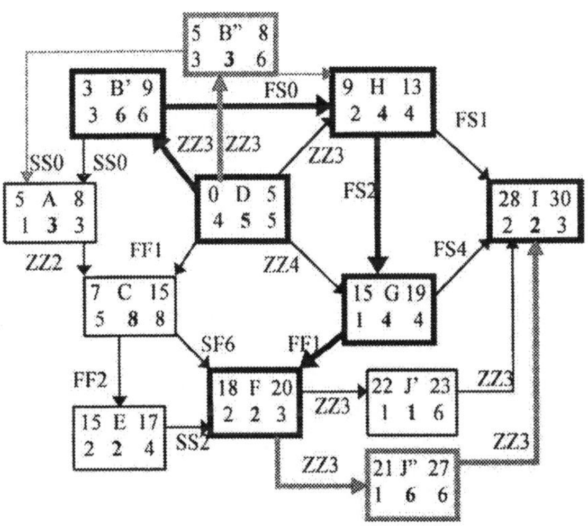

6.5 HISTORICAL REVIEW

Changes of activity durations can have different impacts on project duration. This topic has been comprehensively discussed by Weist [65], who was first to characterize the criticality term in PDM. He has also showed the anomalous behavior of the PDM network, which is due to the reverse critical activities. Reverse critical activities are the reasons why the minimal project duration can be smaller than the project duration calculated by the crash durations, and why the maximal project duration can be greater than the project duration calculated by the normal durations. However, this interesting phenomenon can only arise if activity splitting is not allowed, or if it is allowed, the duration of interruption is bounded. If splitting is allowed and not limited, then p_{min} is always equal to p_a and p_{max} is always equal to p_b. The algorithms given in this chapter originally appeared in Hajdu and Mál振usz's [31] paper.

Chapter 7

PDM LEAST COST SCHEDULING

In Chapter 3: CPM Least Cost Scheduling we discussed that the CPM time-cost trade-off procedure chooses the duration time for each activity so as to minimize the total project costs to a given project duration. Despite that the advantages of this method are apparent, there can hardly be found industrial applications for this method. The main reason for this is that finding proper cost data is very tiresome. This process can sometimes be longer than the whole planning and scheduling phase itself.

The PDM time-cost trade-off can be used for the same purpose as the original CPM, that is, to minimize the total project direct cost and at the same time to satisfy the constraints on the project completion date, the deadline. However, there might be an another important field of usage. In the first chapter of this book we mentioned that planning and scheduling are always iterative processes. The first schedule is rarely satisfactory, the schedule usually extends far beyond the desired project dates. During the iterative process the scheduler can modify the network logic and the activity durations. However, changing an activity duration can unexpectedly change the project duration, as it was comprehensively discussed in the previous chapter. It is almost impossible to predict what changes will emerge if we change a duration by one or two days. Good example for this is Problem 2 in Chapter 6. The maximal project duration is 27 days, and to achieve this, all the activities have to be accomplished in their normal durations. Shortening activity B by one day will shorten the project duration by one day, too. Shortening the activity duration by another day will keep the project duration unaffected. Shortening B by one day will again result in one day increment in project duration. If during the iteration process we change the duration of different activities, may be at the same time, then the changes are unpredictable. Any changes in the duration of an arbitrary activity or modification of the lag time of any precedence relationship can modify the behavior of hundreds of activities.

PDM time cost trade-off can help solving this problem in an efficient way. For this, one only has to produce the crash durations and provide the expected project duration. The PDM time-cost trade-off procedure will result in the desired project duration without unneeded changes in activity durations. Instead of cost slopes, priorities can also be used in the method. We can give a priority one to an activity on a 1 to 10 scale if we want to reduce this activity's duration first in case of necessity, and a priority 10 if we want to avoid reducing the activity duration if possible.

Only PDM time-cost trade-off can help us to achieve the desired project duration in a PDM network efficiently. We discuss this methodology in the present chapter.

In the first part, Introducing The Sample Project, we establish the sample project, which will be used through the chapter.

In the second section, Differences Between CPM and PDM Cost Curves, we discuss the differences between the CPM and PDM time-cost trade-off problems, and we show, why heuristic methods cannot be used in PDM time-cost trade-off.

In the third chapter, Exact Solution, we give a solution based on network flow theory. The knowledge of the CPM time-cost trade off algorithm, which was shown in Chapter 3, makes the PDM time-cost trade-off algorithm easy to understand.

In the following part, Special Cases, we discuss the changes which can modify the algorithm. In the fifth part, Practical Problems and Solutions, we give exercises.

In the last part of this chapter we give a historical review.

7.1 INTRODUCING THE SAMPLE PROJECT

A covered depot has to be built from prefabricated standardized steel elements, in order to protect large industrial equipment, which is used for engine production. The project has been divided into fifteen processes from A to O. The activity interdependencies are the same as in Chapter 4.

The accomplishing time of most processes can vary depending on the number of crew, the power of the applied machines, and the length of the workdays. Project managers usually assume a normal-size crew, normal workdays, and 'owned' equipment in deciding the activity durations. Activity durations that result from this kind of argument are called *normal durations*. The activity duration, which stands for the fastest possible accomplishing time is called the *crash duration*. We use the same crash and normal durations as in Chapter 3.

Changing the activity durations will affect the project duration and the project cost. From the project's point of view, we can distinguish between two types of costs.

The first type of cost is called the *direct cost*, and indicates the cost which can be directly attributed to a task. The sum of activities' direct costs, which are related to the project, is called the *project direct cost*.

The other type of cost is called the *indirect* or *overhead*, sometimes *fixed* cost. This type of cost indicates those cost factors that are not easily attributable to activities, and rather belong to the whole project. The sum of project direct cost and indirect cost is called the *project total cost*.

In the PDM/cost model the emphasis is on the changes of direct costs. The direct cost that belongs to normal activity duration is called the *normal cost*. Shortening the duration of an activity to its crash duration usually results in higher accomplishing cost. Premium for overtime will increase the accomplishing cost of an activity. In order to shorten the activity, material cost can also increase. For example in concrete work the use of expensive rapid setting cement increases the costs too. The cost related to crash duration is called the *crash cost*. The nature of change of the cost can be seen in Figure 7-1.

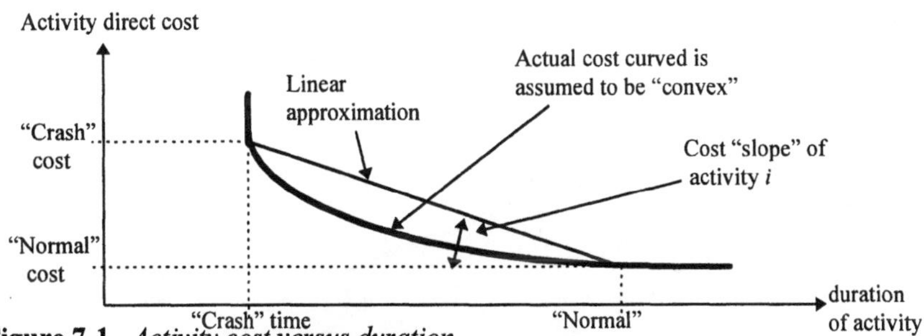

Figure 7-1 *Activity cost versus duration*

In the PDM/cost model a linear approximation[i] is used instead of the convex activity time-cost curve. Using the linear approximation, the so called *cost slope* can be determined. In the example in this chapter, we are going to use the cost data of the CPM/cost model. The input data for the sample project can be seen in Table 7-1.

[i]There are other types of cost curves when the assumption of linearity cannot be used. We discuss these cases in this chapter, in the section "Special Cases".

Act.	a_i Crash duration	b_i Normal duration	Cc_i Crash cost	Cn_i Normal cost	c_i Cost slope	Preceding activities and precedence relations
A	20	30	200	100	10	-
B	15	20	50	30	4	A FS0;
C	20	30	600	400	20	B FS0; E FS0;
D	6	10	50	30	5	A FS0;
E	10	14	230	150	20	A FS0;
F	25	30	195	120	15	E SS0;
G	12	20	154	90	8	D FS0; E SS3;
H	50	80	370	160	7	D FS0; E SS0;
I	50	90	520	280	6	D FS0; E SS0;
J	5	10	720	600	24	F FS0; E FS0;
K	25	30	1500	1200	60	J ZZ5; G ZZ3;
L	15	20	350	150	40	K ZZ10; H ZZ10;
M	5	5	200	200	0	L FS0; C FS0; K ZZ20;
N	12	20	1200	800	50	M FS0; I SS2; I FF0;
O	10	15	210	150	12	N FF0; SS0

Table 7-1 *Input data for the sample project.*

In a network, where activity duration can change between its lower and upper bound, different project durations are available; what is more, the same project duration can be achieved in hundreds of different ways. The set of the feasible solutions can be seen in Figure 7-2. Two points of the feasible solution set come forward automatically. These are the 'all crash' and 'all normal' points.

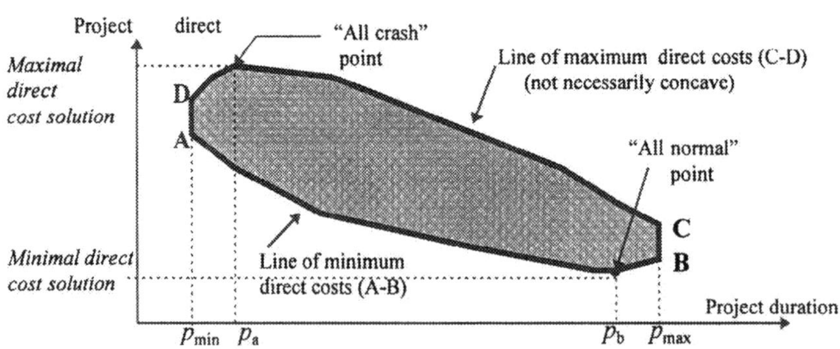

Figure 7-2 *Project duration versus project direct cost in PDM/cost model.*

The contour of the feasible solution set differs from the contour of the solution set of the CPM/cost model. We explain the reasons for this in the following section.

7.2 DIFFERENCES BETWEEN CPM AND PDM COST CURVES

The set of feasible solutions for the PDM network can be seen in Figure 7-2. The set of feasible solutions for the CPM network can be seen in Figure 7-3. Several differences can be observed between the two sets.

First, in the CPM network the maximal project duration is equal to the project duration calculated from the normal durations, that is, $p_{max}=p_b$ and the minimal project duration is also equal to the project duration calculated from the crash durations. In Precedence Diagramming Method, p_{max} is not necessarily equal to p_b and p_{min} is not necessarily equal to p_a. We discussed this phenomenon in the previous chapter, and pointed out that that reverse critical activities are in the behind this anomalous behavior.

The second basic difference is that the shape of the two sets are not identical. The 'all crash' point, which belongs to p_a project duration, gives the highest project direct cost. The pmin project duration can be achieved if reverse critical activities will be lengthened, but this results in a lower project direct cost if the cost slopes of these activities are not equal to zero. Using the same logic, it can be proved that the minimal cost solution that belongs to p_{max} has to be higher than the minimal cost solution that belongs to p_b.

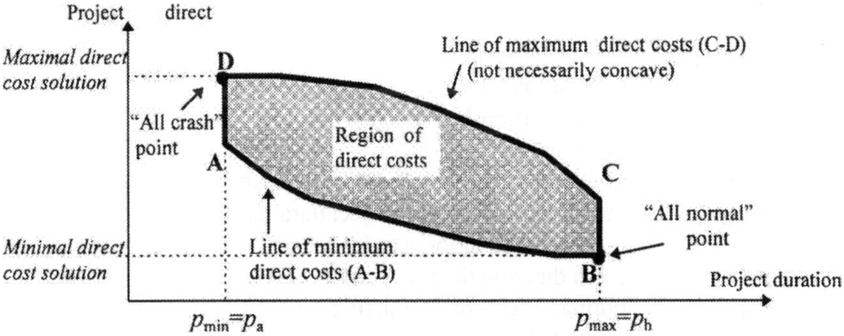

Figure 7-3 *Project duration versus project cost in CPM/cost model.*

The third basic difference between the CPM/cost and the PDM/cost method is the way project duration changes as activity duration changes. Shortening a critical activity in CPM - in case of one critical path - always results in a decreasing project duration. In PDM, depending on the characteristics of the critical activity, the project duration can increase, decrease or remain unaffected. As the characteristics of an activity in question can change if a simple lag factor or activity duration far away from the activity in question changes, the application of heuristic algorithms makes no practical sense.

7.3 EXACT SOLUTION

In the solution of the PDM least cost scheduling problem we follow the same methodology we followed in the solution of the CPM time-cost trade-off model. First we give the mathematical model of the time-cost trade-off problem, which is an optimization problem. After this we formulate another problem using the same network and the same data. The first model will be referred to as the primal, the second as the dual problem. We are going to solve these two models together. In the solution, we follow the methodology which has been followed in the CPM time-cost trade-off problem, but the solutions will differ since the structure of the PDM network differs from that of the CPM network.

We use the following notations in the algorithm:

N	- set of activities
S	- subset of **N** which contains the initial activity
T	- subset of **N** which contains the terminal activity
A	- set of precedence relationships
s, t	- the initial and terminal activities
i	- a discretionary activity
$\{(i,j)k\}$	- precedence relationship from i to j
τ_i	- duration of activity i
π_{iS}/π_{iF}	- start/finish of activity i
ES_i/EF_i	- earliest start/finish of activity i
LS_i/LF_i	- latest start/finish of activity i
p	- project duration
p_a/p_b	- project duration calculated from crash/normal durations
p_{min}/p_{max}	- the minimal/maximal project duration
b_i	normal duration for activity i
a_i	- crash duration time for activity i
Cn_i	- normal direct cost for activity i
Cc_i	- crash direct cost for activity i
c_i	- cost slope for activity i

Let [**N**,**A**] be a directed loopless graph with one source (s) and sink (t). Let **N** be the set nodes. Nodes stand for the activities that have to be carried out continuously with the same intensity. Multiple arcs are allowed between nodes. Arcs stand for the logical relationships between activities. The $\{(ij)k\}$ notation is used to identify a precedence relationship between activity i and activity j. The four basic minimal types of relationships can be used in the model. Let the duration of the ith activity be τ_i which has to be between its normal (b_i) and crash (a_i) durations.

Let a Cn_i normal and a Cc_i crash cost be given to all the activities which show the cost of activity i if its accomplishing duration is b_i or a_i. The assumption of the model is that the crash cost is equal or greater than the normal cost, and the change of the costs are linear between Cn_i and Cc_i.

Knowing all this, we can determine the cost slope c_i which shows how much the normal cost increases when the duration of activity i decreases by one day. All this can be seen in Figure 7-4.

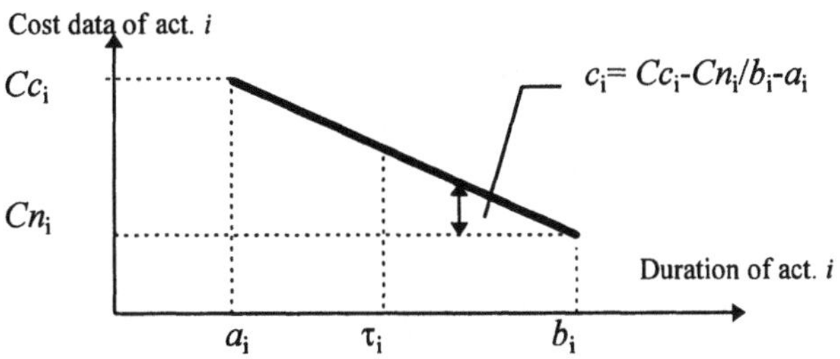

Figure 7-4 *Activity cost versus activity duration assumption in PDM/cost.*

In a network where τ_i is not fixed but can change between its lower and upper bound, different project durations are posssible. What is more, the same project duration can be achieved in several different ways. It is evident that in case of different cost slopes c_i different total costs will correspond to the same project duration. Our goal is to minimize the total cost that corresponds to a given p project duration.

Let us denote the beginning of activity i by π_{iS} and the finishing of it by π_{iF}. Knowing all these, our mathematical model, hereinafter referred to as the PDM/cost primal problem, is the following:

$a_i \le \tau_i \le b_i$	$\forall i \in \mathbf{N}$	**7.1**
$\pi_{iF}-\pi_{iS}=\tau_i$	$\forall i \in \mathbf{L}$	**7.2**
$\pi_{jS}-\pi_{iS} \ge z_{(ij)k}$	$\forall\{(i,j),k\} \in \mathbf{A}$ and $SSz_{(ij)k}$	**7.3**
$\pi_{jF}-\pi_{iS} \ge z_{(ij)k}$	$\forall\{(i,j),k\} \in \mathbf{A}$ and $FSz_{(ij)k}$	**7.4**
$\pi_{jF}-\pi_{iF} \ge z_{(ij)k}$	$\forall\{(i,j),k\} \in \mathbf{A}$ and $FFz_{(ij)k}$	**7.5**
$\pi_{jS}-\pi_{iF} \ge z_{(ij)k}$	$\forall\{(i,j),k\} \in \mathbf{A}$ and $SFz_{(ij)k}$	**7.6**
$\pi_{sS}=0$		**7.7**
$\pi_{tF}=p$		**7.8**
$\min(\sum_{N} \{Cn_i+(b_i-\tau_i)c_i\}$		**7.9**

As Cn_{ij} and $(b_{ij}*c_{ij})$ are constants the above objective function (**7.9**) is equivalent to :

$$\max\left(\sum_{N} \{c_i \tau_i\}\right) \qquad 7.10$$

A dual problem corresponding to the primal problem is established herein. These problems will be solved together.

To establish the dual problem, we have to convert the nodes (activities) into arrows. The starting node of activity i will be denoted by (iS), the finishing node of the activity arrow will be denoted by (iF). After the conversion all the SSz_{ji} FSz_{ji} precedence relationships have to run into (iS) and all the FFz_{ji} and SFz_{ji} precedence relationships have to run into (iF). The SSz_{ij} and SFz_{ij} relationships will run out from (iS) and the FFz_{ij} and FSz_{ij} relations have to run out from (iF). A sample transformation can be seen in Figure 7-5. This transformation does not change the primal problem.

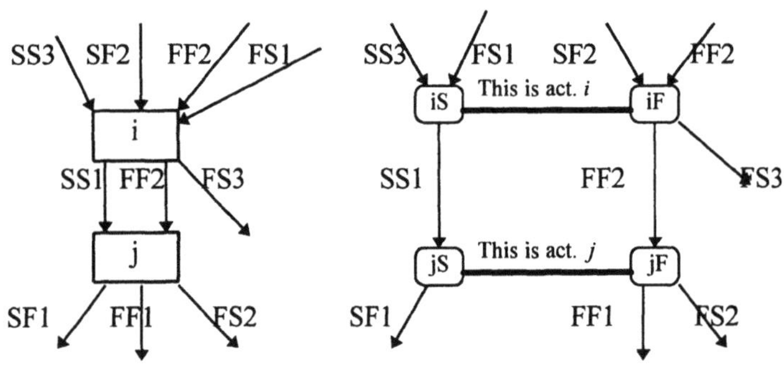

Figure 7-5 *Transformation of PDM network.*

The dual problem can be formulated in the following manner:

$$\sum_{j} \varphi_{(iS,j)k} - \sum_{j} \varphi_{(j,iS)k} + \varphi_{iS,iF} = \begin{cases} 0 & \text{if } i = \mathbf{N} - s \\ \Theta & \text{if } \quad i = s \end{cases} \qquad \forall\, i \in \mathbf{N} \qquad 7.11$$

$$\sum_{j} \varphi_{(iF,j)k} - \sum_{j} \varphi_{(j,iF)k} - \varphi_{iS,iF} = \begin{cases} 0 & \text{if } i \in \mathbf{N} - t \\ -\Theta & \text{if } \quad i = t \end{cases} \qquad \forall\, i \in \mathbf{N} \qquad 7.12$$

$$\varphi_{ijk} \geq 0 \qquad \forall\, \{(ij)k\} \in \mathbf{A} \qquad 7.13$$

$$\varphi_{iS,iF} \geq -\infty \qquad \forall i \in \mathbf{N} \qquad 7.14$$

Our goal is to minimize the following objective function:

$$\min\{p\Theta - \sum_{A} z_{ijk}\varphi_{ijk} + \sum_{\substack{c > \varphi_{iS,iF} \\ N}} [c_i - \varphi_{iS,iF}] b_i + \sum_{\substack{c < \varphi_{iS,iF} \\ N}} [c_i - \varphi_{iS,iF}] a_i \} \qquad 7.15$$

Conditions (**7.11**) and (**7.12**) are the normal flow conditions. They prescribe that the sum of the flows that enter a node is equal to the sum of the flows that leave a node. Condition (**7.14**) tells us that the flow can pass through an activity from both direction.

We are going to solve the primal and dual problem together. First we show that the dual problem is really a dual of the primal task. The following lemma points out the strict connection between the primal and dual problem. This lemma and its consequences are leading to the solution of the problem.

Lemma 7-1: If there exist policies π and τ satisfying the primal (**7.1**)-(**7.8**) conditions and a flow φ satisfying the dual (**7.11**)- (**7.14**) conditions, then (**7.10**)≤(**7.15**), that is:

$$\sum_{N} c_i\tau_i \leq p\Theta - \sum_{A} z_{ijk}\varphi_{ijk} + \sum_{\substack{c>\varphi_{iS,iF} \\ N}}[c_i - \varphi_{iS,iF}]b_i + \sum_{\substack{c<\varphi_{iS,iF} \\ N}}[c_i - \varphi_{iS,iF}]a_i \qquad \mathbf{7.16}$$

and equality holds if and only if the following are satisfied:

$$\left.\begin{array}{llll}
\text{if} \ \pi_{jS}-\pi_{iS}>z_{(ij)k} & \text{then } \varphi_{(ij)k}=0 & \forall & \{(i,j),k\}\in \mathbf{A} \text{ and } SSz_{(ij)k} \\
\text{if} \ \pi_{jF}-\pi_{iS}> z_{(ij)k} & \text{then } \varphi_{(ij)k}=0 & \forall & \{(i,j),k\}\in \mathbf{A} \text{ and } FSz_{(ij)k} \\
\text{if} \ \pi_{jF}-\pi_{iF}> z_{(ij)k} & \text{then } \varphi_{(ij)k}=0 & \forall & \{(i,j),k\}\in \mathbf{A} \text{ and } FFz_{(ij)k} \\
\text{if} \ \pi_{jS}-\pi_{iF}> z_{(ij)k} & \text{then } \varphi_{(ij)k}=0 & \forall & \{(i,j),k\}\in \mathbf{A} \text{ and } SFz_{(ij)k}
\end{array}\right\} \quad 1^{o}$$

$$\begin{array}{lllll}
\text{if} \ b_i > \tau_i & \text{then} & c_i \leq \varphi_{iS,iF} & \forall & i\in\mathbf{N} \qquad 2^{o} \\
\text{if} \ a_i < \tau_i & \text{then} & c_i \geq \varphi_{iS,iF} & \forall & i\in\mathbf{N} \qquad 3^{o}
\end{array}$$

Proof: We replace b_i and a_i by τ_i, z_{ijk} by the left side of the primal conditions (**7.3**)-(**7.6**) and τ_i by the left side of primal condition (**7.2**). In this case the value of the dual objective function will decrease or remain the same. If we can prove for the so modified function that it is greater or equal than the primal objective function, the original statement will be proved.

For the sake of brevity, we suppose that only one precedence relationship can be between two activities, that is, we can use z_{ij} instead of z_{ijk} and φ_{ij} instead of φ_{ijk}.

First we replace b and a by τ. With this the value of the dual objective function will decrease or remain the same.

$$\sum_{N} c_i\tau_i \leq p\Theta - \sum_{A} z_{ij}\varphi_{ij} + \sum_{N}[c_i - \varphi_{iS,iF}]\tau_i$$

After substitution we come to the following expression:

$$0 \leq p\Theta - \sum_{A} z_{ij}\varphi_{ij} - \sum_{N} \varphi_{iS,iF}\tau_i$$

We replace τ_i by the left side of (**7.2**)

$$0 \le p\Theta - \sum_{N} z_{ij}\varphi_{ij} + \sum_{N} \pi_{iS}\varphi_{iS,iF} - \sum_{N} \pi_{iF}\varphi_{iS,iF}$$

We replace z_{ij} by the left side of conditions (7.3)-(7.6), so $-\Sigma z_{ij}\varphi_{ij}$ will change as follows:

$$-\sum_{A_{SS}}(\pi_{jS} - \pi_{iS})\varphi_{ij} - \sum_{A_{FF}}(\pi_{jF} - \pi_{iF})\varphi_{ij} - \sum_{A_{FS}}(\pi_{jS} - \pi_{iF})\varphi_{ij} - \sum_{A_{SF}}(\pi_{jF} - \pi_{iS})\varphi_{ij}$$

after substitution the $-\Sigma z_{ij}\varphi_{ij}$ expression will change to

$$-\sum_{SS,FS}\pi_{jS}\varphi_{ij} - \sum_{FF,SF}\pi_{jF}\varphi_{ij} + \sum_{SS,SF}\pi_{iS}\varphi_{ij} + \sum_{FF,FS}\pi_{iF}\varphi_{ij}$$

so the right side of the equilibrium will look like this:

$$p\Theta - \sum_{A_{SS,FS}}\pi_{jS}\varphi_{ij} - \sum_{A_{FF,SF}}\pi_{jF}\varphi_{ij} + \sum_{A_{SS,SF}}\pi_{iS}\varphi_{ij} + \sum_{A_{FF,FS}}\pi_{iF}\varphi_{ij} + \sum_{i}\pi_{iS}\varphi_{iS,iF} - \sum_{i}\pi_{iF}\varphi_{iS,iF}$$

Now we are going to transform the $p\Theta$ expression in the following way:

$$p\Theta \Rightarrow [\pi_{sS}(-\Theta) + \pi_{sF}0] + [\pi_{2S}0 + \pi_{2F}0] + \ldots + [\pi_{jS}0 + \pi_{jF}0] + \ldots + [\pi_{tS}0 + \pi_{tF}\Theta]$$

$$p\Theta \Rightarrow \sum_{j}\pi_{jS}[\sum_{i_{SS,FS}}\varphi_{ij} - \varphi_{jS,jF} - \sum_{i_{SS,SF}}\varphi_{ji}] + \sum_{j}\pi_{jF}[\sum_{i_{SF,FF}}\varphi_{ij} + \varphi_{jS,jF} - \sum_{i_{FS,FF}}\varphi_{ji}]$$

$$p\Theta \Rightarrow \sum_{A_{SS,FS}}\pi_{jS}\varphi_{ij} - \sum_{j}\pi_{jS}\varphi_{jS,jF} - \sum_{A_{SS,SF}}\pi_{js}\varphi_{ji} + \sum_{A_{SF,FF}}\pi_{jF}\varphi_{ij} + \sum_{j}\pi_{jF}\varphi_{jS,jF} - \sum_{A_{FS,FF}}\pi_{jF}\varphi_{ji}$$

After substitution and simplification we come to the following expression:

$0 \le 0$

With this the inequality is proved, and equality exists if and only if

if	$\varphi_{(ij)k} > 0$ then $\pi_{jS} - \pi_{iS} = z_{(ij)k}$		$\forall$	$\{(i,j),k\} \in \mathbf{A}$ and $SSz_{(ij)k}$	
if	$\varphi_{(ij)k} > 0$ then $\pi_{jF} - \pi_{iS} = z_{(ij)k}$		$\forall$	$\{(i,j),k\} \in \mathbf{A}$ and $FSz_{(ij)k}$	
if	$\varphi_{(ij)k} > 0$ then $\pi_{jF} - \pi_{iF} = z_{(ij)k}$		$\forall$	$\{(i,j),k\} \in \mathbf{A}$ and $FFz_{(ij)k}$	1^{oo}
if	$\varphi_{(ij)k} > 0$ then $\pi_{jS} - \pi_{iF} = z_{(ij)k}$		$\forall$	$\{(i,j),k\} \in \mathbf{A}$ and $SFz_{(ij)k}$	
if	$c_i > \varphi_{iS,iF}$	then $b_i = \tau_i$	$\forall$	$i \in \mathbf{N}$	2^{oo}
if	$c_i < \varphi_{iS,iF}$	then $a_i = \tau_i$	$\forall$	$i \in \mathbf{N}$	3^{oo}

Reversing these,

if	$\pi_{jS} - \pi_{iS} > z_{(ij)k}$	then $\varphi_{(ij)k} = 0$	$\forall$	$\{(i,j),k\} \in \mathbf{A}$ and $SSz_{(ij)k}$	
if	$\pi_{jF} - \pi_{iS} > z_{(ij)k}$	then $\varphi_{(ij)k} = 0$	$\forall$	$\{(i,j),k\} \in \mathbf{A}$ and $FSz_{(ij)k}$	
if	$\pi_{jF} - \pi_{iF} > z_{(ij)k}$	then $\varphi_{(ij)k} = 0$	$\forall$	$\{(i,j),k\} \in \mathbf{A}$ and $FFz_{(ij)k}$	1°
if	$\pi_{jS} - \pi_{iF} > z_{(ij)k}$	then $\varphi_{(ij)k} = 0$	$\forall$	$\{(i,j),k\} \in \mathbf{A}$ and $SFz_{(ij)k}$	
if	$b_i > \tau_i$	then $c_i \le \varphi_{iS,iF}$	$\forall$	$i \in \mathbf{N}$	2°
if	$a_i < \tau_i$	then $c_i \ge \varphi_{iS,iF}$	$\forall$	$i \in \mathbf{N}$	3°

These are exactly the equilibrium conditions stated in the lemma.

Now, we have proved that the primal objective function usually has smaller value than the dual objective function, and that the two objective functions have the same value if certain conditions are satisfied. In the consequence of the lemma we prove that if the values of the objective functions are equal, then both solutions are optimal.

Theorem 7-1 (*weak form of equilibrium*): If there exist policies π and τ to a given project duration p satisfying the primal **(7.1)-(7.8)** conditions and a φ flow satisfying the dual **(7.11)-(7.14)** conditions and the value of both objective functions are equal **(7.10)=(7.15)**, then the solutions of both problems are optimal.

Proof (*indirect way*). Let us denote the value of the primal objective function (5) by P and that of the dual by D.

Let us assume that P=D, but there exists a better primal solution P* where P*>P. In this case P*>P=D but this is a contradiction according to the lemma.

Let us assume that P=D, but there exists a better dual solution D* where D*<D. In this case D*<D=P but this is a contradiction according to the lemma.

In this way the theorem is proved.

Theorem 7-2 (duality theorem): If there are feasible systems π, τ according to a given project duration p where p is bounded ($p_{min} \leq p \leq p_b$), then there are systems π, τ and φ where the values of both objective functions are equal, that is, they are optimal.

(The proof of the theorem is constructive, that is, it gives the algorithm too.)

Proof: To start a trivial solution, let $\pi_{iS}=ES_i$ and $\pi_{iF}=EF_i$ be the time policy calculated from $\tau_i=b_i$ values and let $\varphi_{(ij)k}=0$ $\forall\{(ij)k\}\in\mathbf{A}$ and $\varphi_{iS,iF}=0$ $\forall$ $i\in\mathbf{N}$. This is an optimal solution as they satisfy **(7.1)-(7.8)**, **(7.11)-(7.14)** conditions and the value of the two objective functions are equal. The project duration is $p_b=\pi_{tF}$.

So in the initial step (Step 0.) we have to make a simple time analysis, and we get the "all normal" point as starting trivial solution.

If we know optimal solutions π, τ and φ corresponding to any project duration p, then we can give systems p^*,π^*, τ^* and φ^* , which also satisfy conditions **(7.1)-(7.8)**, **(7.11)-(7.14)** and equilibrium conditions 1°, 2°, 3°, that is, are also optimal and $p^*<p$.

This statement says that if there exists an optimal solution to a project duration p, then we can move on to another optimal solution which corresponds to a smaller project duration. As we know the optimal solution (step 0.) corresponding to a project duration p_b, we can give the optimal solutions to all project durations where $p<p_b$.

We comprehend this through a two step construction.

In the first step we increase the flow from φ to φ^* so that the duality conditions formulated in the lemma continue to be true, that is, the solution remains optimal. In the second step we decrease the project duration p in a way that the duality conditions remain fulfilled, that is, the solution remains optimal. After finishing Step 2. we go back to Step 1. and start increasing the flow again. We have to repeat Step 1. and Step 2. until we reach the desired project duration or the minimum project duration.

Step 0. (Initial Step): In this step a simple PDM network calculation has to be applied. This is a two phase calculation, the first phase, which is a kind of longest path calculation from the source node, is called forward calculation, the second phase, which is called backward calculation, is the same but in this case calculation has to be started from the finishing node. At the initial step we need the result of forward calculation, so the backward calculation phase can be omitted.

The calculation rules can be found in Chapter 4.

Step 1. (increasing of flow): Let us examine an arrow (precedence relationship) and a node (activity) from an optimal solution. Which duality conditions can be accomplished in what combination? These are shown in the first three columns of Table 7-2 and Table 7-3. The '+' sign indicates if a condition is satisfied, the '-' sign indicates if it is not.

Depending on which conditions are accomplished on a certain arrow or node -- which are parts of an optimal solution --we can get some information on the φ flow passing through arrows, and some information on $\varphi_{iS,iF}$ passing through nodes. These pieces of information can be found in the fourth column of Table 7-2. and Table 7-3.

We have classified the arrows and nodes of an optimal solution for further references. These notations can be found in the fifth column.

In the first step the flow has to be increased so that the solution remains optimal, that is, the dual conditions **(7.11)-(7.14)** and equilibrium conditions 1°, 2°,3° are satisfied for all arrows and nodes. We increase the flow in a way that both arrows and nodes remain in their own classes, that is, the solution remains optimal. With the help of the data on flows, we can give a value on how much the flow $\varphi_{(ij)k}$ or $\varphi_{iS,iF}$ can be increased or decreased. The 6th and 7th columns of Table 7-2. and Table 7-3. show the capacities which have come about as a result of this kind of argument.

These capacities will ensure that the solution remains optimal.

It can be read out from the equilibrium conditions that activities in class **N1** have the duration of a_i, in class **N2** a duration of b_i, in class **N3** the duration is between the crash and normal duration, in class **N4** the crash and normal durations are equal, therefore $\tau_i=a_i=b_i$.

Equilibrium conditions 1°	2°	3°	φ information	Classing of arrows	$\kappa_{(ij)k}$	$\kappa_{(ji)k}$
+	not defined on		$\varphi_{(ij)k}=0$	**A1**	0	0
-	the arrows		$\varphi_{(ij)k}\geq 0$	**A2**	∞	$\varphi_{(ij)k}$

Table 7-2 *Arrows in optimal solution.*

Equilibrium conditions 1°	2°	3°	$\varphi_{iS,iF}$ information	Classing of activities	$\kappa_{iS,iF}$	$\kappa_{iF,iS}$
Not	+	-	$\varphi_{iS,iF}\geq c_i$	**N1**	∞	$\varphi_{iS,iF}-c_i$
defined	-	+	$\varphi_{iS,iF}\leq c_i$	**N2**	$c_i - \varphi_{iS,iF}$	∞
on	+	+	$\varphi_{iS,iF}=ci$	**N3**	0	0
activities	-	-	$\varphi_{iS,iF}>-\infty$	**N4**	∞	∞

Table 7-3 *Activities in optimal solution*

After the initial step precedence relationships can be definitive or non definitive, so they can be in class **A1** or **A2.** As their duration is b_i, after the initial step, activities can be in class **N2** or **N4**. Searching for maximal flow on this network with the above given capacities(κ), we get a $\psi_{(ij)k}$ $\psi_{iS,iF}$ flow. Adding this to the flow φ the new flow on the arrows and the new flow on the nodes from their start to the finish will be:

$$\overset{*}{\varphi}_{(ij)k}=\varphi_{(ij)k}+\psi_{(ij)k} \qquad \overset{*}{\varphi}_{iS,iF}=\varphi_{iS,iF}+\psi_{iS,iF}$$

During this step both arrows and activities have remained in their own classes due to the applied capacities. The equilibrium conditions and the conditions of the primal and dual task are still satisfied, that is, the solution is still optimal.

<u>*Second step:*</u> In the network we are to find a system ($p^*<p$, π^*, τ^*) corresponding to the increased flow φ^* where the equilibrium and the primal dual conditions are accomplished, that is, the solution remains optimal. In the first step we were looking for a maximal flow. In this case, there exists a cut **(S,T)** in the network the arrows (precedence or activity) of which are saturated and $s\in$**S** and $\tau\in$**T**. (**S**$\cap$**T**=0, **S**$\cup$**T**=**N**) In the cut there can be **A1** type arrows and **N2** and **N3** type of activities. Arrows of **A2** and **N1, N4** type of activities cannot be in the cut because it would mean an infinite flow from s to t.

In the cut backwards there can be **A1**, **A2** type arrows and **N1, N3** type nodes. To determine the new values of p^*, π^*, τ^* we give a δ value. With this δ we decrease the values of π in the set **T**. During this process the equilibrium conditions of the lemma remain fulfilled, so the solution is still optimal. This means that the solution we have is still an optimal solution and $p^*<p$. The value of δ can be determined in such a way that all the activities and arrows in or backward in the cut have to remain part of the optimal solution. It means that if they change their own class, these changes have to correspond to the increased (decreased) flow. The determination of δ comes about in the following way:

$$\delta=\min\{\delta_{A1}, \delta_{N2}, \delta_{N3}, \delta_{A1*}, \delta_{A2*}, \delta_{N1*}, \delta_{N3*}\}$$

where,

in case of precedence relations and activities in the cut:

$$\begin{aligned}
\delta_{A1}:=\min\Big[&\pi_{jS}-\pi_{iS}-z_{(ij)k} && : \forall\{(ij)k\}\in\mathbf{A1} \text{ and } \{(ij)k\}\in(\mathbf{S},\mathbf{T}) \text{ and } SSz_{(ij)k};\\
&\pi_{jF}-\pi_{iF}-z_{(ij)k} && : \forall\{(ij)k\}\in\mathbf{A1} \text{ and } \{(ij)k\}\in(\mathbf{S},\mathbf{T}) \text{ and } FFz_{(ij)k};\\
&\pi_{jF}-\pi_{iS}-z_{(ij)k} && : \forall\{(ij)k\}\in\mathbf{A1} \text{ and } \{(ij)k\}\in(\mathbf{S},\mathbf{T}) \text{ and } SFz_{(ij)k};\\
&\pi_{jS}-\pi_{iF}-z_{(ij)k} && : \forall\{(ij)k\}\in\mathbf{A1} \text{ and } \{(ij)k\}\in(\mathbf{S},\mathbf{T}) \text{ and } FSz_{(ij)k}\Big]\\
\delta_{N2}:=\min[&\tau_i-a_i && : \forall\ i\in\mathbf{N2} \text{ and } (i_S, i_F)\in(\mathbf{S},\mathbf{T})]\\
\delta_{N3}:=\min[&\tau_i-a_i && : \forall\ i\in\mathbf{N3} \text{ and } (i_S, i_F)\in(\mathbf{S},\mathbf{T})]
\end{aligned}$$

in case of precedence relations and activities backwards in the cut:

$$\begin{aligned}
&\delta_{A1*}=\infty && : \forall\ \{(ij)k\}\in\mathbf{A1} \text{ and } \{(i,j)k\}\in(\mathbf{T},\mathbf{S})\\
&\delta_{A2*}=\infty && : \forall\ \{(ij)k\}\in\mathbf{A2} \text{ and } \{(i,j)k\}\in(\mathbf{T},\mathbf{S})\\
&\delta_{N1*}:=\min[b_i-\tau_i && : \forall\ i\in\mathbf{N1} \text{ and } (i_S\, i_F)\in(\mathbf{T},\mathbf{S})\\
&\delta_{N3*}:=\min[b_i-\tau_i && : \forall\ i\in\mathbf{N3} \text{ and } (i_S\, i_F)\in(\mathbf{T},\mathbf{S})
\end{aligned}$$

The δ value thus defined is definitely greater than zero. Let the new system of π^* in the transformed network be determined in the following way:

$$\pi_i^* = \begin{cases} \pi_i & \text{if } i\in\mathbf{S} \\ \pi_i-\delta & \text{if } i\in\mathbf{T} \end{cases}$$

Thus p will become $p^*=p-\delta$ and the duration of activity i can be determined from the following formula:

$$\tau_i^*=\pi^*_{iF}-\pi^*_{iS} \qquad \forall i\in\mathbf{N}$$

The δ value had been constructed so that the duality conditions of the precedence relationships and of the activities continue to be accomplished.

After these modifications the following, below described changes take place:

In the case of precedence relationships or activities of which both nodes fall into set **S** or **T**, they remain in the same class.

In the case of precedence relationships and activities in the cut, the following changes take place corresponding to the flow.

- **A1** type of arrows remain **A1** type or become **A2** type.
- **N2** type of activities become **N1** or **N3** type.
- **N3** type of activities remain **N3** type or become **N1** type.

*In the case of relations and activities **backwards** in the cut,* the following changes take place corresponding to the flow.

- **A1** type of precedence relationships remain **A1** type.
- **A2** type of precedence relationships become **A1** type.
- **N1** type of activities become **N2** or **N3** type.
- **N3** type of activities remain **N3** or become **N2** type.

These changes in types correspond to the flow, that is, the solution remains optimal.

Thus we completed step 2.

After all this we have to go back to Step 1 and increase the flow again and then in the Step 2 the project duration can be decreased. These steps must be alternately repeated until we reach the desired project duration or the flow becomes infinitely great. Let us examine the case of infinite flow. In this case there exists a $P(s \to t)$ path along which all the capacities are infinite. This means that precedence relationships on this path determine the beginning/finishing of the activities on the path, so this is the longest the critical path. Activities of classes **N1** and **N4** can be on this path which means that they are normal critical, that is, shortening their duration will shorten the critical path, too. As $\tau_i = a_i$ for these activities, they cannot be shortened anymore. Activities of classes **N2, N4** can be backward in this path, so they are reverse critical activities, i.e., the longer they are, the shorter the project duration. As $\tau_i = b_i$ for these activities, the project duration cannot be shortened furthermore.

This means that the infinite flow indicates that the project duration is the minimal and no further compression is possible.

With this procedure the optimal direct cost curve can be determined in the p_b, p_{min} interval.

An important consequence of this theorem is presented hereby.

Theorem 7-3 (strong form of equilibrium). If there exists an optimal primal solution of which the value of its objective function is (P), and an optimal dual solution of which the value of its objective function is (D), then these two values are equal.

Proof: According to the duality theorem there exists a maximal primal solution (P*) and a minimal dual solution (D*), which are optimal, that is (P*)=(D*). As (P) and (D) are also optimal, (P*)=(P) and (D*)=(D), but then (P)=(D)

We illustrate the algorithm using a very small example. The network and the input data can be seen in Figure 7-6. Activities cannot be interrupted.

We are to find the optimal cost curve in the p_b-p_{min} interval. Let us start from the policy π calculated from normal durations (b_i) as a starting trivial solution. All the flows are equal to zero. The initial policies (π, τ) are also shown in Figure 7-6.

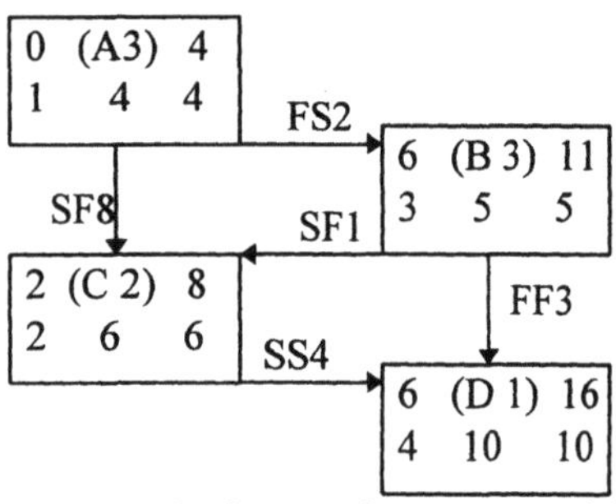

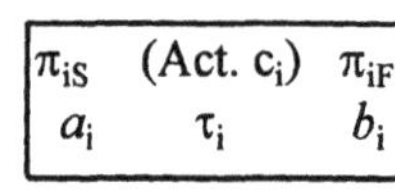

Figure 7-6 *The sample project and the initial solution*

Now the iteration procedure can start. Each of the following figures will show a complete iteration procedure. (Step 1 and Step 2) Each figure consist of 5 parts. The first shows the classification of the arrows and activities, the second the capacities with which the flows can be increased or decreased (first and second number on the arrows), the third shows the maximal flow ($\psi_{(ij)k}$) calculated from these capacities and the cut, the fourth shows the new flow, and the fifth the new system (π, τ) .

We start our first iteration (Figure 7-7) from the trivial solution calculated from normal durations. The value of the flow is Θ=1. Observe that the flow goes backward on activity C. Activity D and the precedence relationship between C and D are in the cut. As δ_{N2}=min[τ_i-a_i]=6 for activity D and δ_{A1}=min{π_{DF}-π_{BF}-z_{BD} }=2 for precedence relationship B-D, therefore δ=min{δ_{A1}, δ_{N2}}=2.

By decreasing the values π in the set **T** we come to the new system (π, τ) as it can be seen in the figure (Figure 7-7e). The new project duration is 14 days. The increment of project cost is $\delta * c_D$=2 units.

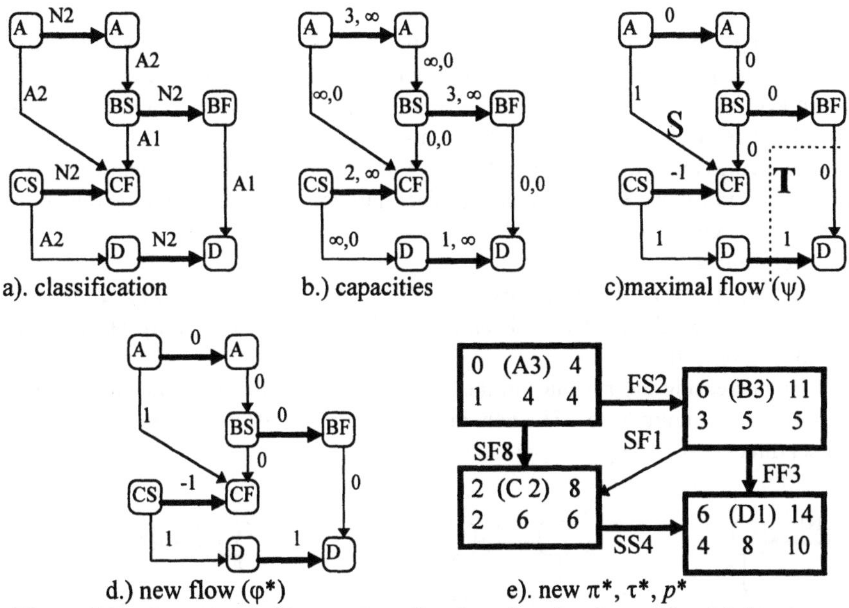

Figure 7-7 *Iteration I. (Decreasing of project duration from 16 to 14 days.)*

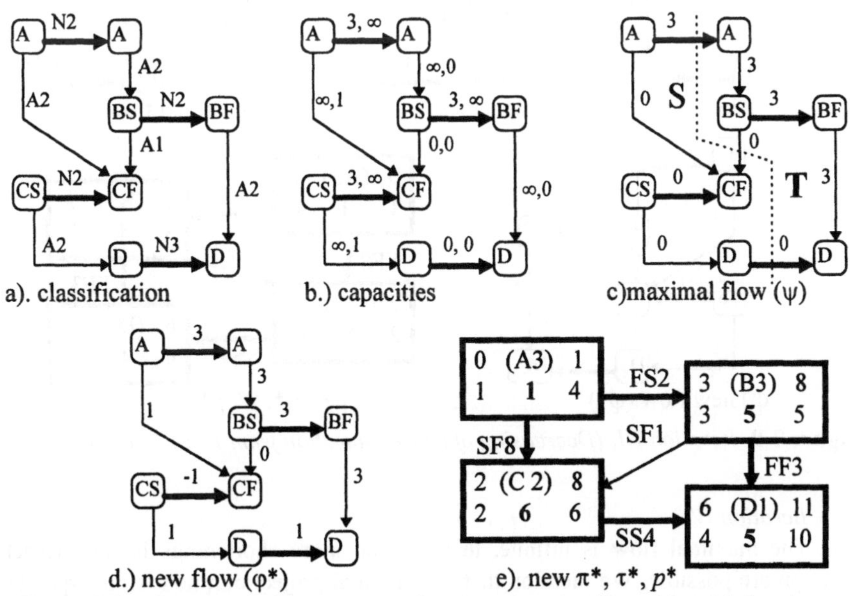

Figure 7-8 *Iteration I. (Decreasing of project duration from 14 to 11 days.)*

Iteration II.

In Step 1 we increased the value of the flow by 3. Activities A and D are in the cut. Precedence relationship SF1 between B and C goes backward in the cut. As $\delta_{N2}=\min[\tau_i-a_i]=3$ for activity A and $\delta_{N3}=\min[\tau_i-a_i]=4$ for activity D and $\delta_{A1*}=\infty$ for precedence relationships B-C therefore $\delta=\min\{\delta_{N2}, \delta_{N3}, \delta_{A1*}\}=3$.

By decreasing the π values in the set **T** we come to the new system (π,τ) as it can be seen in Figure 8-8e. The new project duration is 11 days. The increment of project direct cost is $\delta*(c_D+c_A)=12$ units.

Iteration III.

In Step 1 we could not increase the flow. Activities B and D are in the cut. As $\delta_{N2}=\min[\tau_i-a_i]=2$ for activity B and $\delta_{N3}=\min[\tau_i-a_i]=1$ for activity D, therefore $\delta=\min\{\delta_{N2}, \delta_{N3}\}=1$.

By decreasing the π values in the set **T** we come to the new system (π,τ) as it can be seen in Figure 7-9e. The new project duration is 10 days. The increment of project direct cost is $\delta*(c_D+c_B)=4$ units.

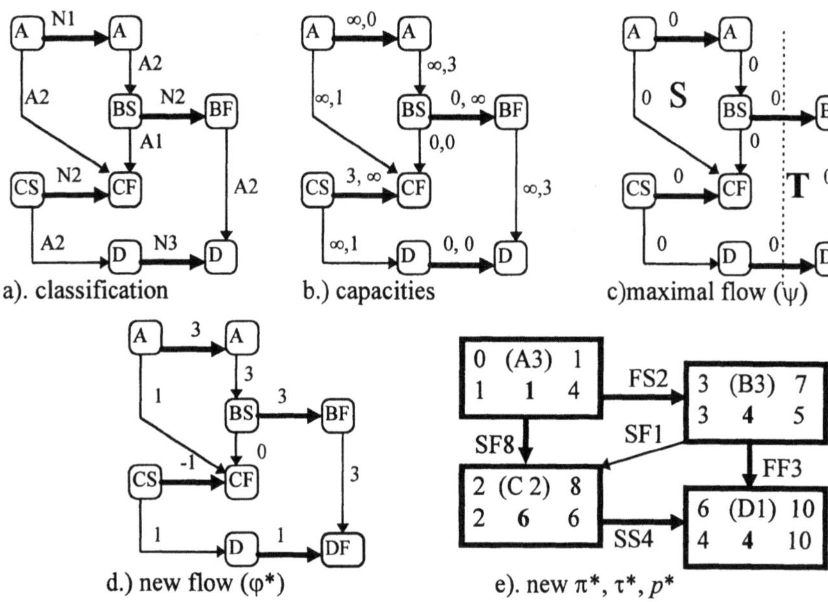

Figure 7-9 *Iteration III. (Decreasing of project duration from 11 to 10 days*

Iteration IV.

The maximal flow is infinite, therefore no more shortenings in the project duration are possible. We mention that the minimal project duration is not equal to the “all crash” project duration which is $p_a=14$ days.

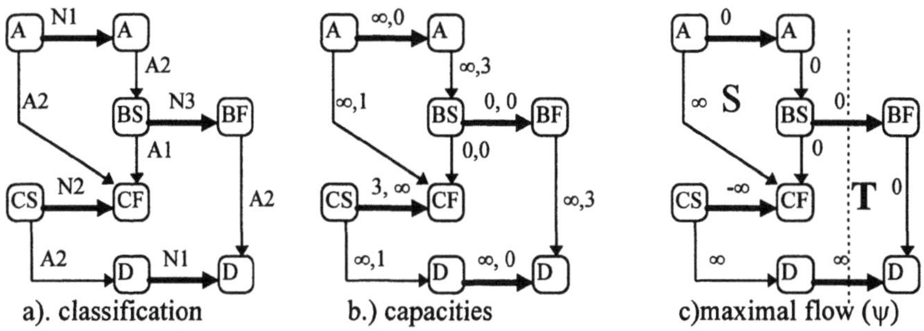

Figure 7-10 *Iteration IV. (No more possibility for decreasing the project duration)*

Remarks to the algorithm:

The algorithm starts from a trivial solution, which is the "all normal" solution. This means that we do not have the optimal curve in the pb pmax-interval. Finding this part of the curve is especially troublesome, because our model 7.1-7.8 and 7.10 is not complete. There is one more condition, which states that there should be a critical path from s to t. This is a non-linear condition, therefore it cannot be handled by linear programming and related methods such as maximal flow minimal cut. Fortunately, this condition is automatically fulfilled in the p_b-p_{min} *interval so it can be deleted from the model. But solving the model with other algorithms, e.g., linear programming, would result in a horizontal curve at the height of the "all normal" cost point, because this is the absolute minimum cost solution.*

Any kind of maximal flow algorithm can be used in Step 1.

Finally, we give the solution of the depot building project. The results can be seen in Table 7-4, and the optimal cost curve in Figure 7-11.

P	A	B	C	D	E	F	G	H	I	J	K	L	M	N	O	cost increment
155	30	20	30	10	14	30	20	80	90	10	30	20	5	20	15	0
151	30	20	30	***6***	14	30	20	80	90	10	30	20	5	20	15	4* 5= 20
135	30	20	30	6	14	30	20	***64***	90	10	30	20	5	20	15	16* 7=112
125	***20***	20	30	6	14	30	20	64	90	10	30	20	5	20	15	10*10=100
120	20	20	30	6	14	***25***	20	***59***	90	10	30	20	5	20	15	5*22=110
112	20	20	30	6	14	25	20	59	90	10	30	20	5	***12***	15	8*50=400
107	20	20	30	6	14	25	20	***54***	90	10	***25***	20	5	12	15	5*67=335

Table 7-4 *Optimal solutions for the depot building project.*

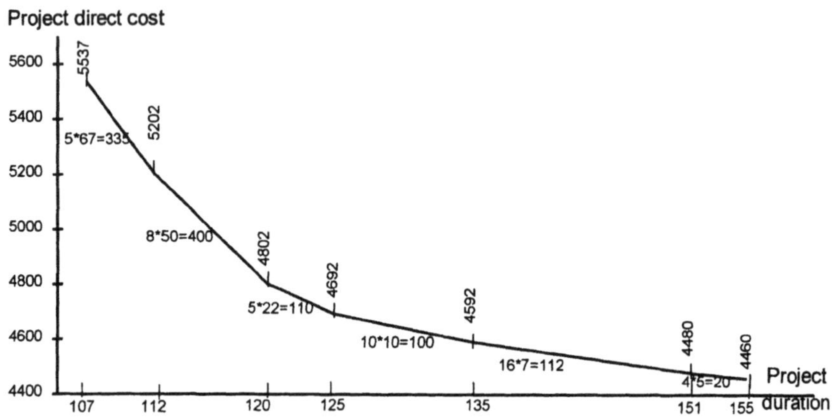

Figure 7-11 *Optimal direct cost curve for the sample project*

7.4 SPECIAL CASES

We talk about special cases if changes take place in the original (7.1)-(7.8) conditions, or the structure of the network is being modified, or the activity cost curves differ from Figure 7-4. Changes in the activity cost curves have been discussed in Chapter 3: Special Cases. In Figure 3-11 we have been showed how the activity direct cost versus time curve can vary and how they can be solved with the modification of the algorithm and the network or sometimes with application of another optimization procedures. In the PDM/cost method we can handle these activity cost curves in the same way. The only difference is that we cannot handle those activities for which the crash cost is less than the normal cost. This was also a problem with the CPM/cost model, and we suggested to choose the crash duration in any circumstances when the crash cost is the smallest of all. From a practical point of view, this seemed to be a satisfactory solution. In the PDM/cost model if the activity is reverse critical then it is sometimes necessary to lengthen the activity duration in order to decrease the project duration.

Another main group of special cases is when changes in the model arise. For example, if splitting is allowed for all the activities, if splitting is allowed for some of the activities, if splitting is allowed but limited in time, if maximal types of precedence relationships are applied, if lead/lag factors are given in the percentage of work, all these are considered special cases. All these special cases can be handled with the maximal flow methodology, but some changes are necessary. For example, if splitting is allowed for some of the activities, then activities have to be classified into twelve different classes instead of four. Special considerations are necessary if maximal type of precedence relationships are applied in the network. In this case there is no trivial optimum solution. We do not discuss these modifications in this text. Our aim was to introduce the reader to the solution of the basic PDM/cost problem. In the Historical Review of this chapter we give a survey of the existing algorithms.

7.5 PRACTICAL PROBLEMS AND SOLUTIONS

Problems

1. Find the optimal cost curve for the below given network in the p_b-p_{min} interval!

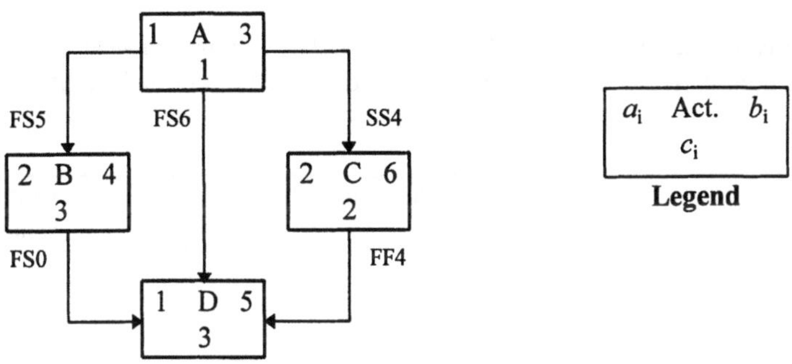

2. Find the optimal cost curve for the below given network in the p_b-p_{min} interval!

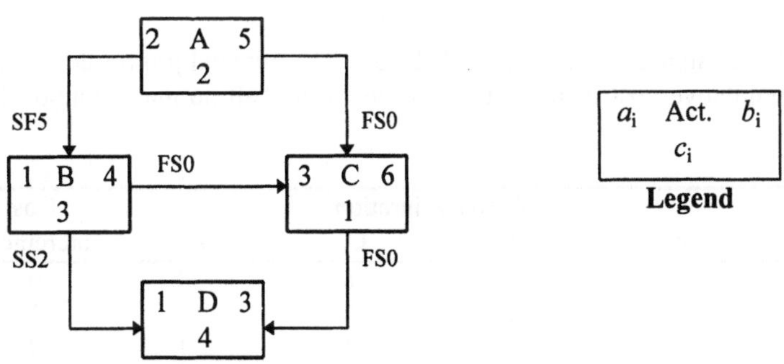

3. Find the optimal cost curve for the below given network in the p_b-p_{min} interval!

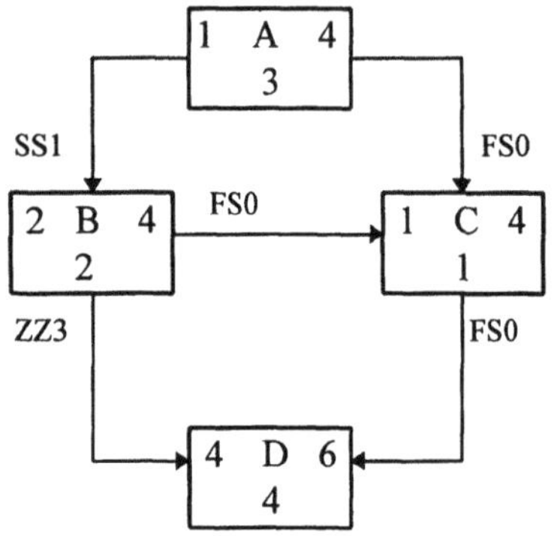

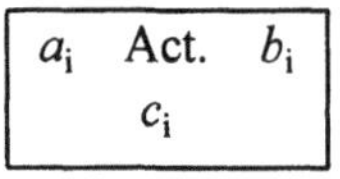

Legend

Solutions

1. The results are summarized in the table below. The minimal project duration is 10 days, the cumulative cost increment compared to the "all normal" solution is 25 units.

Project duration (p)	Activity duration A	B	C	D	Cost increment
17	3	4	6	5	0
15	***1***	4	6	5	2
14	1	***3***	6	5	3
13	1	***2***	***5***	5	5
10	1	2	***2***	***2***	15

2. The results are summarized in the table below. The minimal project duration is 9 days, the cumulative cost increment compared to the "all normal" solution is 11 units.

Project duration (p)	Activity duration A	B	C	D	Cost increment
14	5	4	6	3	0
11	5	4	3	3	3
9	5	4	3	1	8

3. The results are summarized in the table below. The minimal project duration is 8 days, the cumulative cost increment compared to the "all normal" solution is 18 units.

Project duration (p)	A	Activity duration B	C	D	Cost increment
15	4	4	4	6	0
12	4	4	*1*	6	2
11	4	*3*	1	6	3
9	4	3	1	*4*	5
8	*3*	*2*	1	4	15

7.6 HISTORICAL REVIEW

The PDM time-cost trade-off problem such as the CPM time-cost trade-off can be formulated as a linear programming model, therefore all the existing methodology for solving the CPM/cost problem can be applied to the PDM/cost model. However, the same methodologies will result in different algorithms, as the two models differ from each other. The PDM/cost algorithm, which is given in this book has first appeared in Hajdu's [32],[33],[34] works. In this works the original PDM assumptions have been used, that is, the splitting is not allowed approach. An algorithm when splitting is allowed for all the activities and an algorithm when some of the activities are splittable and the rest of them are non-splittable is also a result of Hajdu's [27],[28] work.

Chapter 8

RESOURCES IN SCHEDULING

During the time analysis we obtained the early and late start and finish times of the activities. During these procedures no resources were assigned to the activities. However, in the practice the completion of individual tasks requires different resources, e.g., manpower, machines and materials. The limitations of these resources can influence the start and the finish of activities and sometimes the project duration. On the other hand, even if the resource requirements for a projects are far below the limitations, a good arrangement of the activities can cut the peak resource requirement and can make the project more efficient. No doubt, systematic and optimal planning of resource demand can cut the project cost, therefore it has growing importance.

In the first section of this chapter the possibility of using operational research models in resource planning is discussed.

The heuristic procedures that optimize resource handling procedures can be divided into two groups. One, when the assumption of limited resources is being used. This means that the schedule has to be such that the limitations in resources have to be satisfied. Procedures that arrange activities in this way can lengthen the project duration that has been gained from the ordinary time analysis. These kinds of procedures are called resource allocation procedures.

Procedures of the second group cannot change the project duration. In these cases a more efficient resource utilization is the main goal. These methods are known as resource leveling methods. The second section, Resource Leveling (Fixed Project Duration) discusses a procedure that can be used for resource leveling. Procedures that allow the change of project duration are the topic of the third section, Resource Allocation (Limited Resources).

In the fourth and fifth sections the changes caused by the application of maximal precedence relationships are discussed.

Finally, we mention that the terms of allocation and leveling are widely used in the above mentioned sense, but other definitions are also used in the practice. An often used approach is that the term leveling covers all the resource optimization procedures and the term of resource allocation means the assignment of resources to the tasks of the project. In this text we use the term resource allocation as a method that helps to utilize the resources on the most efficient level in a way that allows the lengthening of the project duration. However, to avoid confusion the term allocation with limited resources or leveling by keeping the project duration will always be indicated in the text.

8.1 USING OPTIMAL PROCEDURES

Both types of allocation methods are optimization problems in mathematical sense. The planner wants to make a schedule that satisfies all the resource limitations but causes the least increase in project duration in case of resource allocation (limited resources). When the project duration is fixed he or she wants to level the plan so that the peak resource requirements will be as small as possible. Operational research offers a wide variety of different optimization techniques. Mathematical models for allocation and leveling can be created that can be solved by Linear Programming, Integer Programming or other methods, e.g., different kinds of enumeration. However, the application of these exact methods cannot be found in the practice. The reason for this is that these models work with so many variables -- even in the case of a relatively small networks -- that it makes the computational process very slow. The other side of the coin is that engineers know well that due to the bias in duration and resource estimation the optimal solution is just a fiction, and a close optimal solution from practical point of view can be as good as a real optimal from mathematical point of view. According to the above mentioned reasons instead of exact optimization procedures algorithms based on heuristic rules are used in the practice.

8.2 RESOURCE LEVELING (FIXED PROJECT DURATION)

Resource leveling also known as leveling within the float or smoothing is used when resource limits are far beyond the requirements, therefore they do not have any influences on project duration. It is also used when the amounts of required resources are based on estimation therefore a resource profile that is close to the planned resource profile can be satisfactory. The efficient utilization of resources, that is, a better distribution of them can make the project more successful. The ideal resource profile is gradually built up to the peak and has a downward tendency towards the end. Ideal resource profiles can be seen in Figure 8-1. Generally it can be stated that the closer the resource profile is to the ideal form the less the overall cost will be. Every sudden rise in the resource histogram, especially after the peak, is usually solved by hiring extra personnel which results in extra costs.

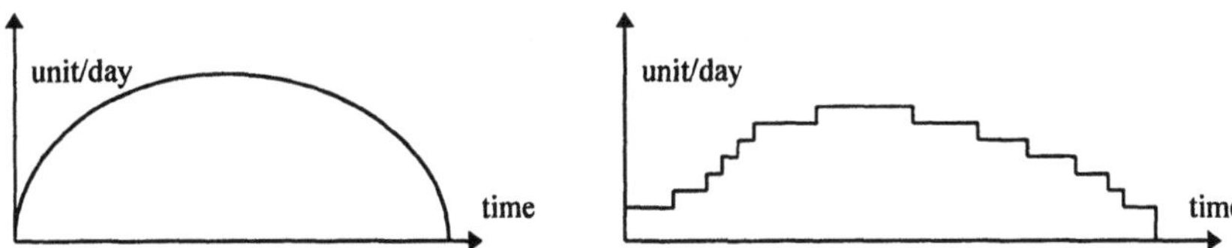

Figure 8-1 *Ideal resource profiles*

Resource leveling procedures are based on the idea of rescheduling activities within the limits of the early start and late finish times. A heuristic approach that uses the same thought will be presented below.

For the sake of simplicity it is assumed that there are no maximal type of precedence relationships in the network, the network contains one start and one finish activity and only one resource is applied in the network for which leveling is required. Finally, it has to be mentioned that the desired resource profile to which the resources will be smoothed can take any form. This depends on the planner.

During the algorithm the following notations are used:

N - set of activities
A - set of precedence relationships
n - number of activities in the network
i - notation of an activity, i=1..n
s - start activity; no precedence relation runs into this activity
t - finish activity; no precedence relationships runs out of this activity
τ_i - duration time of activity i
f - the fth day of the project
R_i - required amount of resource per time unit for activity i
SR_f - value of the resource histogram on the day f.

ES/EF_i - earliest start/finish of activity *i*
LS/LF_i - latest start/finish of activity *i*
G_f - the desired amount of resource on the fth day
V - $\Sigma\,(G_f\text{-}SR_f)^2$ $(f{:=}0 \ldots p)$ the value of deviation between the desired and the actual resource histograms
S_i/F_i - scheduled start/finish of activity *i*
D_i - the maximal possible shift for activity *i*

The algorithm is based on the observation that activities have to be rescheduled within the limits of their total or free float. The reverse order of the time analysis has to be followed. An activity can be taken into account if all the successor activities have been rescheduled. This implies that the algorithm has to be started with the finish activity. The activity in question has to be placed within its boundaries to where the deviation from the desired histogram is the least. This can be defined by using the least squares method. The activity has to be shifted to where it gives the smallest value of V. The iteration is finished when all the activities have been taken into account. In case of changes of any start time the whole procedure can be repeated. To start the leveling, time analysis has to be performed and the resource histogram arranged for the early starts has to be done.

Resource Leveling (Fixed Duration)

Step 0. Perform a time analysis. Let $S_i{=}ES_i$ and $F_i{=}EF_i$ $\forall\ i \in \mathbf{N}$
Step 1.
REPEAT
there_were_changes=false
REPEAT
Take an activity *i* of which all the successors are being rescheduled
Determine the amount of the possible shift (D_i) for activity *i*
{ $S_j\text{-}SSz_{(ij)k}\text{-}S_i$;
$F_j\text{-}SFz_{(ij)k}\text{-}S_i$;
$S_j\text{-}FSz_{(ij)k}\text{-}F_i$;
$F_j\text{-}FFz_{(ij)k}\text{-}F_i$} $\forall\ \{(ij)k\} \in \mathbf{A}$
Shift *i* by day to day in the defined interval and calculate V after each shift. Where V takes its minimum value defines the start of *i*.
IF the start of *i* has been changed THEN *there_were_changes*=true
UNTIL *i=s*
UNTIL *there_were_changes*=false

To illustrate the algorithm let us solve the resource leveling problem depicted in Figure 8-2. The result of the time analysis and the resource histogram can be seen in Figure 8-3. Let the desired resource profile to which the resource has to be smoothed be a constant number during the whole project, that is, the average number of personnel.

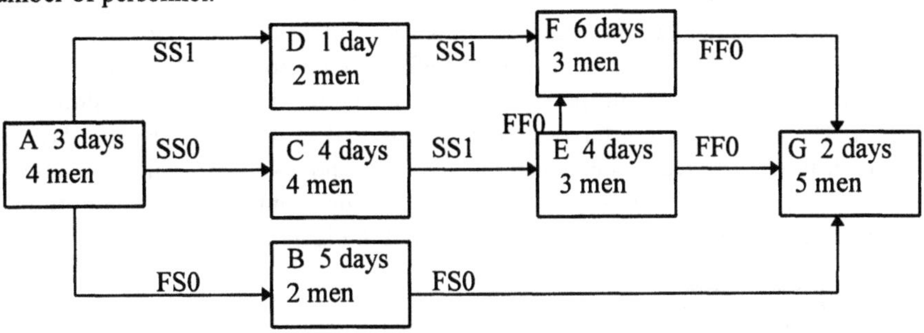

Figure 8-2 *Example project to illustrate the resource leveling procedure*

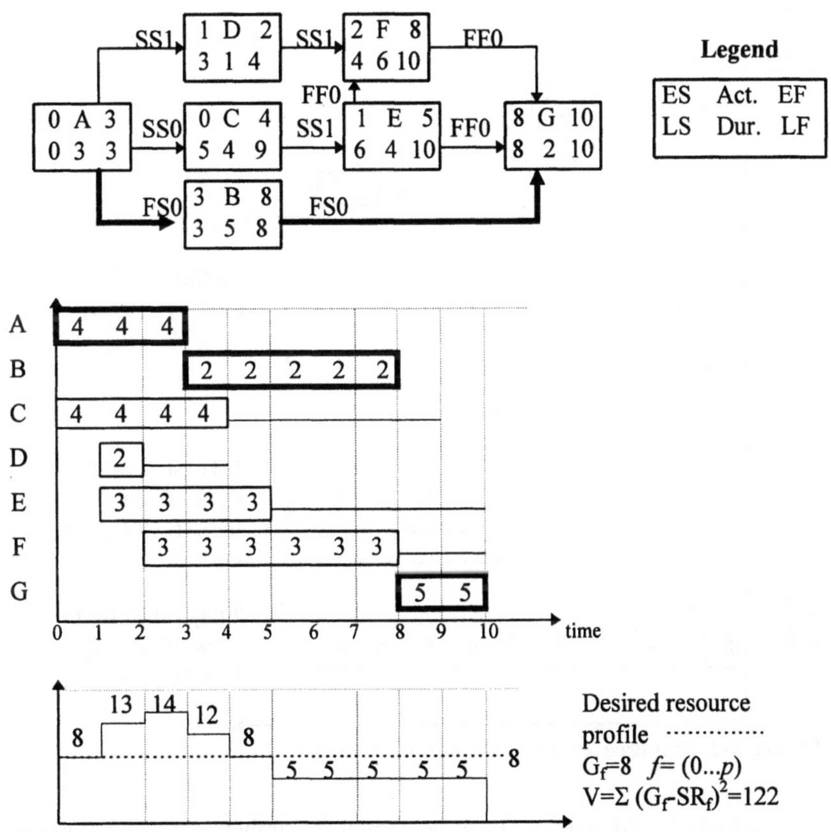

Figure 8-3 *Results of the time analysis and the resource histogram*

In the first step activity G is the only activity of which all the successors are known. This activity is critical, therefore it cannot be shifted. A new activity has to be chosen. Activities B and F can be chosen for further computations. Any of them can be chosen. Activity B is critical, therefore we do not have to deal with it and activity F is chosen for further computations. The maximum possible shift is two days. The deviation from the desired profile is V=122 if the shift is zero day. One day shift will decrease the deviation to V=86 and two days shift decrease the deviation to V=62. This is the best variant from our point of view, therefore F has to be shifted by two days. This can be followed in Figure 8-4

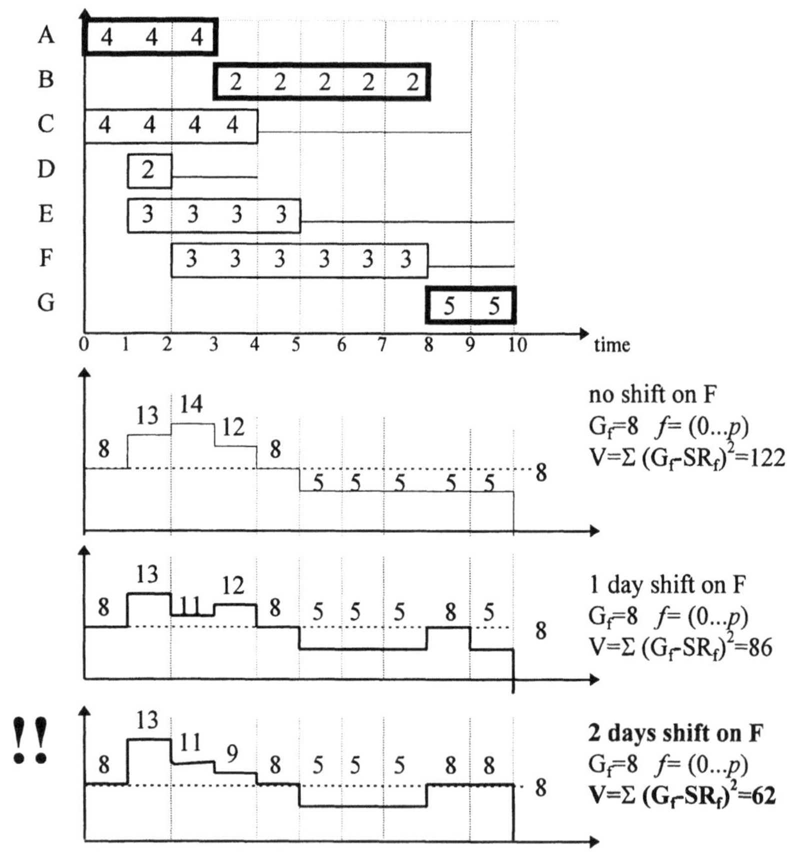

Figure 8-4 *Resource leveling: examining activity F*

Activity E and activity D can be chosen for further computation as all their successors have been rescheduled. Let us choose activity E. The maximum possible shift is five days. The calculation can be followed in Figure 8-5

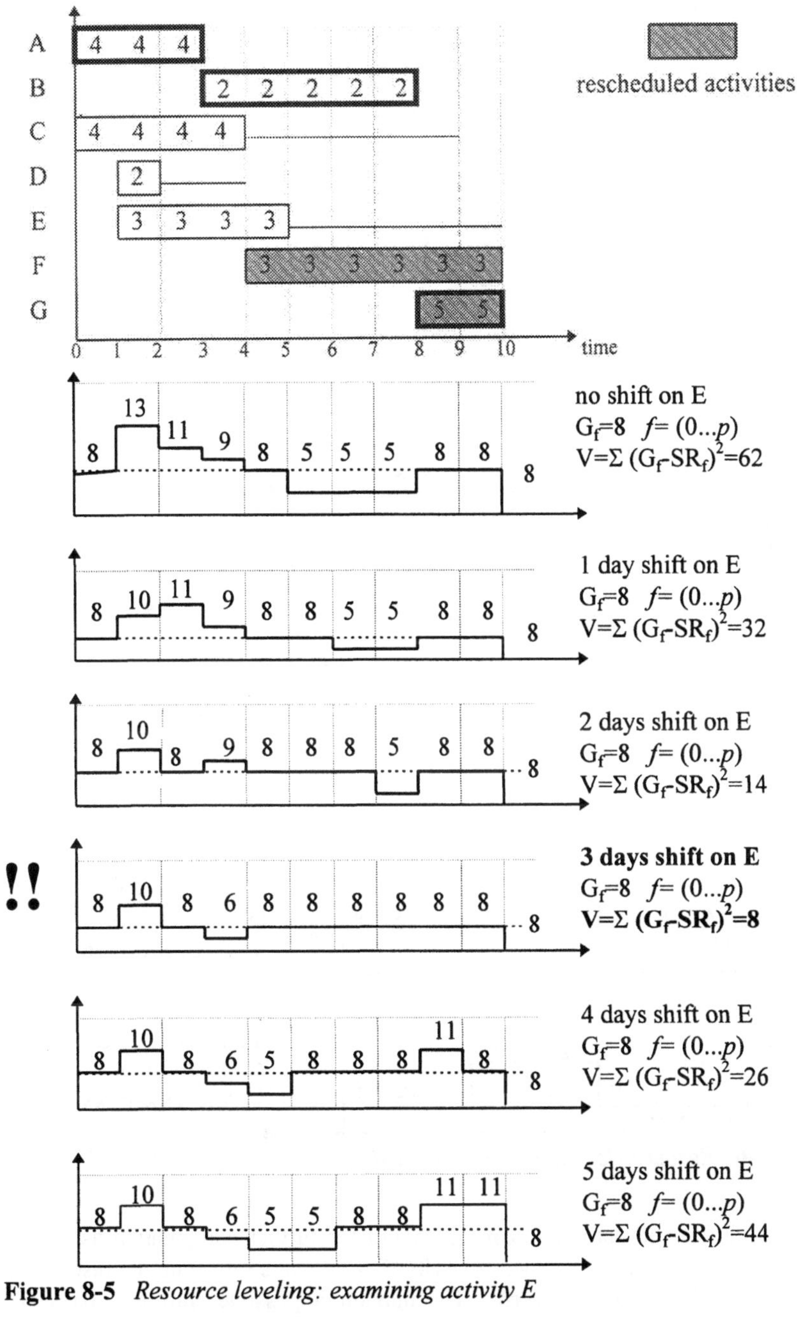

Figure 8-5 *Resource leveling: examining activity E*

Activity C and activity D can be chosen for further computation as all their successors have been rescheduled. Let us choose activity C. The maximum possible shift is 3 days. The calculation can be followed in Figure 8-6. No shift on C has to be performed.

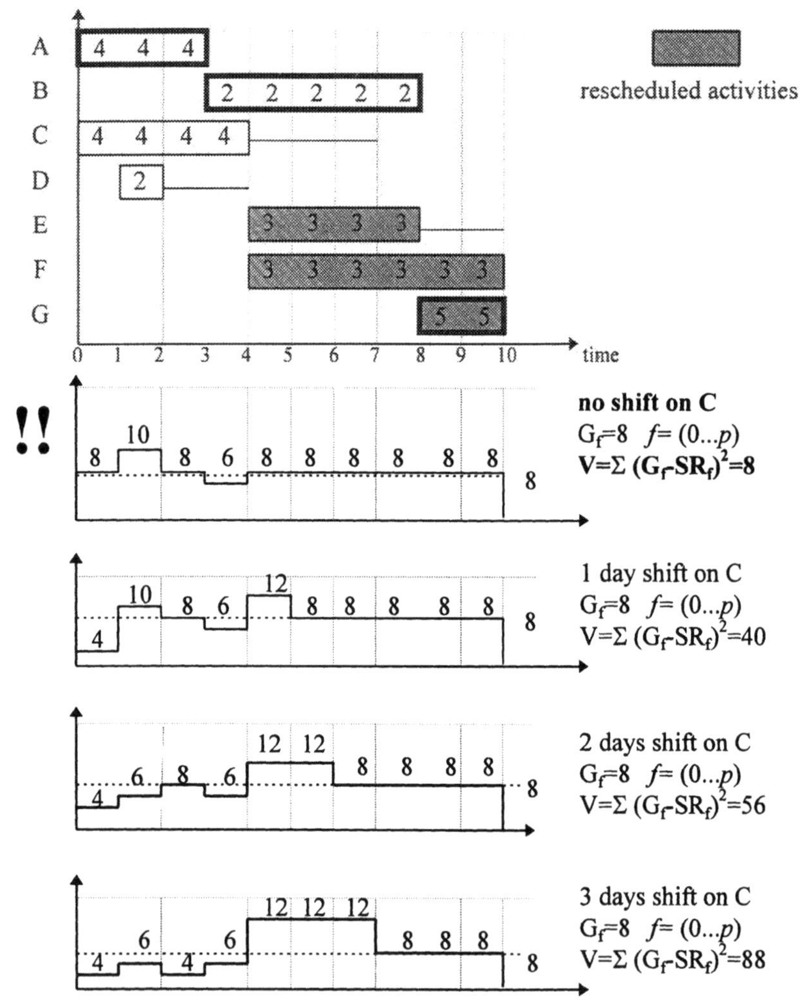

Figure 8-6 *Resource leveling: examining activity C*

Activity D and activity B can be chosen for further computation as all their successors have been rescheduled. Activity B is critical so it cannot be rescheduled and activity C has to be chosen. The maximum possible shift is 2 days. The calculation can be followed in Figure 8-7. To achieve the best result two days shift on D has to be performed. The deviation calculated by the least squares method is zero in this case which means that the desired and the leveled resource histograms perfectly coincide. This result almost never appears in real projects. It was simply our intention to prepare an example where the optimal solution perfectly coincides with the desired resource profile and this solution can be achieved by using heuristic methods. In average cases the iteration has to be started again after every activity has been taken into account, and when it stops it usually does not agree with the desired resource profile. The final results can be seen in Figure 8-8.

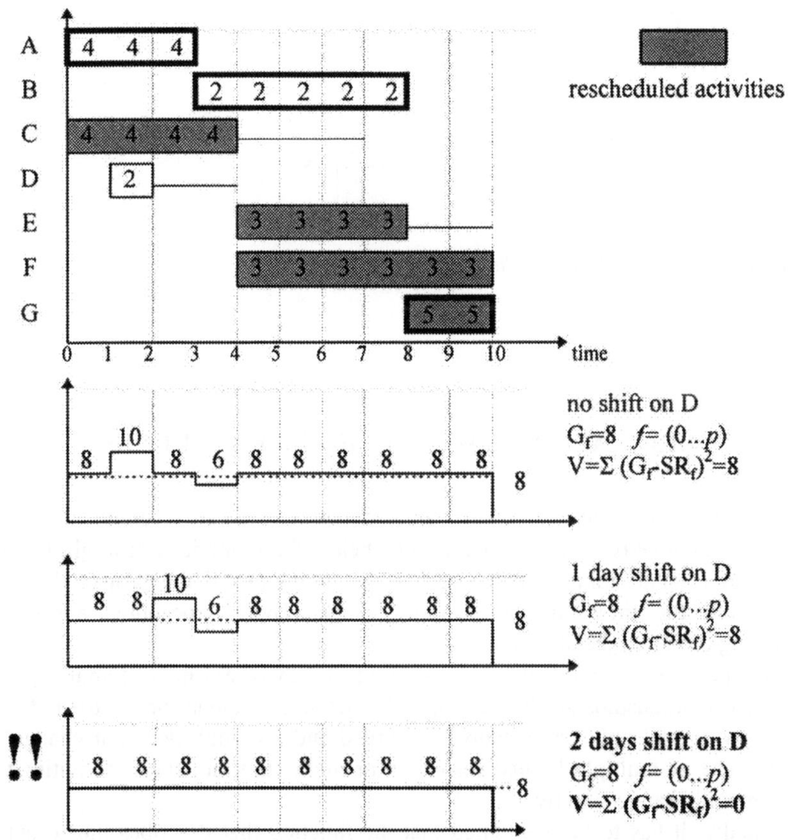

Figure 8-7 *Resource leveling: examining activity D*

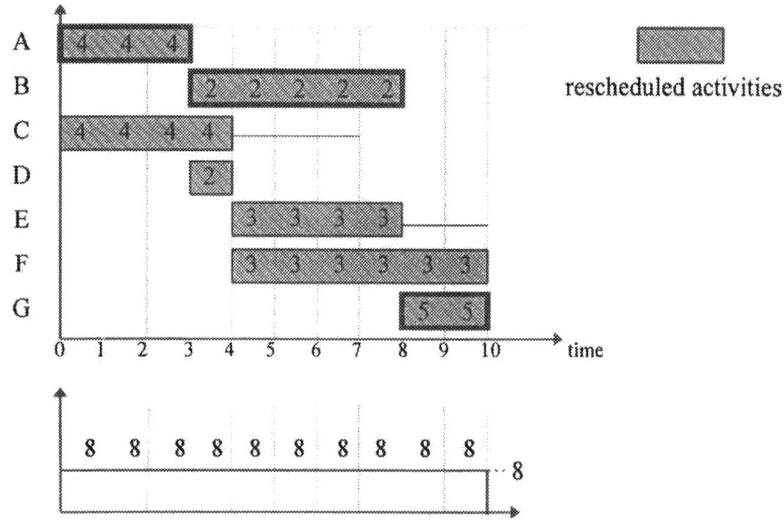

Figure 8-8 *Resource leveling: final result*

8.3 RESOURCE ALLOCATION (LIMITED RESOURCES)

Resource allocation also known as resource constrained leveling, has to be used when resource requirements have to be below the limitations. It might be that there is no feasible solution that satisfies the limitations if the project duration gained from the time analysis cannot change. These procedures allow the lengthening of the project duration.

For the sake of simplicity it is assumed that there are no maximal type of precedence relationships in the network, the network contains one start and one finish activity, and only one resource is limited and this limit does not change in time. In case of multiple limited resources or more start and finish activities the algorithm will change slightly.

Finally, it has to be mentioned that the function that describes the resource limitation can take any form, especially if it is used in the company level in multiproject environment. Subtracting the resource requirements of a project from the company's resource pool will result in resource limitations that can change from day to day for the rest of the projects.

During the algorithm we use the following new notations:

L - resource limit
pp - project duration (result of the allocation)

A possible solution to the resource allocation problem can be when *resource leveling* (leveling for fixed duration) is used to satisfy the resource limits. If the limits cannot be kept within the given project duration, then a fictive SF(p+1) relationship has to be added to the network between the start and finish activities. The application of this precedence relationship will increase the project duration by one day. Now, another leveling has to be performed for this increased project duration. If the resource requirements are still above the limits the project duration has to be increased by another day and new leveling must be performed. This procedure has to be continued until the resulted resource profile will be below the limits.

The above mentioned procedure is rather time consuming, a faster algorithm is going to be presented here. The algorithm is similar to the algorithm used for the simple time analysis, it is just being completed with a resource handling step. It is based on the observation that an activity can be arranged on the time scale if its predecessor activities are arranged too. This implies that the algorithm has to be started with the start activity. Once an activity has been chosen its start has to be calculated by its predecessors. Adding its resources to the resource histogram, it has to be examined if the resource histogram was below the resource limit or not. If not, the activity has to be shifted until the resource limits are satisfied. Knowing these, the following algorithm can be applied for resource allocation.

Resource Allocation (Resource Limited Allocation): Forward Calculations

Step1. Let $ES_s=0$; $EF_s=ES_s+\tau_s$; $SR_f=0$ ($f:=1$ to ∞);
$SR_f:=SR_f+R_s$ ($f:=ES_i$ to EF_i)

Step 2.
REPEAT
$k:=k+1$
Take an activity j of which all the predecessors are calculated
IF there is no such an activity, then *GO TO* step 3.

$$ES_j=\max\{\ ES_i+SSz_{(ij)k};\ EF_i+FSz_{(ij)k};\ ES_i+SFz_{(ij)k}-\tau_j;\ EF_i+FFz_{(ij)k}-\tau_j\}\quad \forall\ \{(ij)k\}\in \mathbf{A}$$

$EF_j=ES_j+\tau_j$

$SR_f:=SR_f+R_j$ ($f:=ES_j$ to EF_j)
IF $SR_f>L$ THEN
BEGIN
Shift j in the time ahead until $SR_f \leq L$ will satisfies for all f
IF $SR_f \leq L$ satisfies then ES_j= start of j and $EF_j=ES_j+\tau_j$
END
UNTIL $k=n$

Step 3
IF $f<n$ *THEN* STOP! (There is a loop in the network) *ELSE* $p=EF_t$

Backward calculations based on the reverse order of forward calculations can be performed as well. The procedure has to be started from the time policy and the resource histogram (SR_f) gained from the forward calculations. An activity can be chosen if all the successor activities have been arranged for the latest start/finish times. Once an activity is chosen, it simply has to be shifted to the latest possible occurrence where the histogram is still below the limit. Knowing the latest and earliest activity times the critical activities can be defined. However, it does not make much sense. Different heuristic rules lead to different early and late activity times and different project durations. It is hard to believe that an activity that has been indicated as critical according to some heuristic computations will not be critical according to another heuristic calculation where the project duration is less than in the first case. Application of exact optimization methods can help a lot to solve these theoretical problems.

To illustrate the algorithm let us solve the resource limited scheduling problem depicted in Figure 8-9. The availability of the resource is limited to 4 men. The result of the time analysis and the resource histogram can be seen in Figure 8-10. It can be seen that the resource limit makes this time policy non-feasible. The result of the resource allocation (allocating for limited resources) can be seen in Figure 8-11.

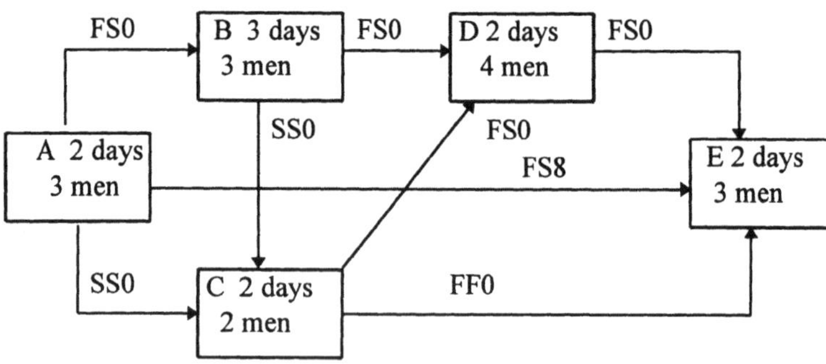

Figure 8-9 *Example project for resource allocation*

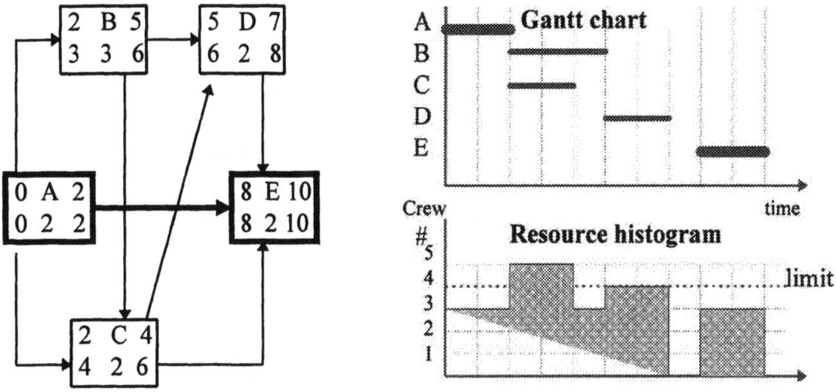

Figure 8-10 *Results of the time analysis*

The calculation starts with activity A and with an empty resource histogram. Activity A is scheduled from day 0 to day 2. The resource requirement of the activity has to be added to the resource histogram. The only activity for which all the predecessors are scheduled is activity B, therefore activity B is chosen for further computations. Based on its predecessors, it can be started on day 2 and finish on day 5. Activity C is the next in the queue. Based on its predecessors, it can be started on day 2, but a conflict arises when adding its resources to the histogram. To solve this resource conflict, its start has to be shifted to day 5 and it will be finished on day 7. Activity D can be chosen for further computations. According to the previous shift in the start and finish of activity C, activity D can start on day 7. Finally, activity E can start on day 9. The new project duration that satisfies the resource limitation is 11 days.

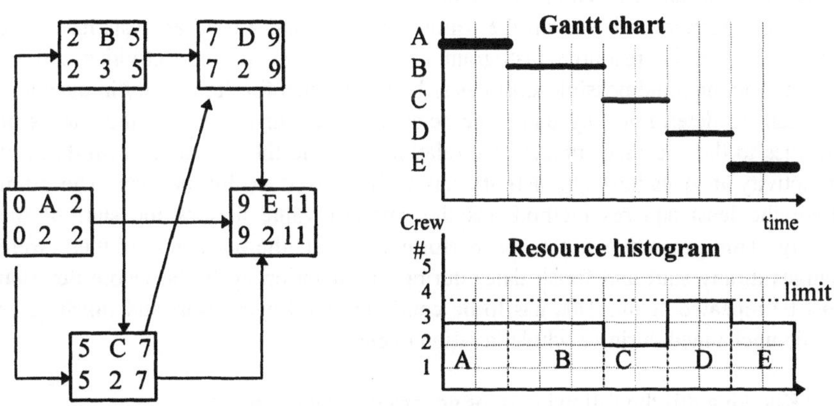

Figure 8-11 *Results of the time analysis*

Remarks to the algorithm:

- *It is possible that more activities can be taken into account at the same time. In this case the scheduler can decide what activity should be the next. The following rules can be applied for choosing among the activities:*

  Choose the activity with the greatest resource demand.
  Choose the activity with the greatest total float.
  Choose the activity that has the minimum early start time.
  Choose the activity that has the maximum late start time.
  Choose the activity with the smallest priority (priorities are given by the user, activities with the smallest priorities are the most important)

- One can never know which rule gives the better solution. Rules that perform well in some cases can perform poor in others.

8.4 MAXIMAL PRECEDENCE RELATIONSHIPS IN LEVELING

When maximal type of precedence relationships are applied in the network transformations have to be done. During the transformations maximal type of precedence relationships are transformed into minimal type of precedence relationships and these transformations cause transformation loops. The scheduler faces the same problem that he or she had in time analysis if maximal type of precedence relationships had been added to the network. The solution is similar to when activities can be taken into account in any order.

The algorithm is based on the following idea. First a time analysis has to be performed and the resource histogram calculated. Any of the activities can be chosen. The maximal possible shift has to be determined for the activity in question. This can be determined by using the scheduled start and finish of the successor activities to the successor precedence relationships and the start and finish dates of the activity in question. The activity has to be moved in this interval. The point where the least squares method has the smallest value defines the start of the activity. This procedure has to be performed for all the activities. If there were changes in any start and finish dates during this iteration cycle the whole iteration must be repeated again. This has to be continued until every start and finish dates remain unchanged during a whole iteration cycle.

Knowing this the following rules govern the algorithm:

Resource Leveling: Maximal Type Of Precedence Relationships Are Allowed

Step 0.

Perform a time analysis. Let S_i=ES_i and F_i=EF_i $\forall$ $i \in \mathbf{N}$

Step 1.

REPEAT

there_were_changes=false; *i*:=0

REPEAT

i:=*i*+*1* (Take an arbitrary activity *i*)

Determine the amount of the possible shift (D_i) for activity *i*

D_i=min{ S_j-$SSz_{(ij)k}$-S_i;

F_j-$SFz_{(ij)k}$-S_i;

S_j-$FSz_{(ij)k}$-F_i;

F_j-$FFz_{(ij)k}$-F_i} $\forall$ $\{(ij)k\} \in \mathbf{A}$

Shift *i* by day to day in the defined interval and calculate V after each shift. Where V takes its minimum value defines the start of *i*. (S_i)

IF the start of *i* has been changed THEN *there_were_changes*=true

UNTIL *i*=*n*

UNTIL *there_were_changes*=false

8.5 MAXIMAL PRECEDENCE RELATIONSHIPS IN ALLOCATION

The application of maximal type of precedence relationships in the network makes the allocation much more difficult. Remember, allocation is used when certain resource limits have to be kept, and these limits can cause increases in time gap between the activities and increases in the project duration. The procedure described in Chapter 8.3 always leads to a feasible solution if the resource requirement of individual resources are below the resource limits. However, the existence of a solution is not guaranteed if maximal precedence relationships are used in the network. The reason for this is that maximal type of precedence relationships keep the activities or activity chain together and do not allow the increase of the time gap beyond a given value. If a solution does not exist in the time interval defined by the maximal precedence relationships then resource limits cannot be satisfied. A small sample for this can be seen in Figure 8-12. To maintain the resource limit (3 men), it would be necessary to shift the activity by four days. However, the maxSS3 relationship allows only three days.

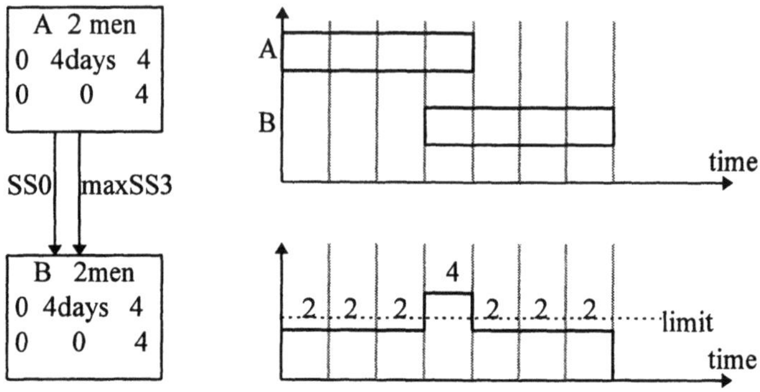

Figure 8-12 *Resource allocation: No solution exists*

Maximal type of precedence relationships make the allocation so difficult that it is very hard to produce good heuristic rules. The procedure that has been mentioned in Chapter 8.3 can be applied to solve this problem. Resource allocation (resource limited allocation) can be solved by leveling for fixed duration. Resource leveling has to be performed for the original project duration. If the limits could not be kept then the project duration has to be increased by adding a SF(p+1) relationship between the start and the finish and the leveling has to be performed again. If the resource profile is still above the limits then the project duration has to be increased again. This has to be repeated until the scheduler gets satisfactory results.

Chapter 9

ART OF SCHEDULING

So far we have treated network techniques as pure mathematical structures, directed graphs on which different computations have to be performed in order to achieve the results. To make network techniques more understandable, skeletonization has been done in previous chapters. We got rid of all the components that are used in practice but make the understanding more difficult on the beginner level. Problems of structured planning, special activity types, tracking the project, loops, calendar and multiproject scheduling are discussed in this chapter.

The first section, Work Breakdown Structure, deals with the problems that can arise in the scheduling phase if structured planning was applied in the planning phase. Milestones and hammock activities are the topics of the next section.

In the third section, Cash Flow Forecasting, Monitoring and Control, the use of network techniques for cost analysis, cash flow forecasting and tracking are discussed.

In the next part, Loops in Networks, we deal with the problem of finding the loops in the network. Multiproject scheduling is the topic of the fifth section.

The last section, Calendar Versus Workday Schedule covers some of the problems of using calendars in scheduling. Changing to calendar from workday schedule causes so many, so far solved and unsolved, problems that can confuse even the most experienced experts of network techniques.

9.1 WORK BREAKDOWN STRUCTURE

It was discussed in the introductory chapter that the first step of network-based project management is the determination of activities in the necessary details. For the first try it is almost impossible to create the activity lists detailed in the necessary depth especially in case of complex projects consisting of thousand of activities. Planners use a special technique called 'top to down' or the inverse of it called the 'bottom up' technique. In the first case the project is first divided into its main phases, then the main phases are divided into smaller phases, etc., until the planner reaches the desired depth, that is, the lowest practical level. Cost, duration, resources can be applied to the activities on this level. The result of this planning method is the so called Work Breakdown Structure (WBS). The WBS is a tree-like structure that has only one element on the top level which represents the project itself.

A possible WBS of the depot construction sample can be seen in Figure 9-1. Activities that have successors in the WBS are called summary tasks. For example, Foundation is a summary task, its subordinate tasks are activities J and K.

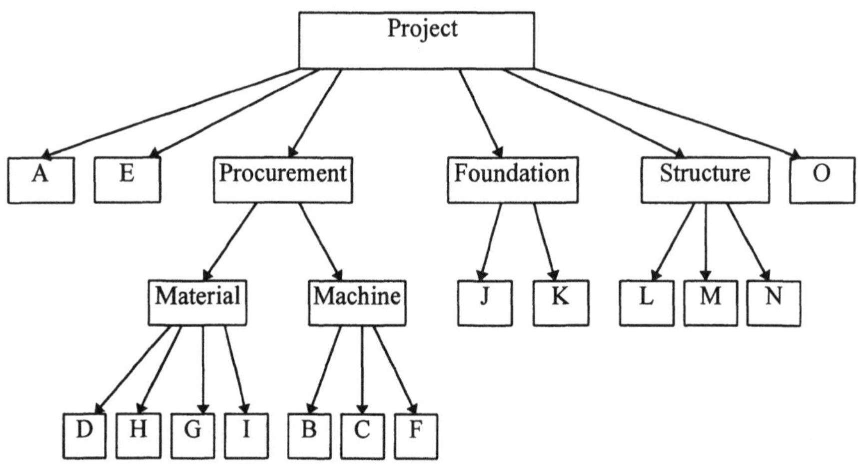

Figure 9-1 *Possible WBS of the depot construction sample.*

Activity duration of a summary task is not given by the planner. It is determined by the time spanned by their subordinates. The start of a summary task can be defined as the minimum amongst the starts of its immediate subordinates. The finish of a summary task is the maximum amongst the finishes of its immediate subordinates. Labor, materials, machines and costs cannot be directly ordered to a summary task, these measures have to be derived from the subordinates.

Different levels of management need information on project in different details. Other important use of WBS is that it can supply every participant of the project with information in the necessary depth. Crew members or foremen need information from the lowest level. Top management, especially in a multiproject environment, needs information from the uppermost level, usually the project start, finish and project total cost.

Using WBS in practice can lead some problems in time analysis. The situation is the simplest if summary tasks are *only served as tools for grouping activities and they are not part of the network logic (PA list).* This implies that only those activities can be found in the PA/SA lists that have no subordinate tasks. In this case one has to perform the time analysis in the usual way then define the start and finish times of the summary tasks from their subordinates going from bottom to up.

The situation is more complicated if summary tasks can also be part of the network logic, that is, if they can have immediate predecessor or successor activities from all level of the WBS. Let us examine this case now! Making a network logic that allows interdependencies between activities of any level, one must keep in mind the following two rules.

Rule 1. Activities that lie on the same branches of the WBS tree cannot be connected by precedence relationships. For example, precedence relationships between activities A and A11 or A and A22 are not allowed (see Figure 9-2). However, they are allowed, for example, between activities A and B, A and B1, A1 and A21 or A11 and B1.

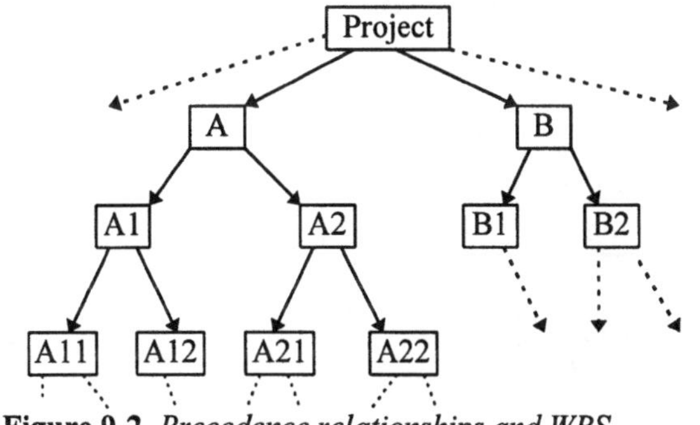

Figure 9-2 *Precedence relationships and WBS*

Rule 2. A summary task can have only FS, SS, maxSF and maxFF predecessor relationships and FS, FF, maxSS and maxSF successor relationships.

This rule also implies that only FS and maxSF relationships can exist between two summary tasks.

This is not a rule that has to be obeyed under all circumstances, but rather a recommended, well established methodological concept. Let us see what happens if one omits this rule and adds a SF predecessor to a summary task (see Figure 9-3). Suppose that the successor relationships of activity X define the start of activities A1 and A2 and the duration time of activities A1-A6 is 5 days each. In this case the start of A1 defines the start of the summary task. The finish of the summary task is defined by the SF90 precedence relationship. This relationship pushes the early finish of the summary task to day 100. However, the length of the A1-A4 chain is only 20 days and the length of the A5-A6 chain is only 10 days. Now there is a confusion how to display the subordinate tasks on the Gantt chart for the earliest start dates. The only thing known is that no subordinate activity can start before day 10 and no subordinate activity can finish later then day 100. On the other hand, one of the subordinates has to finish on day 100 according to the logic. But which one of them? Activity A4 or A6 or both? The scheduler faces the same problem that he or she had when using the splitting allowed concept. Just like with splittable activities, activities now also have to be arranged manually. The situation is more difficult if some of the subordinates tasks are splittable. The extensive use of the manual correction of the plan eliminates the greatest advantage of the network techniques, that is, the automatic calculations and automatic displaying of the results. To avoid this it is advisable that a summary task cannot have an SF preceding relationships.

The same argument excludes the application of FF preceding relationships, and the use of maxFF and maxFS as successor relationships of a summary task. This argument has to be applied for the latest activity times and this excludes the application of SF and FF as successor relationships and the use of maxSS and maxFS relationships as predecessors of a summary task.

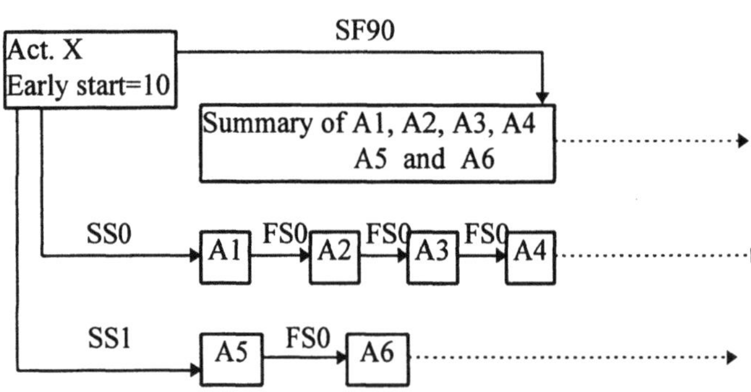

Figure 9-3 *Part of a network with FF precedence relationship to a summary task*

Keeping Rule 1 and Rule 2, the following method can be used for calculations. All the predecessor and successor relationships of a summary task have to be transformed to the lowest level and the calculations have to be performed on this level the same way as in a network with no summary tasks. To finish the calculations, the start and finish times of the summary tasks have to be determined by a 'bottom up' calculation. The start of a summary task is the minimum of the starts of its immediate subordinates. The finish of a summary is the maximum of the finishes of its immediate subordinates.

To illustrate this let us see the WBS structure of a sample network shown in Figure 9-4. The network itself can be seen in Figure 9-5. The transformed network can be seen in Figure 9-6. **It is not necessary to make the transformation physically** but the calculation has to be performed as though the transformation has happened. The calculations can be followed in Figure 9-7. The results of the original sample are displayed in Figure 9-8.

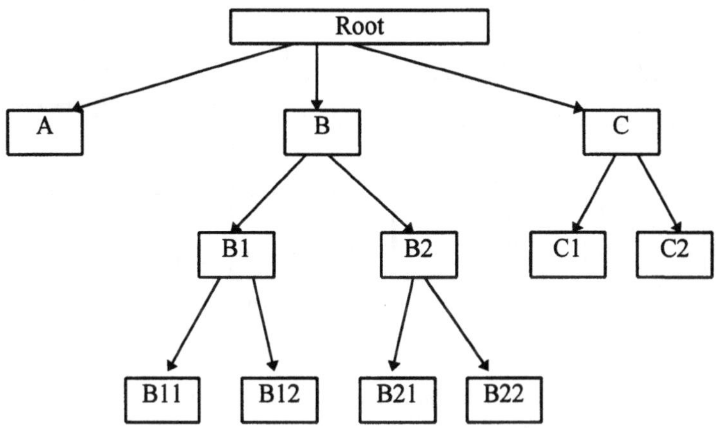

Figure 9-4 *Network analysis: Work Breakdown Structure*

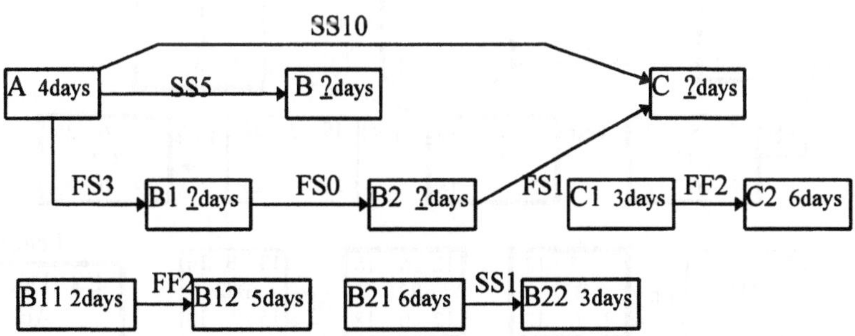

Figure 9-5 *Network analysis: Activity interdependencies*

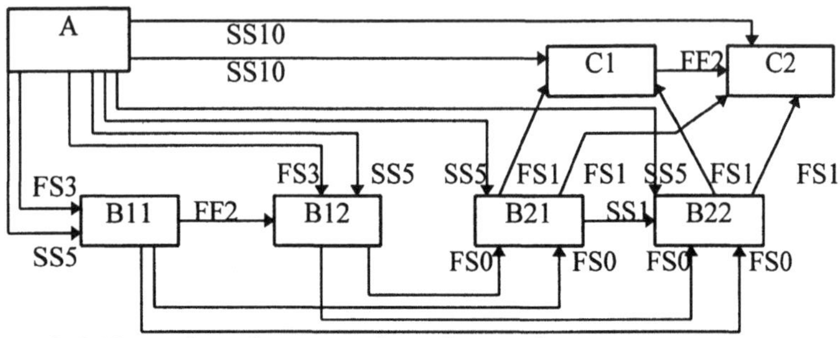

Figure 9-6 *Network analysis: transformed network*

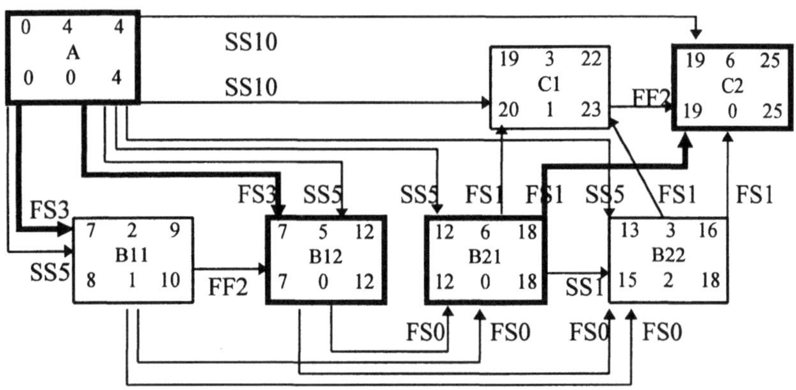

Figure 9-7 *Network analysis: calculations on the transformed network.*

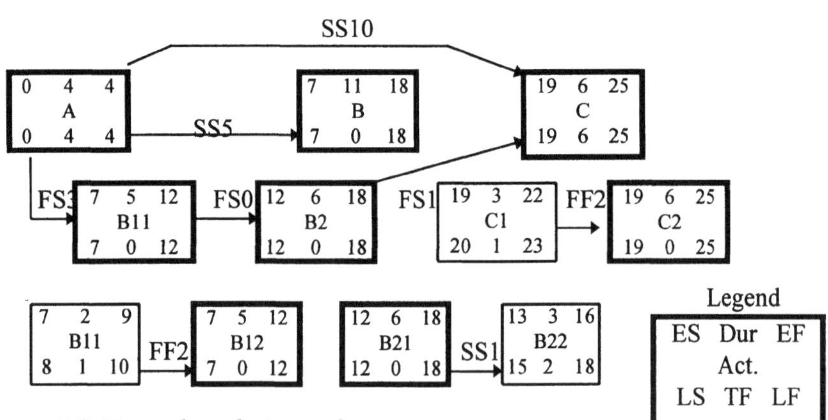

Figure 9-8 *Network analysis: results.*

The early start and late start of B1 come as the minimum of the early/late starts of B11 and B12. The early/late finish of B1 comes as the maximum of the early/late finishes of B11 and B12. Early and late times for B2 and C have to be determined in the same way. Duration times for summary activities come as the difference between their early finish and early start times. Start and finish times for activity B as well as its duration can be determined after these data are known for its immediate subordinates, that is, for B1 and B2.

Note that critical activities do not necessarily form a critical path. Summary tasks can be critical while one or more of their subordinates are not, and vice versa, critical subordinates not necessarily make their summary task critical.

Different calculation rules can be applied to handle WBS, but one has to be very careful in dealing with it. Transformations that leave summary tasks in the network will always require modifications in the original calculation rules.

Finally, we have to mention that different breakdown structures can be used in project management. Resource Breakdown Structures (RBS), Cost Breakdown Structures (CBS) or Organizational Breakdown Structures (OBS) are frequently used in the process of project management. The application of these breakdown structures do not modify any of the applied algorithms. They have special importance in producing reports and supplying different levels of management with necessarily detailed information. The use of CBS and RBS in computer implementation makes the software easy to integrate with finance and other applications.

9.2 SPECIAL ACTIVITY TYPES

Management is primarily interested in the attainment of project objectives -- usually the start and finish dates of the main phases -- rather than in the accomplishment of individual tasks. To satisfy the management requirements a special activity type has been introduced in network techniques, the milestone activities. Milestone activities are similar to the events of the activity on arrow networks, in fact they are events applied in activity on node networks. The difference between ordinary events and milestones is that milestones have special importance from the project point of view. Milestones have no duration; they signal the start or end of project phases or certain group of activities. Starting or finishing of project main phases can be emphasized by start and finish milestones. Finish milestones usually have FF0 predecessors start milestones have SS0 successors. The user has to write in the plan what type of milestone is the milestone in question. Milestones do not cause any problem in calculations. They have to be treated as activities with zero duration. However, start milestones and finish milestones are treated differently in calendar schedule and they have to be treated differently from the normal activities. The reason for this is that calendar schedules have a special characteristics. An activity which has one day duration will start and finish on the same day, and an activity with two days duration will finish one day after its start (see Fig. 9-9). This means that one has to subtract one day from the EF value and calculate the finish calendar date from this new value. However, this problem does not arise if start milestones have only SS successors and finish milestones have only FF predecessors.

Planners often apply constraints, especially Must Start On or Must Finish On constraints to the milestones, and make them critical.

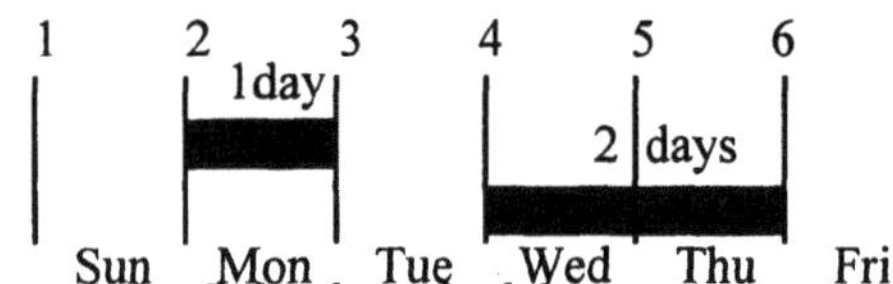

Figure 9-9 *Calendar schedule*

During the planning phase there are activities with unknown duration. A good example for this is dewatering. During a pipe laying project dewatering might be necessary from the start of trenchning till the finish of backfill (Figure 9-10).

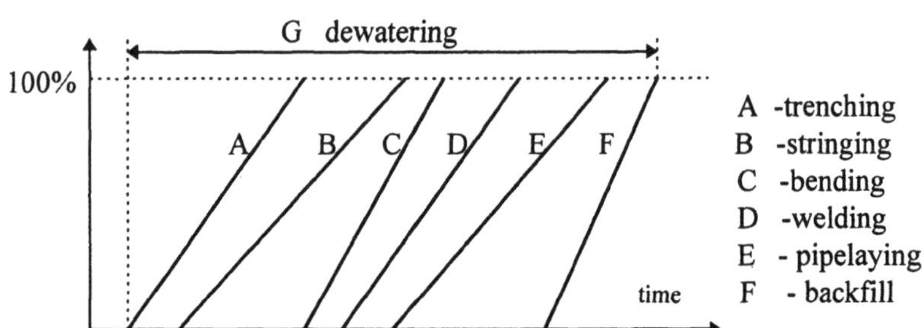

Figure 9-10 *Determining the duration of dewatering using LSM*

In this case the duration of dewatering cannot be determined in the planning phase, as it depends on the duration of certain activities and the precedence relationships between these activities. To handle this problem in networks a new activity type has to be introduced. This type of activity is called the hammock activity. The hammock activity is an activity that spans a group of activities. The hammock does not change the schedule dates for any activities. It includes the start date of its predecessor and the finish date of its successor. The network representation of the pipelaying project can be seen in Figure 9-11. Two different interpretation of hammock activities exist.

One, when a hammock has to have SS0 and FF0 predecessors. In this case the start of the hammock is equivalent to the minimum of the start dates determined by the SS0 relationships and the finish date is equivalent to the maximum of the finish dates determined by the FF0 relationships. The first interpretation of hammock activities can be seen in Figure 9-11a.

Another possible interpretation of hammock is when it has only SS0 predecessors and only FF0 successors (see Figure 9-11b). In this case the start of the hammock is the minimum of the start dates of its predecessors and the finish date is the maximum of the finish dates of its successors. Both approaches are acceptable, they give the same results. The second interpretation is not so correct as it can imply -- following the nature of the original network logic -- that backfilling logically depends on dewatering, that is, backfilling has to be finished as dewatering has been finished. The real interdependency, however, is the opposite of the above mentioned.

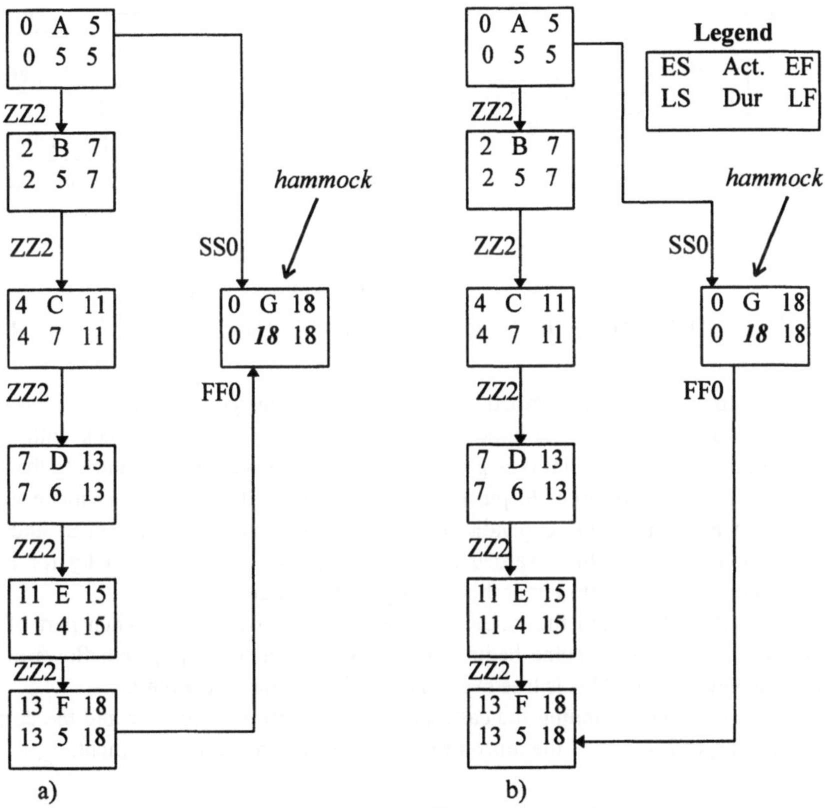

Figure 9-11 *Two existing interpretations of hammock activities*

Hammock activities and WBS summary activities practically are handled in the same way during the calculation. Both activity types span a group of activities and their durations are determined during the calculations. The difference between them that hammock activities can have their own resource demand and they are on the same level of WBS as the activities they span.

Some computer applications can handle hammocks, others cannot. There is another way of handling activities with unknown duration. This can be seen in Figure 9-12. In this case dewatering has to be splitted into two activities with zero duration. One stands for the start of dewatering, the other signals the finish of it. The difference between the two occurrence gives the duration of the dewatering. The disadvantage of this approach is that it is awkward to apply resources to these kind of activities.

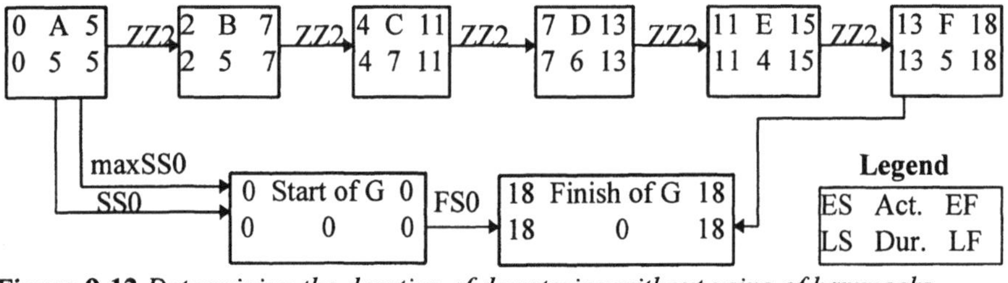

Figure 9-12 *Determining the duration of dewatering without using of hammocks*

9.3 Cash Flow Forecasting, Monitoring And Control

In today's highly competitive market when the competition for getting a project is ever stronger, the importance of keeping the project on track within the cost, time and resource limits is greater than ever. Owners award contracts through bidding processes in order to push down prices. Awarded contractors make every effort to minimize their expenditures. If their endeavors are successful and the expenditure is below the awarded price then they realize profit. Penalty for tardy completeness is the main factor that usually decreases the profit.

On the other hand the value of money changes over time. When part of the project is completed income begins to flow in. The later the payment flows in the less the value it has. This is true even in a non-inflationary economy.

Nevertheless, planning the cash income and outflow over time and the correct monitoring of it is one of the most important tasks of project management.

Cash Flow

In order to determine the financial feasibility of the project, a financial analysis must be undertaken. Cash flow analysis is one of the most important tools for planning the project financial life cycle.

Cash flow diagrams may be defined as the graphical portrayal of the inflow (payments) and the outflow (expenditures) of cash during a project. Progress payments received from the owner represent the inflow; payments made to subcontractors and suppliers represent the outflow. Cash flow diagram can be seen in Figure 9-13. From the contractors point of view, at the early stages of a project the sum of expenditures is usually higher than the sum of the incomes. There are different reasons for this. Owners like to hold back a certain percentage of the total value of completed work as security. With this sum they ensure that the contractors will finish the work to their satisfaction. Both income and outflow have to be planned. Expenditures are relatively higher in the earlier stage if activities start at their earliest start and relatively lower if activities start as late as possible. Contractors also have to plan the income, that is, the payment received for the work completed. Negotiations between contractors and owners will result in the planned inflow. This negotiation process is more difficult if subcontractors are also involved in the construction. One of the contractors' main objectives during the cash flow planning is to minimize funding, that is, keep the sum of inflow (payments received) as close as it possible to the outflow (expenditures). This planning requires cumulative cash flow diagrams, also called progress or S curves.

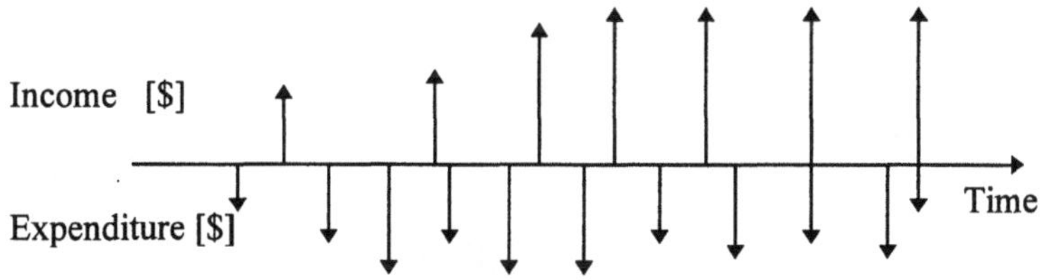

Figure 9-13 *Cash flow diagram*

Progress Curves

Progress curves, also called 'S' curves express some aspects of a project. They graphically depict some measures of cumulative progress on the vertical axis, while horizontal axis plots time. Progress can be measured in terms of money expended or any other measure which makes sense. The shape of a typical S curve can be seen on Figure 9-14. This comes from the fact that expenditures usually tend to start slowly, they are the highest in the middle of the construction and then taper off near the end. A daily expenditure diagram and the S curve gained from it can be seen in Figure 9-14.

Once the project is underway actual progress can be compared with the planned progress. The two outmost planned progress curves are the S curves that come from the early start and late start times. Scheduled progress should be plotted between these two possibilities (Figure 9-15).

These three progress curves start and finish in a common point. In ideal cases reported progress curves fall between the two extremes, what is more, they coincide with the actual planned S curve.

S curves that measure expenditures are the most commonly used in practice. One can draw different S curves that give important information about the project. An S curve that is based on BCWS (Budgeted Cost of Work Scheduled) shows the planned cost for the original schedule. Budgeted Cost of Work Scheduled (BCWS) is the amount of the budget which should have been earned by now, based upon original (baseline) dates and durations.

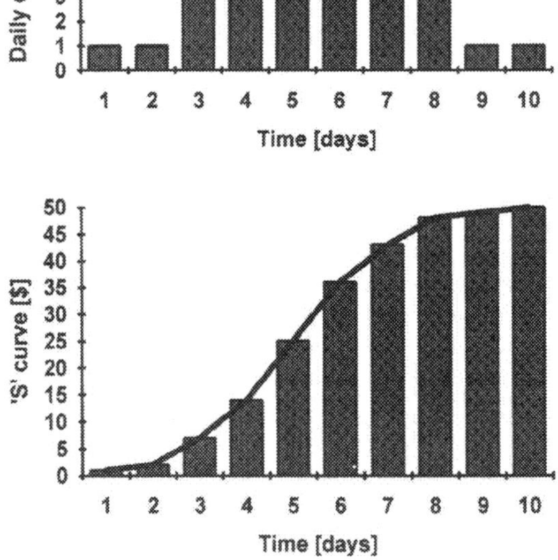

Figure 9-14 *Development of a progress curve.*

This can be compared with the S curve based on BCWP (Budgeted Cost of Work Performed). Budgeted Cost of Work Performed is the amount of the budget earned as measured by the percentage of current total scheduled work actually completed.

The difference between these to curves is measured by Schedule Variance (SV). Schedule Variance (SV) is the variance between what should have been earned by now (BCWS) and what has actually been earned as measured by earned value (BCWP). SV=BCWP-BCWS. A negative value indicates being over budget or behind schedule.

To S curve based on Actual Cost of Work Performed (ACWP) is used to decide whether the project is over budget or behind schedule. It shows the expenditures incurred up to now.

Cost Variance (CV) shows the difference, in currency, between actual costs and costs earned by earned value. CV=ACWP-BCWP. A positive value indicates the amount over budget.

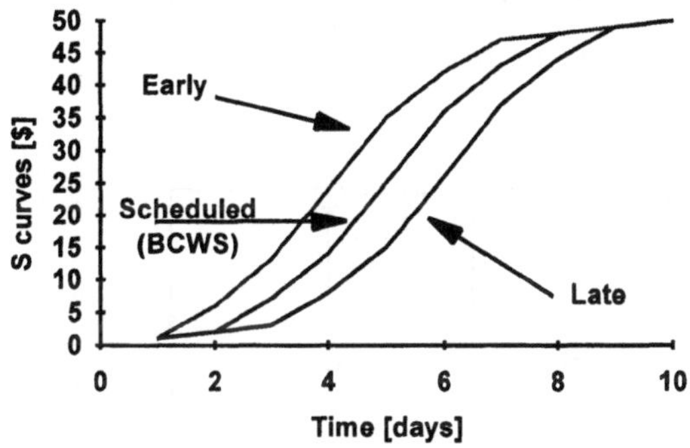

Figure 9-15 *Early late and scheduled S curves*

9.4 LOOPS IN NETWORKS

We have pointed out in Chapter 4: Loops in PDM Network that different interpretations of loops are used in practice. One is -- and we follow this approach -- when leaving an activity and going through some activities and precedence relationships the activity in question will be reached again. According to this definition Figure 9-16 shows a loop. In the second approach activities are replaced by arcs directed from the start to the finish of the activities and the start and finish points are represented by individual nodes (see Figure 9-17). In this approach a loop is a directed path that leaves a node and following the directed arcs and precedence relationships will reach the node in question again. What has been considered a loop according to the first approach is not a loop according to the second approach (see Figure 9-17). Again, we follow the first approach in the remaining part of this section but all the rules and algorithms introduced here can be applied to the second approach.

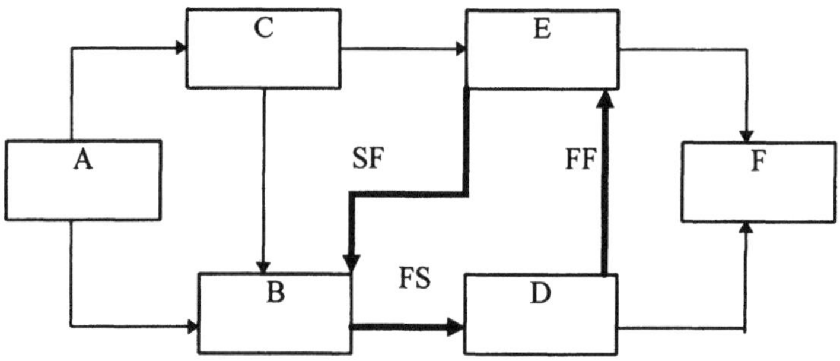

Figure 9-16 *Loop in PDM network B-FS-D-FF-E-SF*

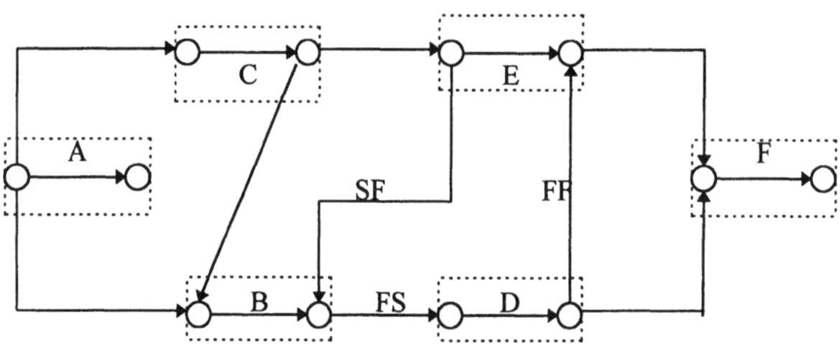

Figure 9-17 *B-FS-D-FF-E-SF do not form loop*

In the earliest versions of CPM the so called *i<j rule was implemented* to prevent the network from occurring loops. This rule means that events must be numbered so that the head event number must be greater than the tail event number. This can be seen Figure 9-18. This rule excludes the possibility of creating loops in a network. Most of the users are not aware of the fact that their computer application uses this rule to prevent the creation of loops. The reason for this is that this numbering process is taking place in the background. This rule can be implemented in PDM technique as well. However, this rule has two serious shortcomings. First, there is a big chance for renumbering the nodes as new activities are added to the network. This can be really tiresome in case of big and frequently modified networks.

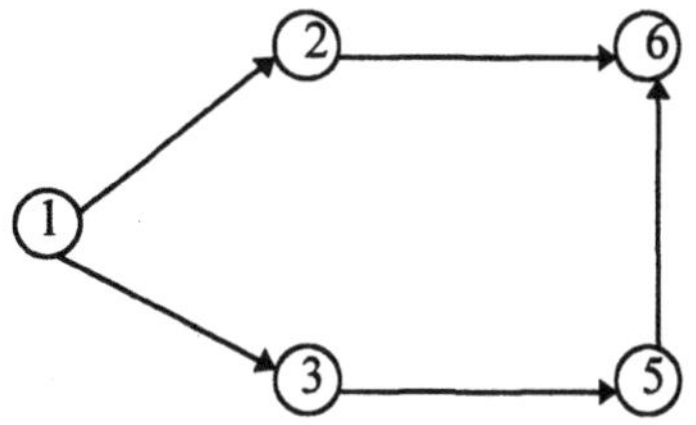

Figure 9-18 *The $i<j$ rule in CPM*

Imagine that a new activity is being added to the network of Figure 9-18 from event 3 to event 2. According to the $i<j$ rule a renumbering process has to be applied. A possible result of it can be seen in Figure 9-19. This shortcoming has virtually disappeared since computers are used in project management.

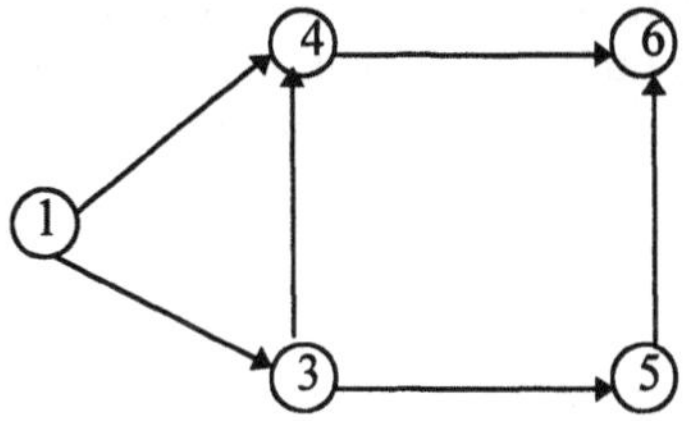

Figure 9-19 *Renumbering is necessary after modification*

If this renumbering cannot be performed then a loop has occurred in the network. This can be seen in Figure 9-20 where a new activity form node '6' to node '3' has been tried to be added to the network.

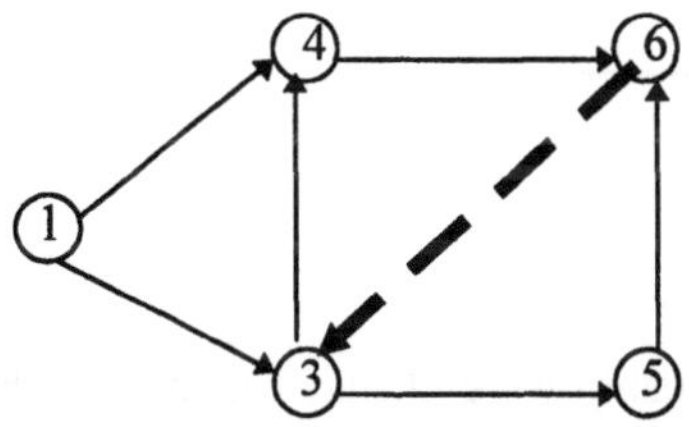

Figure 9-20 *The $i<j$ rule cannot be applied*

The $i<j$ rule cannot be implemented and this indicates that the newly added activity together with other activities form a loop in the network. The second shortcoming of the $i<j$ rule can be found in this phenomenon. Suppose that the

activity that was intended to be added describes existing logical interdependencies and the reason for the loop can be found in a previously typed in activity. The $i<j$ rule signals the loop when it is arisen and does not allow the last activity to be added that would close this loop even if it is a good and existing logical relationship.

To detect what activity or activities in CPM or what precedence relationship or relationships in PDM are guilty for the emerging of the loops all the partakers of the loop have to be known. Before introducing the loop detecting algorithm let us examine in detail the PDM network from the loop's point of view. It has become evident that loops are not allowed in PDM technique either. However, in case of application of maximal type of precedence relationships loops can arise in the network during the time analysis. Figure 9-21 shows a very small PDM network which consist of two precedence relationships. During the time analysis all the maximal type of precedence relationships have to be transformed to minimal type of precedence relationships. The transformation rule is shown in Figure 5-10 and in Table 5-2. According to this rule the direction of the maximal type of precedence relationships has to turn back and this process will lead to the emergence of loops (Figure 9-21). Loops that appear as a result of this transformation have to be treated differently because they are not logical loops just a result of the necessary transformation process.

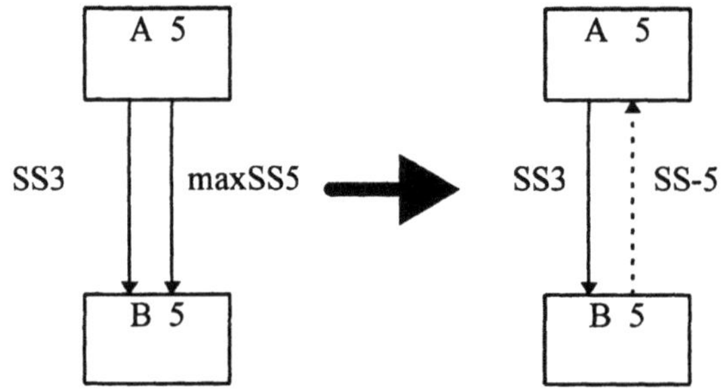

Figure 9-21 *Transformation loop*

If the lag factor of the maxSS relationship were changed to 2 days then the network could not be solved. Both relationships define a half line for the start of B in the LSM chart (Figure 9-22) and the common part of the two half lines defines the set where activity B can start. In case of maxSS2 relationship the two half lines defined by the two precedence relationships have no common part, therefore no schedule exists that satisfies these relationships.

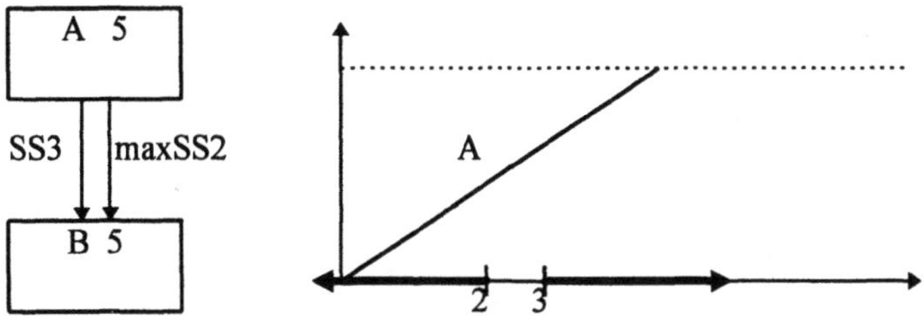

Figure 9-22 *The two precedence relationships have no common part*

The difference between the two cases is that in the case of maxSS5 relationship the transformation loop has a negative value, that is, 3+(-5)=-2 and in the case of maxSS2 relationship the transformation loop has a positive value, that is, 3+(-2)=1.

If a loop contains transformed maximal type of precedence relationships one has examine the value of the loop. If this value is less or equal to zero then the network is solvable, if this value is positive then the network cannot be solved, that is, no schedule exists that satisfies the precedence relationships.

The necessary steps of loop detecting can be summarized as follows:

- Detect all the loops in the network.
- Examine each loop. If the loop in question contains minimal type of relationships only then the network has to be corrected. If the loop contains transformed maximal type of precedence relationships then examine the value of the loop. If it is less or equal to zero then the network analysis can be performed. If the loop has positive value then the network cannot be solved.

In the following we present a simple algorithm that can be used for loop detecting. The algorithm will stop as a loop has been found. For detecting all the loops some kind of enumeration has to be used.

During the algorithm we divide the set of nodes (**N**) into two sets **S** and **T**. Set **S** includes the nodes that do not form loop or loops. Set **T** contains the rest of the nodes. If this set is empty then no loops have been included in the network. The disjunction of the nodes can be done by the application of an ordinary time analysis algorithm (see, e.g., Chapter 2.6 or Chapter 4.5) but simpler algorithms can also be used.

After **T** has been detached a new procedure has to be followed. During this procedure labels will be ordered to the nodes. A label will show that the node in question can be reached from the node indicated by the label. If a labeled node will be labeled again, then a loop has been detected.

Loop detecting: Phase I (Defining set S)

Step 1.

Let **S**=*s* **T**=**N**-*s*; *f*:=0; *loop_found*=*FALSE*

Step 2.

REPEAT

Take a *j* activity of which all the predecessors are in **S**. If there is no such an activity then Go To Step 3. **S**:=**S**+*j* **T**:=**T**-*j*

f:=*f*+1;

UNTIL f=n

IF *f*=*n* *THEN* Stop (There are no loops in the network)

Step 3.

Set **T** contains a loop. Continue with Phase II.

Loop detecting: Phase II. (finding the loop in T)

Step 1.

Take an arbitrary node *j*.

Step 2.

REPEAT

Label node j by node i, if i is one of j's immediate predecessor nodes

If j is labeled second times THEN loop_found=TRUE

else i:=j

UNTIL loop_found=TRUE

The algorithm will be illustrated on the following example. Let us find the loop in the below given network (Figure 9-23). Phase I and Phase II can be followed in Table 9-1 and Table 9-2.

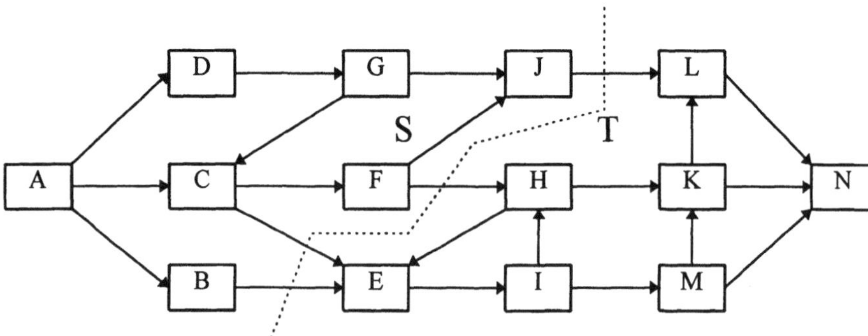

Figure 9-23 *Sample network for loop detecting*

	A	B	C	D	E	F	G	H	I	J	K	L	M	N
A		*	*	*										
B					*									
C					*	*								
D							*							
E									*					
F								*		*				
G			*							*				
H					*						*			
I								*					*	
J												*		
K												*		*
L														*
M											*			*
N														

+	+	+	+		+	+			+				

Table 9-1 *Phase I: Separating* **S** *and* **T**

Phase I can be followed in Table 9-1. Labeled nodes are in **S** the rest of the nodes are in **T**. At the start of Phase I only *s,* that is, the start node is in **S**. It is labeled by '+'. Let the rows and columns of **S** be covered. Set **S** can be enlarged by the nodes of which all the predecessor nodes can be found in S, that is, by the nodes the column of their uncovered part are empty.

In Iteration I set **S** can be enlarged by node D and B. The rows and columns of node D and B have to be covered. **S**= {A,B,D,}

In Iteration II set **S** can be enlarged by node G. The row and column of node G have to be covered. **S**= {A,B,D,G}

In Iteration III set **S** can be enlarged by node C. The row and column of node C have to be covered. **S** ={A,B,D,G,C}

In Iteration IV set S can be enlarged by node F. The row and column of node F have to be covered. **S**= {A,B,D,G,C,F}

In Iteration V set **S** can be enlarged by node J. The rows and column of node D and B has to be covered. **S**= {A,B,D,G,C,F,J}

Set **S** cannot be further enlarged. At least one loop exists in the network, precisely in set **T**. **T**={E,H,I,K,L,M,N}

Any of the nodes can be chosen from set **T** to start Phase II. Let us choose node N. The node in question has to be labeled by one of its predecessors. Node N has three predecessors (K,L,M). Any of them can be chosen. Let us label node N by K. Now we have to move on node K and label it by one of its predecessors. We can label it by M. Then node M can be labeled by I, node I can be labeled by node E, node E can be labeled by node H and node H can be labeled by node I. As node I has already been labeled, the loop has been detected. It can be read out from the labels. The label of I is E, the label of E is H, the label of H is I (I-E-H-I). Phase II can be followed in Table 9-2.

	E	H	I	K	L	M	N
E			*				
H				*			
I		*				*	
K					*		*
L							*
M					*		*
N							
			!!				
	H	I	E	M		I	K

Table 9-2 *Finding the loop in set T*

9.5 MULTIPROJECT SCHEDULING

Large companies usually have more than one project in progress at a time. It might be that these projects have no physical relation to each other. The nature of these projects can be different and they can take place thousands of miles away from the others. However, there is one common characteristic in these projects, they use, at least partially, the same resources. This common resource can be the same top management or the stakeholders' money, etc. What is more, these projects can employ the same labor and machines. Multiproject scheduling can help to utilize the company's resources in the most efficient way. Thinking in multiproject environment requires different approach to the project management. Planners must think not only about the most efficient delivery of one project, but rather on behalf of the whole company. The contradiction between the success of a single project and the common success of multiple projects can cause conflicts between the different levels of management. Project managers in charge of the success of a single project and top managers have different interests and these are usually in contradiction with the other party's interests. How to solve these kinds of problems is an important task, but is not be the objective of this book. Here, we are going to present the tools and methods that can be used in multiproject environment.

Multiproject management requires more preparatory work than single project management. It is advised to use common coding systems for the activities, costs and resources. If all the projects are planned and scheduled by the same team it is easy to satisfy this requirement. In this case all the projects can be integrated into a big one by simply adding a new activity, that is, a new root, above the root of the single projects (Figure 9-24). In this case all the algorithms and methods can be applied, just as for a single project. Precedence relationships can be added from a project to others, which can be a great deal of help in resource planning.

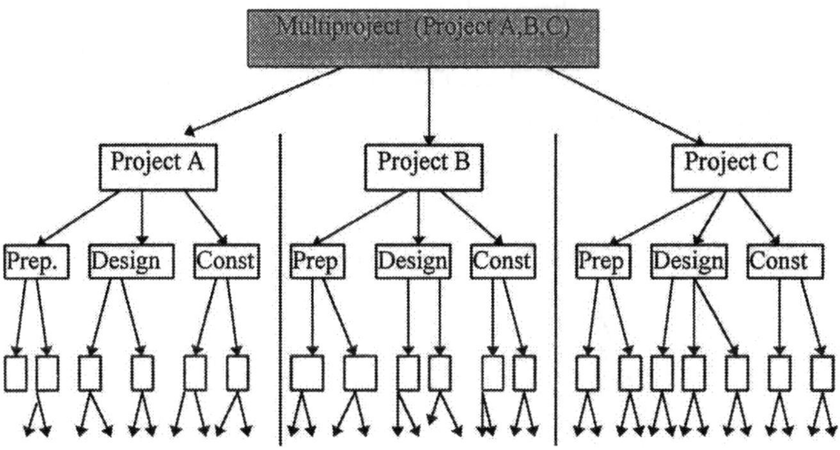

Figure 9-24 *Multiproject WBS*

Due to computer limitations and some other organizational reasons this method cannot be applied every time. Often, individual projects are planned and scheduled in individual networks. For efficient project management it is necessary to develop a tool that is able to connect networks if the planner does not want to merge these projects into a big network. One of the existing solutions to this problem is the use of interfaces that can connect activities of different networks. An example multiproject, which is connected by interfaces can be seen in Figure 9-25.

The simplest solution of handling more projects is when skeleton network is used. In this network individual nodes can represent individual projects. The disadvantege of this solution is that interrelationships between activities cannot be described (Figure 9-26) as this method makes connections between the start and the finish dates of the projects.

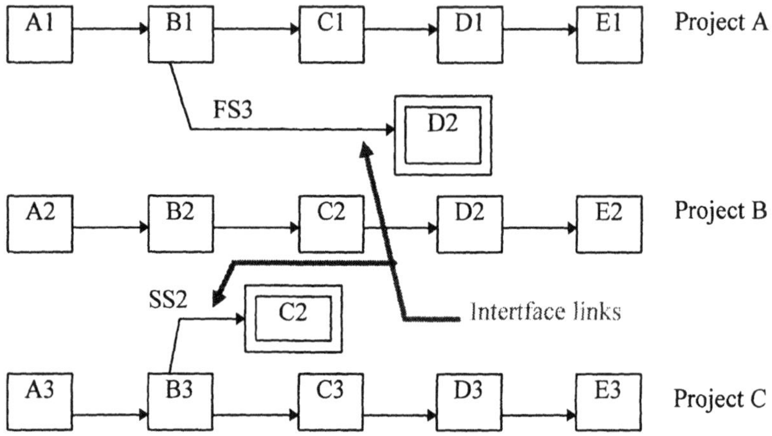

Figure 9-25 *Network interfaces*

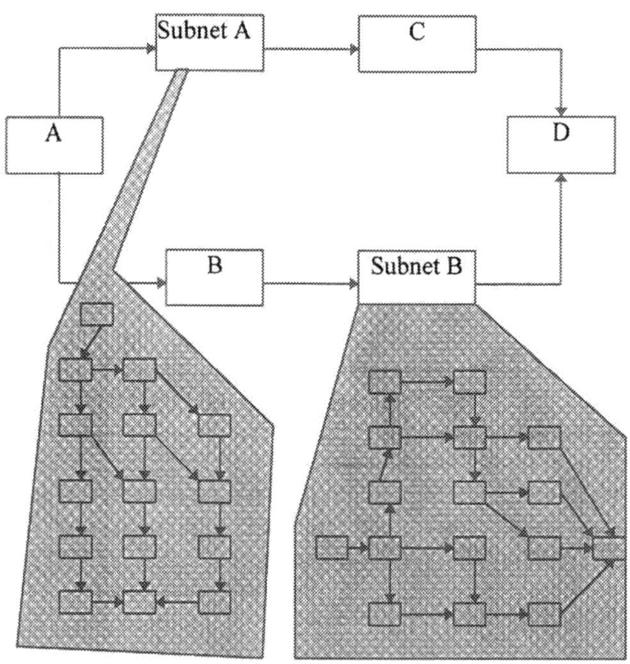

Figure 9-26 *Including subnets into the main network*

9.6 CALENDAR VERSUS WORKDAY SCHEDULE

Calendar schedule causes so many problems to the schedulers that it is necessary to discuss it here. In case of one calendar the scheduler will not face problems. He or she just has to transform the results into a calendar format. However, the application of different calendars, which is a natural demand on the schedulers part, can confuse even the experts.

A good example for this can be seen in Figure 9-27. Suppose that the duration times of activities A and B are given in elapsed duration, which means that weekends and public holidays have to be considered as workdays in case of these activities. The same machine is used in carrying out both activities. The FS2 relationship describes that after finishing task A the machine has to be moved to the location of task B. According to the traffic rules this delivery can take place on workdays only and it takes two days.

Figure 9-27a). shows the results of the usual time analysis and Figure 9-27b) shows the result of the calendar scheduling.

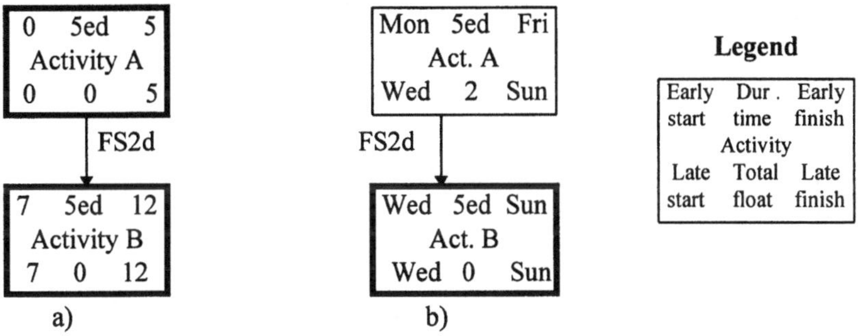

Figure 9-27 *Calendar versus workday scheduling*

The basic difference between the two schedules is that as a result of the application of different calendars (normal and elapsed times) the project duration can vary, that is, in this case the project duration calculated from the early activity times is greater than the project duration calculated from the late activity times. The change from workday scheduling into calendar scheduling can change the critical path or can make critical activities non critical. What is more, it is possible that there is no feasible solution to the project that was solved without problems using the workday scheduling. An example for this can be seen in Figure 9-28. Suppose that two kinds of resources are necessary to carry out activity B, resource 1 (R1) and resource 2 (R2). For example, let R1 represent the concrete forms and R2 the crew who is able to assembly it. However crew members are available in the odd months only, and concrete forms in the even months. Certainly, there is no feasible solution to this schedule, concrete forms will never be erected. However, if the above mentioned availability is valid in the first half of the year and both of the resources are available from July, then the project, which originally has a two month duration, can be executed in seven months (Figure 9-28). This phenomenon points out the biggest problem of calendar scheduling: one can never know if it really has no feasible solution or the calculations just have not gone far ahead in the calendar.

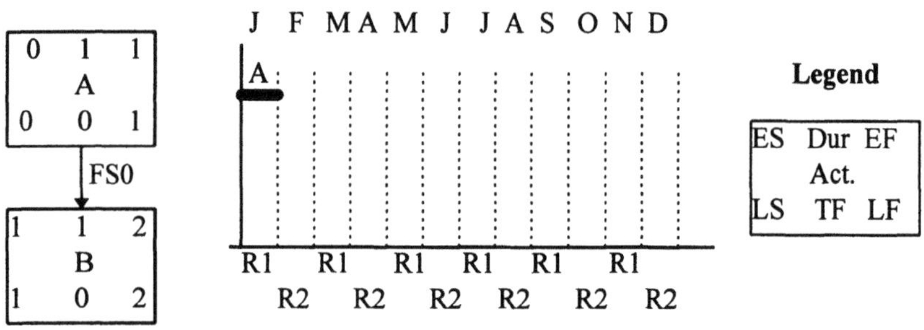

Figure 9-28 *There is no feasible schedule in calendar scheduling*

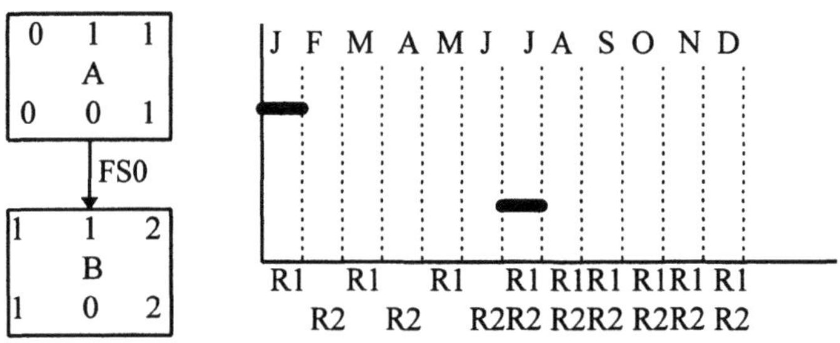

Figure 9-29 *Schedule depends on the availability of resources*

The application of maximal precedence relationships can cause schedule conflicts as well. We have discussed in Chapter 9.4 that transformation loops arise in the calculation process if maximal precedence relationships have been added to the plan. If the value of a transformation loop is equal or less than zero then a solution exists. Unfortunately, the originally negative value of a transformation loop can be positive when changing from workday to calendar schedule, and what is the worst this value can change depending on the actual place of the activities in question in the time scale. A small example for this can be seen in Figure 9-30. Every second day of the calendar has to be considered as holiday in the first ten days. As there is a maxSS1 elapsed day precedence relationship between the two activities, activity A can start only on the eleventh day.

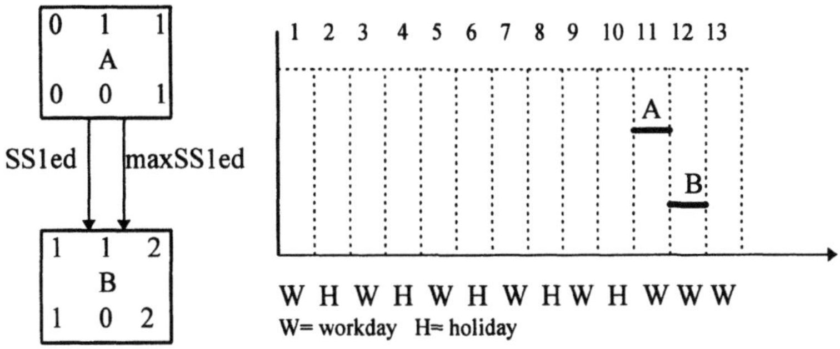

Figure 9-30 *The value of transformation loops can change in the calendar*

Scheduling by using different calendars requires different algorithms, and this is the point when one has to leave the calculations to the computer. Keeping sometimes dozens of different calendars in mind is almost impossible and the chance for making mistakes in hand calculations is almost hundred percent. Computer algorithms usually follow procedures for time analysis that is similar to the resource allocation procedures. For every calendar a fictive resource has to be supposed. There is no resource limit in workdays, however, the number of available resources are zero on the days when the resource in question is not available, that is, on weekends, public holidays, etc. Activities that use the same calendar can use the same fictive resource as well.

Appendix A

MATHEMATICAL BASIS

1. DIGRAPH

There are n nodes given where n is a finite positive integer. The set of the nodes is denoted by **N**. (**N**=$\{1,2.....i,...j,.....n\}$) From these nodes ordered pairs of nodes will be constructed which will be called directed arcs. The set of these arcs is denoted by **A**. The set of nodes and the joining arcs form a structure, which is called *digraph* and denoted by the [**N**,**A**] symbol. An arc is identified by its tail and head nodes. Direction of arcs is indicated by arrows at the end of the arcs. Multiple arrows are not allowed, looping arcs are, however, allowed. Digraphs can be seen in Figure A-1.

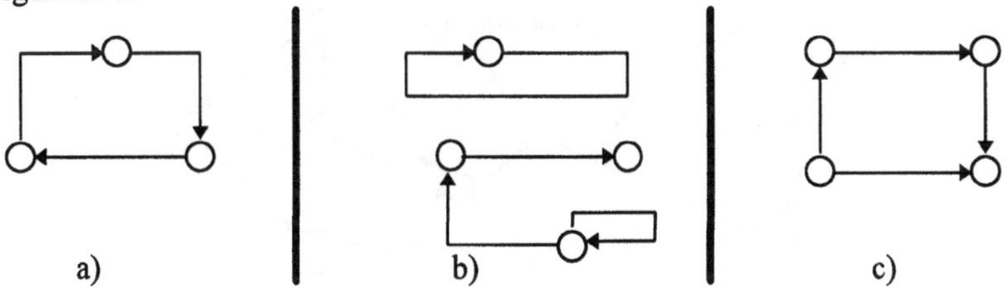

Figure A-1 *These are digraphs*

The structures of Figure A-2 cannot be considered as digraphs, because they have multiple arcs.

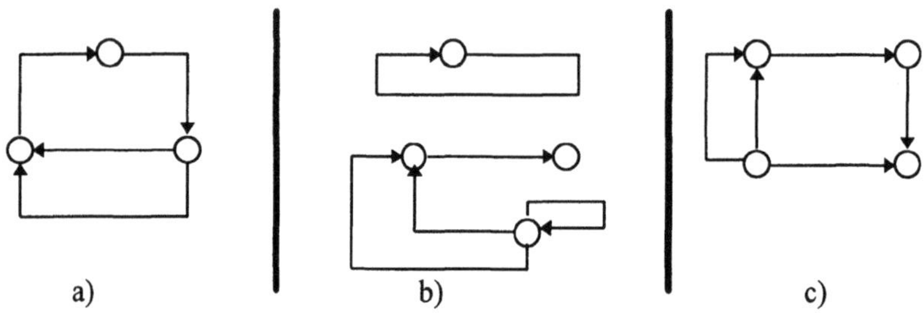

Figure A-2 *These are not digraphs.*

To indicate which pair of nodes of set **N** belong to set **A**, that is, to indicate the arcs, the so called *structure matrix* can be used. The rows and columns of this matrix represent the nodes of the digraph. A sign in the cell of the ith row and j^{th} column indicates an arrow from node i to j.

There are two important definitions in connection with digraphs which the reader should be familiar with.

Definition A-1 *(Definition of cut)*
Let s and t be two distinguished points of digraph **[N,A]**. *Let* **N** *be divided into two nonempty disjoint subsets, that is,* **S∩T=0** and **S∪T=N**. *Let* $s \in$ **S** *and* $t \in$ **T**. *The arcs originating from* **S** *and designating in* **T** *are called a cut separating s and t and denoted by* **(S,T)**.

A cut can be seen in Figure A-3. (The arcs in the cut are depicted in bold line.) Observe that the arcs originating from set **T** and finishing in set **S** do not belong to cut **(S,T)**. Those arcs are the elements of cut **(T,S)** . The separating cut **(T,S)** can be seen in Figure A-4.

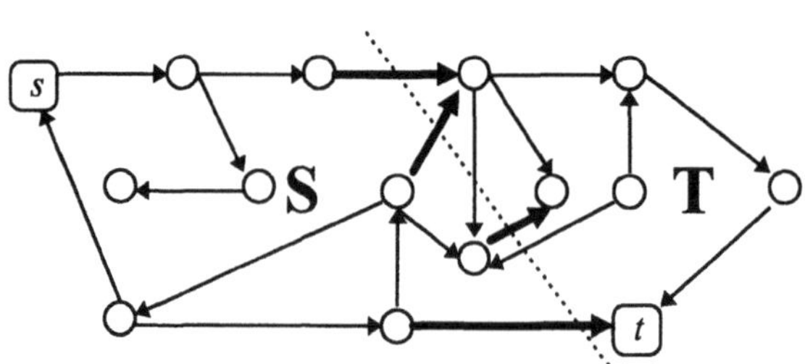

Figure A-3 *Separating cut (***S,T***).*

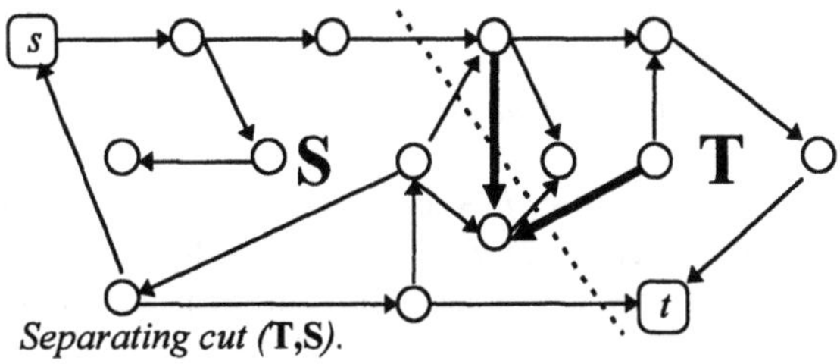

Figure A-4 *Separating cut (***T**,**S***).*

Definition A-2 *(Definition of path)*
Let s and t be two distinguished points of digraph [**N**,**A**]. *Let the set of* $s \equiv x_0, x_1,\ldots x_i,\ldots x_m \equiv t$ *be an ordered sequence of nodes for which* $x_0\ldots x_m \in \mathbf{N}$ *and* $(x_{i-1},x_i) \in \mathbf{A}$ *for all* $i=1,2\ldots m$. *The sequence of* $\{(x_{i-1},x_i),\ i=1..m\} \in \mathbf{A}$ *is called the path leading from s* to *t and denoted by the symbols* $P(s,t)$ or $P(s \equiv x_0, x_1,\ldots x_m \equiv t)$.

A path from point *s* to *t* is depicted in Figure A-5.

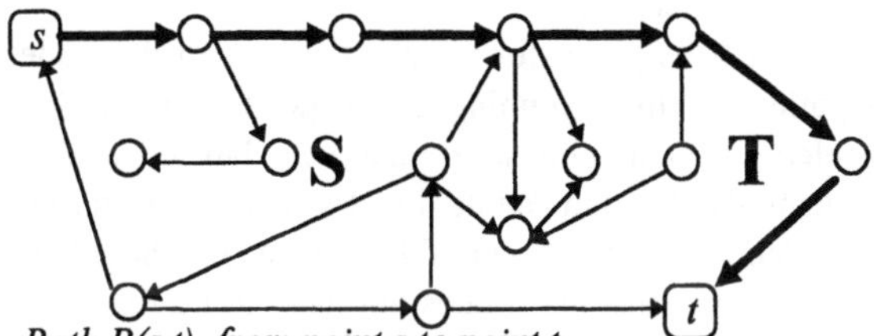

Figure A-5 *Path P(s,t) from point s to point t.*

2. Duality Theorem of Path and Cut

In connection with a digraph the following basic questions can be asked: Is there a path from an $s \in \mathbf{N}$ point to a $t \in \mathbf{N}$ point on a given digraph?

This problem can be answered using the duality theorem between path and cut.

Theorem A-1 (*Duality theorem of path and cut*)
Let [**N**,**A**] be a digraph and $s,t \in \mathbf{N}$. One and only one of the following two statements holds:

- at least one path exists from *s* to *t*,
- there exists a cut separating nodes *s* and *t*, which is empty.

Proof: The given proof is constructive, that is, it gives either a P(s,t) path or an empty cut separating nodes s and t.

Let us construct set **S** in such a way that

- let s∈**S**
- let all j∈**S** if there exist an i∈**S** and (ij)∈**A**, that is, let all the nodes be elements of **S** if they can be reached from s by paths.

If, in this construction, t∈**S**, then at least one path exists from s to t.

If t∉**S**, then let us use symbol **T** to denote the set of nodes that cannot be reached from point s. In this case, due to the construction rules of **S**, there are no arcs between **S** and **T**, that is, the separating cut (**S**,**T**) is empty. With this the theorem is proved. The technique used to construct set **S** is called labeling technique and shown in the following.

Labels and addresses will be used during the algorithm. A '-' label indicates that the labeled node is element of **S** but has to be further investigated. A '+' label indicates that the labeled node is element of **S** and no further investigation is necessary. Addresses are used together for labeled nodes. An address i is assigned for node j if j has been reached from i.

Let an arbitrary set **S** be given for which s∈**S** and all the nodes of **S** can be reached from s by paths. (At the beginning the **S**={s} set will do. In this case node s is negatively labeled.) Let us choose a negatively labeled i∈**S** node. All the j nodes that are not elements of **S** but can be directly reached from i will be elements of **S** and have to be labeled with '-' and addressed by i. The label of i has to be changed into '+', which indicates that i does not have to be further investigated. If node t has been reached, that is, t∈**S**, then the procedure has been finished. In opposite case, an another negatively labeled node has to be chosen and the above given step has to be repeated.

The procedure can end in two ways:

- node t will be an element of **S**
- set **S** cannot be enlarged and t∉**S**, that is, the separating cut (**S**,**T**) is empty.

In the first case the nodes of the path leading from s to t can be traced back from t using the address of t then the address of the node addresses node t, etc., the whole path can be determined by using the addresses of nodes.

To illustrate the labeling technique, let us find a path from s=1 to t=6 points. The procedure will be illustrated step by step both by the digraph and the structure matrix. Figure A-6 and Table A-1 shows the digraph and its structure matrix, the algorithm can be followed in Figure A-7, A-10 and in Table A-21, A-5. The labels and addresses are displayed below the structure matrix and nearby the nodes. At the beginning s is labeled with '-', and s is the only labeled node.

	1	2	3	4	5	6
1			*			
2	*			*		
3		*			*	
4					*	*
5						
6					*	
	-					

Table A-1 *Structure matrix*

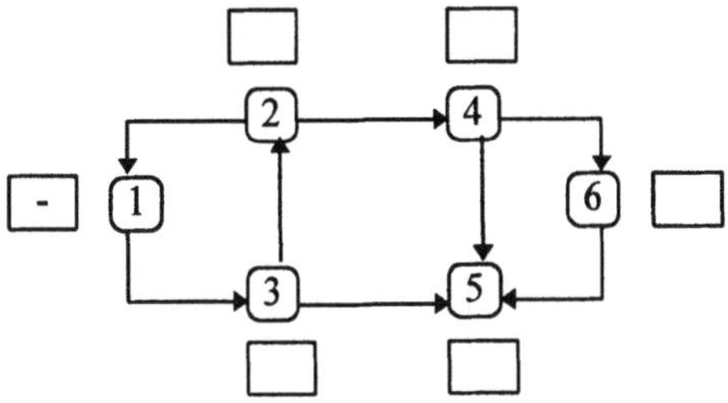

S={1} T={2,3,4,5,6}

Figure A-6 *The sample digraph*

	1	2	3	4	5	6
1			*			
2	*			*		
3		*			*	
4					*	*
5						
6					*	
	+		-1			

Table A-2 *Step 1*

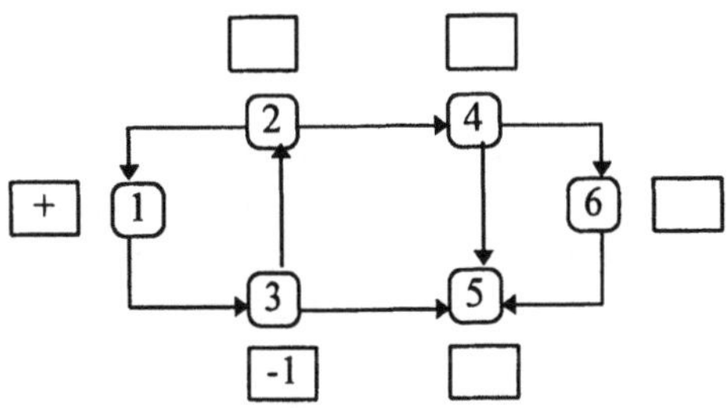

S={1,3} T={2,4,5,6}

Figure A-7 *Step 1*

In step 1 node '3' can be labeled as this is the only node which can be directly reached from *s*. Node *s* does not have to further investigated, therefore it can be labeled with '+'. Node '3' is labeled with '-1' which indicates that it has been reached from node '1' and this node has to be further investigated.

	1	2	3	4	5	6
1			*			
2	*			*		
3		*			*	
4					*	*
5						
6					*	
	+	-3	+1		-3	

Table A-3 *Step 2.*

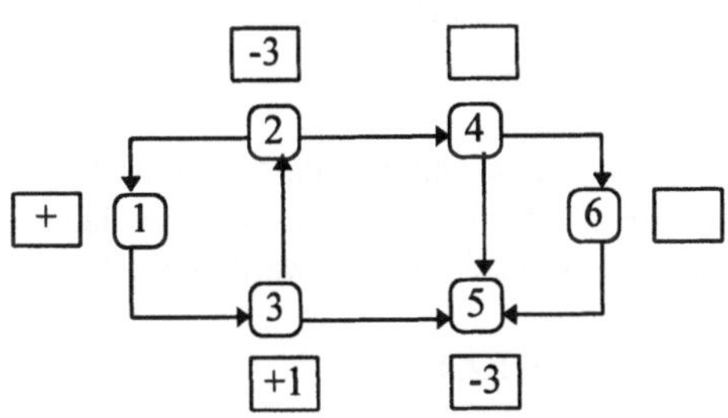

S={1,3,2,5} T={4,6}

Figure A-8 *Step 2.*

In Step 2 nodes '2' and '5' will be labeled with '-3' as they can be directly reached from node '3'. Node '3' will be labeled positively.

	1	2	3	4	5	6
1			*			
2	*			*		
3		*			*	
4					*	*
5						
6					*	
	+	+3	+1	-2	-3	

Table A-4 *Step 3.*

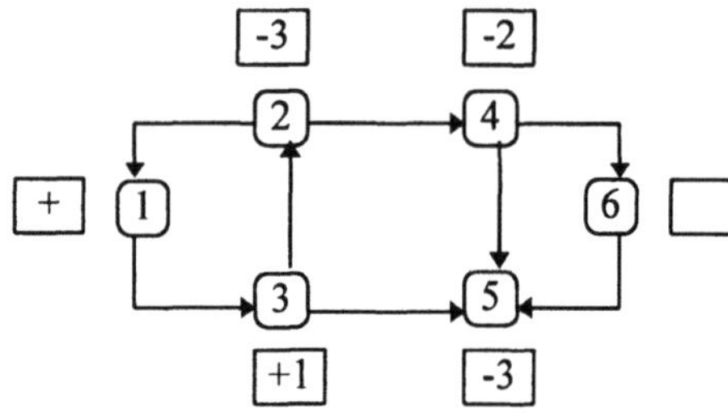

S={1,3,2,5,4} T={6}

Figure A-9 *Step 3.*

In Step 3 either node '2' or node '4' can be chosen for investigation. We follow the 'left to right' rule, that is, in case of more negatively labeled activities the leftmost activity in the structure matrix is chosen. In this step we choose node '2'. Node '4' will be labeled negatively node '2' will be labeled positively.

	1	2	3	4	5	6
1			*			
2	*			*		
3		*			*	
4					*	*
5						
6					*	
	+	+3	+1	+2	-3	-4

Table A-5 *Step 4.*

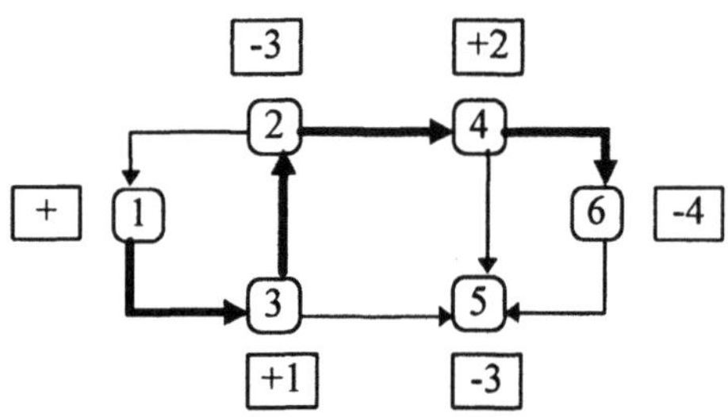

S={1,3,2,5,4,6} T={}

Figure A-10 *Step 4.*

In Step 4 either node '4' or node '5' can be chosen for investigation. We choose node '4'. Node '6' will be labeled with '-4'. The procedure has been finished. Proceeding backward on the labels, the following path is obtained:

P={s≡1,3,2,4,6≡t}

3. MINIMAL PATH - MAXIMAL POTENTIAL PROBLEM

A digraph [N,A] is given. Let us assign $\tau_{ij}\geq 0$ integers to all arcs $(i,j)\in$ **A**. The digraph together with the assigned τ_{ij} values is a called network and denoted by the symbol [N,A,τ]. Let $s\in$ **N** and $t\in$ **N** be two distinguished points of the digraph. One of the basic tasks in the network is to find a path from s to t along which the sum of the τ values is minimal, that is, to find the shortest $P(s,t)$ path. Mathematically, a path

$$P(s\equiv x_0, x_1,\ldots\ldots x_m\equiv t) \qquad \textbf{A-1}$$

has to be found for which the value of the following objective function is minimal:

$$\tau(P)=\sum_{i=1}^{m} \tau_{i-1,i} \rightarrow \min \qquad \textbf{A-2}$$

This problem is called the primal problem.

Let us assign a π_i non-negative integer number to all nodes, which are called potentials. Let the following constraints be prescribed for π_i :

$$\pi_s=0 \qquad \textbf{A-3}$$

$$\pi_j-\pi_i\leq\tau_{ij} \qquad \forall(i,j)\in\mathbf{A} \qquad \textbf{A-4}$$

We are to find the maximum of the following objective function:

$$\pi_t \rightarrow \max \qquad \textbf{A-5}$$

The problem given in **A-3** to **A-5** is called the dual problem.

The first and second problem is called the minimal path - maximal potential problem. We are going to solve the two problems together.

There is a strong connection between the objective functions of the two problems as it is shown by the following lemma.

Lemma A-1 If $P(s,t)$ is an arbitrary path from s to t and π is an arbitrary potential system satisfying **A-3** and **A-4**, then the value of the primal objective function is greater than or equal to the dual objective function, that is,

$$\tau(P)\geq\pi_t \qquad \textbf{A-6}$$

and equality holds if and only if the following satisfies:

$$\pi_j-\pi_i=\tau_{ij} \qquad \forall(i,j)\in P(s,t) \qquad \textbf{A-7}$$

Proof: $\tau(P)=\Sigma\tau_{i-1,i} \geq \Sigma(\pi_i-\pi_{i-1})=(\pi_1-\pi_0)+(\pi_2-\pi_1)+....+(\pi_m-\pi_{m-1})=\pi_{m\equiv t}-\pi_{0\equiv s}=\pi_t$
With this **A-6** is proved and equality holds if $\pi_j-\pi_i=\tau_{ij}$ along the path.

An important consequence of the lemma is presented hereby.

Corollary A-1 *(weak form of equilibrium)* If a path $P(s,t)$ exists and the potential system satisfies **A-3, A-4**, and $\tau(P)=\pi_t$ then they are optimal, that is, $\tau(P)$ is minimal and π_t is maximal.

Proof: (*indirect way*) Let us denote the value of the primal objective function by P and that of the dual by D.

Let us assume that P=D, but the path P is not optimal, that is, a better primal solution P* exist. In this case P*<P=D but this is a contradiction according to the lemma.

Let us assume that P=D, but there exists a better dual solution D* where D*>D. In this case D*>D=P but this is a contradiction according to the lemma.

Now it is proved that the two models form a primal-dual pair. In the next theorem we prove that there exists a solution where the value of the two objective functions are the same, that is, both solutions are optimal.

Theorem A-2 (*duality theorem*) In a network $[\mathbf{N},\mathbf{A},\tau]$ there exist such a $P(s,t)$ path and a π_i potential satisfying **A-3, A-4** that $\tau(P)=\pi_t$, that is, $\tau(P)$ is minimal and π_t is maximal.

Proof: Let us take an arbitrary potential system satisfying **A-3** and **A-4**. At the beginning this can be the $\pi_i=0 \ \forall \ i\in\mathbf{N}$ system.

Let us construct digraph **[N,B]**. Let set **B** be a subset of **A** and those arcs will be elements of **B** for which $\pi_j-\pi_i=\tau_{ij}$ holds. Let us find a path from s to t on digraph **[N,B]**. According to Theorem A-,1 there exists either a path or an empty cut.

If there exists a path $P(s,t)$ on digraph **[N,B]**, then, due to the construction of **[N,B]**, $\pi_j-\pi_i=\tau_{ij}$ along the path, that is, the length of the path is minimal and the potential π_t is maximal as it follows from the consequence of the lemma.

If there is no path, then there exist an empty cut **(S,T)** on **[N,B]** but $\pi_j-\pi_i<\tau_{ij}$ holds for arcs of set **A**. We can define an ε_{ij} value for all arcs *(ij)* in the cut in the following way: $\varepsilon_{ij}:=\{\tau_{ij}+\pi_i-\pi_j\}$

Let us construct a new potential system with the help of the below defined ε value: $\varepsilon:=\min\limits_{\substack{i\in S\\ j\in T\\ (ij)\in A}} \varepsilon_{ij}$

Let the new potential system π^* be constructed in the following way:

$$\pi_i^* = \begin{cases} \pi_i & \text{if } i \in \mathbf{S} \\ \pi_i + \varepsilon & \text{if } i \in \mathbf{T} \end{cases}$$

After this step set **B** will be enlarged by, at least, one arc. The new π^* values still satisfy **A-3**, **A-4**. We prove this in the following:

If $i \in \mathbf{S}$ and $j \in \mathbf{S}$ then $\pi_j^*-\pi_i^*=\pi_j-\pi_i \leq \tau_{ij}$, therefore **A-4** has been satisfied.
If $i \in \mathbf{T}$ and $j \in \mathbf{T}$ then $\pi_j^*-\pi_i^*=\pi_j+\varepsilon-\pi_i-\varepsilon_i \leq \tau_{ij}$, therefore **A-4** has been satisfied .
If $i \in \mathbf{T}$ and $j \in \mathbf{S}$ then $\pi_j^*-\pi_i^*=\pi_j-\pi_i-\varepsilon \leq \tau_{ij}$, therefore **A-4** has been satisfied.
If $i \in \mathbf{S}$ and $j \in \mathbf{T}$ then $\pi_j^*-\pi_i^*=\pi_j+\varepsilon-\pi_i \leq \tau_{ij}$, therefore **A-4** has been satisfied, because of the definition of ε. At least one of these arcs will become element of **B**.

For the new potential system the above procedure has to be repeated. Since the number of arcs is a finite number and set **B** is enlarged by, at least, one arc in each step, the procedure will end after a finite number of steps. If the empty cut on [**N**,**B**] is also empty on [**N**,**A**], then no path exists from *s* to *t*.

The theorem is thus proved.

The following corollary of the theorem tells us that in case of optimal solutions the value of the two objective functions is equal.

Corollary A-2: (*strong form of equilibrium*) If the path $P(s,t)$ and the potential system π are optimal then their values are the same, that is, $\min\{\tau(P)\}=\max\{\pi_t\}$

Proof: It follows from Theorem A-2 that there exists such a system π^* and path P^* that are optimal, that is, $\pi_t^*=\tau(P^*)$. Since P and π are also optimal, therefore $\pi_t^*=\pi_t$ and $\tau(P^*)=\tau(P)$, but then $\pi_t=\tau(P)$.

Let us solve the following example to illustrate the procedure!

Example: Let us find the shortest path from node '1' to node '6'.

Table A-6 shows the structure matrix with two additional rows below the matrix. The first of the extra rows will be used for path searching the last row for storing the potentials. The algorithm can start from any potential system satisfying A-3, A-4. We start from the potential system $\pi_1=0$, $\pi_i=2$ $i=\{2,3,4,5,6\}$, but any system can be used which satisfies the dual conditions. Figure A-11 shows the sample network. The τ_{ij} values are displayed on or near by the arcs. The boxes above or below the nodes contain the actual potential values.

	1	2	3	4	5	6
1		3	2			
2				4		
3		0			4	
4			2			1
5						4
6						
	0	2	2	2	2	2

Table A-6 *Structure matrix*

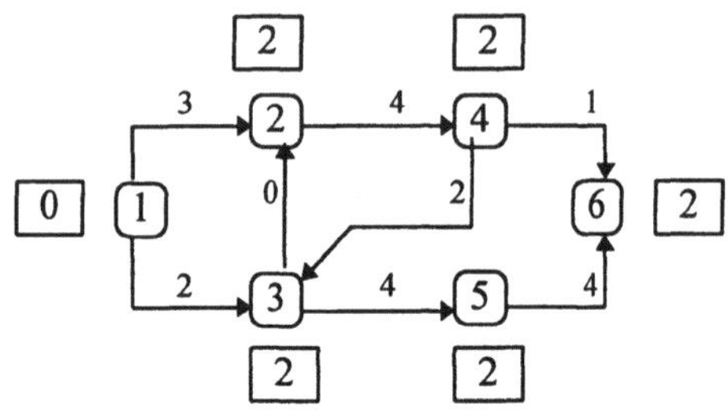

$\pi_1=0$ $\pi_i=2$ $i=\{1,2,3,4,5\}$

Figure A-11 *The sample digraph*

As it derives from the proof of the duality theorem, a digraph **[N,B]** has to be constructed and enlarged during the algorithm.

If there is a path that exists in the digraph **[N,B]** from node '1' to node '6', then, due to the construction of **[N,B]** the solution is optimal. Remember that an arc (i,j) is an element of the **[N,B]** if $\pi_j-\pi_i=\tau_{ij}$, that is, if $\varepsilon_{ij}=0$.

For making the calculation easier we transform the structure matrix to a matrix ε_{ij}. Those cells are the elements of the digraph **[N,B]** in which zero values are written.

Step 1.

The matrix ε_{ij} can be seen in Table A-7. Digraph **[N,B]** has only two arcs. Searching for a path from node '1' to node '6' in this digraph, node '1', '2' and '3' can be labeled. The labeled nodes form set **S**, the unlabeled set **T**. Let us cover the columns of **S** and the rows of **T**. With the help of this methodology, the arcs in the cut can be found in the uncovered part of the ε_{ij} matrix. Table A-7 shows that arcs (2,4) and (3,5) are in the cut. The ε will be the minimum of the values ε_{ij} of the uncovered part. The potentials in set **T** will be increased by ε. The new potentials are shown in the very last row of Table A-7 and in the boxes of Figure A-12. The new matrix ε_{ij} can easily be produced by using the covering methodology. The ε_{ij} values remain the same in the cells covered vertically or horizontally. They increased by ε in the cells covered both vertically and horizontally, and they decreased by ε in the uncovered cells.

The new matrix ε_{ij} can be seen in Table A-8.

	1	2	3	4	5	6
1		1	0			
2				4		
3		0			4	
4			2			1
5						4
6						
	+	+3	+1			
	0	2	2	6	6	6

ε=4

Table A-7 *Step 1. The starting matrix* ε_{ij}

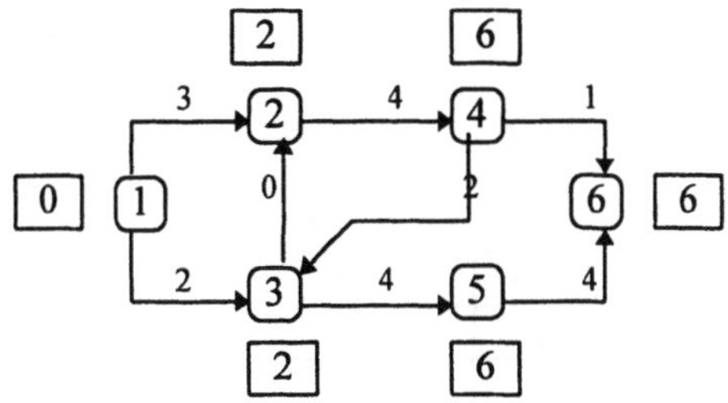

Figure A-12 *Step 1.*

Step 2.

Now, digraph **[N,B]** consists of four arcs. Repeating the procedure on the new digraph **[N,B]** , all the nodes but node '6' will be labeled. Covering the columns of set **S** and the rows of set **T**, the ε can be determined as a minimum of the values of the uncovered cells, as these cells represent the arcs in the cut of digraph **[N,A]**. The values of system π_i remain unchanged in set **S**. The new potential for node '6', which is the only node of set **T**, is $\pi_6=7$. Step 2 can be followed in Table A-8 and Fig. A-13. The new matrix ε_{ij} can be seen in Table A-9.

	1	2	3	4	5	6
1		1	0			
2				0		
3		0			0	
4			6			1
5						4
6						
	+	+3	+1	+2	+ 3	
	0	2	2	6	6	7

ε=1

Table A-8 *Step 2.*

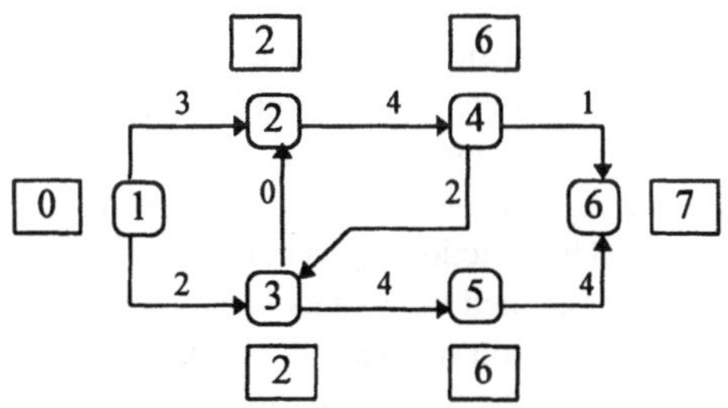

Figure A-13 *Step 2.*

Step 3 can be followed in Table A-9 and Figure A-14. Searching for a path in **[N,B]**, node '6' will be labeled, that is, a path exist from *s* to *t* in **[N,B]**. According to the lemma and its consequence, this P(*s*,*t*) path is the minimal length path and the potential π_t is the maximal, that is, both solutions are optimal. The path can be traced back using the labels. The length of the path can be determined by the sum of the τ values along the path or simply by the use of the $\min\tau(P)=\max\pi_t$ relation.

P $\{s\equiv1\rightarrow3\rightarrow2\rightarrow4\rightarrow t\equiv6\}$ $\tau(P)=2+0+4+1=7$

	1	2	3	4	5	6
1		1	0			
2				0		
3		0			0	
4			6			0
5						3
6						

+	+3	+1	+2	- 3	-4

0	2	2	6	6	7

Table A-9 *Step 3.*

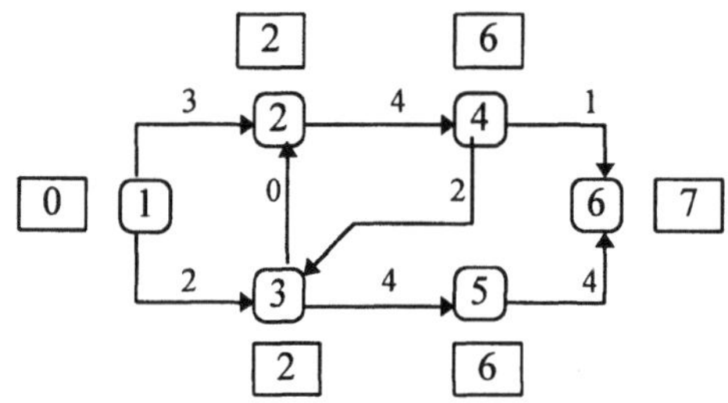

Figure A-14 *Step 3.*

4. MAXIMAL FLOW MINIMAL CUT

In this section we give a solution to the maximal flow minimal cut problem, which is the second basic problem in the network. Before formulating the problem, we give some new definitions.

A digraph **[N,A]** is given. Let us assign $\kappa_{ij}\geq0$ integer function to all arcs $(i,j)\in\mathbf{A}$. The function κ_{ij} is called the capacity of the arc (i,j). The digraph together with the assigned κ_{ij} values is called network and denoted by the symbol $[\mathbf{N},\mathbf{A},\kappa]$. (The κ_{ij} function mathematically is the same as the τ_{ij} in the previous section) Let $s\in\mathbf{N}$ and $t\in\mathbf{N}$ be two distinguished points of the digraph.

Definition A-3 *The integer valued function φ_{ij} interpreted on the arcs of the network* $[\mathbf{N},\mathbf{A},\kappa]$ *is called flow function if*

$\varphi_{ij}\leq\kappa_{ij}$ $\quad\forall\,(i,j)\in\mathbf{A}$ that is, the flow is bounded **A-8**

$\varphi_{ij}\leq-\varphi_{ji}$ $\quad\forall\,(i,j)\in\mathbf{A}$ that is, the flow is skew symmetric. **A-9**

The interpretation of the second condition can be seen in Figure A-15. According to Definition A-3, the following sentences are equivalent:

The value of the flow from i to j is 3.

The value of the flow from j to i is -3.

Figure A-15 *Skew symmetricity of flow*

The assumption of skew symmetricity requires the existance of an arc (j,i) if an arc (i,j) exists. If these arcs do not exist, then the network has to be completed with these backward going arcs, for which the capacities $\kappa_{ji}=0$ has to be assumed.

Skew simmetricity is not compulsory to assume. We need it for making the algorithm, which is based on labeling technique, easier.

Let us now introduce the following notations: if **A**,**B** and **C** are subsets of **N** then

$\kappa_{i\mathbf{B}}=\Sigma\kappa_{ij}$ for all $j\in\mathbf{B}$ and $\varphi_{i\mathbf{B}}=\Sigma\varphi_{ij}$ for all $j\in\mathbf{B}$

$\kappa_{\mathbf{A}j}=\Sigma\kappa_{\mathbf{A}j}$ for all $i\in\mathbf{A}$ and $\varphi_{\mathbf{A}j}=\Sigma\varphi_{\mathbf{A}j}$ for all $i\in\mathbf{A}$

$\kappa_{\mathbf{A},\mathbf{B}}=\Sigma\kappa_{ij}$ for all $i\in\mathbf{A}$ $j\in\mathbf{B}$ and $\varphi_{\mathbf{A},\mathbf{B}}=\Sigma\varphi_{ij}$ for all $i\in\mathbf{A}$ $j\in\mathbf{B}$

If $A\cap B=0$, that is, empty, the following addition relation holds:

$\kappa_{\mathbf{A},\mathbf{B}}=\kappa_{A,C}+\kappa_{\mathbf{B},\mathbf{C}}$

Definition A-4 *The vertex* $s\in\mathbf{N}$ *is called a source concerning flow* φ *in the network* $[\mathbf{N},\mathbf{A},\kappa]$ *if* $\varphi_{sN}>0$.

Definition A-5 *The vertex* $t\in\mathbf{N}$ *is called a sink concerning flow* φ *in the network* $[\mathbf{N},\mathbf{A},\kappa]$ *if* $\varphi_{tN}<0$, *that is,* $\varphi_{Nt}>0$.

Definition A-6 *The vertex* $i\in\mathbf{N}$ *is called an intermediate point concerning flow* φ *in the network* $[\mathbf{N},\mathbf{A},\kappa]$ *if* $\varphi_{iN}=0$.

The interpretation of the definitions can be seen in Figure A-16.

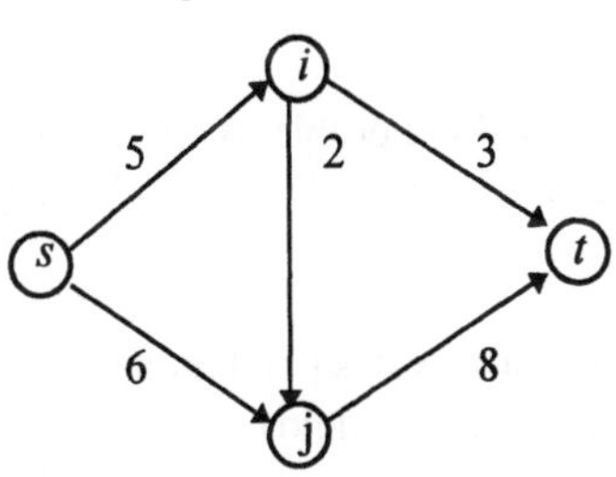

$\varphi_{sN}=5+6=11$ (*source*)

$\varphi_{iN}=2+3+(-5)=0$ (*intermediate*)

$\varphi_{jN}=8+(-2)+(-6)=0$ (*intermediate*)

$\varphi_{tN}=(-8)+(-3)=-11$ (*sink*)

Figure A-16 *Source, sink and intermediate points.*

Definition A-7 *Flow* φ *on the network* [N,A,κ] *is called the flow from s to t if s is the source t is the sink and i is an intermediate point for all* $i \in \mathbf{N}$*-s-t.*

Definition A-8 *The value of* φ_{sN} *is called the value of the flow in the network in case of a flow from s to t.*

Definition A-9 *The capacity of an arbitrary cut* **(S,T)** *in a network* [N,A,κ] *can be defined by the following formula:* $\kappa_{ST}=\Sigma\kappa_{ij}$ *for all* $i \in \mathbf{S}$ $j \in \mathbf{T}$.

The following basic tasks can be formulated concerning the flow in the network:

Determine the maximal flow, that is, φ_{sN} from s to t in the network [N,A,κ] .
Determine the minimal capacity cut separating nodes s and t.

Mathematically a flow φ_{ij} has to be found which satisfies **A-8**, **A-9** and the following condition:

$$\varphi_{iN} = \begin{cases} \varphi_{sN} & \text{if } i = s \\ 0 & \text{if } i \in \mathbf{N} - s - t \\ -\varphi_{sN} & \text{if } i = t \end{cases} \qquad \textbf{A-10}$$

and maximize

$$\varphi_{sN} \rightarrow \max \qquad \textbf{A-11}$$

Conditions **A-8** - **A-10** describe the primal problem together with the objective function A-11.

The dual problem can be formulated in the following way:

A separating cut **(S,T)** has to be found in the network, for which the

$$\kappa_{ST} \rightarrow \min \qquad \textbf{A-12}$$

objective function is minimal.

These problems are called the maximal flow minimal cut problems. There is a strong connection between the flow and the cut as it is shown by the following lemma.

Lemma A-2 For an arbitrary flow from s to t in network [N,A,κ] and an arbitrary cut (S,T) separating node s and node t, the following expression holds:

$$\varphi_{s,N} \leq \kappa_{S,T} \qquad \textbf{A-13}$$

Equality holds in A-10 if and only if

$$\varphi_{ij} = \kappa_{ij} \quad \forall\, (i,j) \in (\mathbf{S},\mathbf{T}) \qquad \textbf{A-14}$$

Proof: It comes from the addition relation that

$\varphi_{sN}=\varphi_{s,N}+\varphi_{S-s,N}=\varphi_{S,N}=\varphi_{S,S\cup T}=\varphi_{S,S}+\varphi_{S,T}=\varphi_{S,T}\leq\kappa_{S,T}$

With this A-10 is proved. The relation $\varphi_{S,T}\leq\kappa_{S,T}$ can be written in the form:

$$\sum_{\substack{i\in S\\ j\in T}}\varphi_{ij}\leq\sum_{\substack{i\in S\\ j\in T}}\kappa_{ij}$$

Here equality holds if **A-14** satisfies. With this the lemma is proved.

An important consequence of the lemma is presented hereby.

Corollary A-3: (*weak form of equilibrium*)
If the flow from *s* to *t* and the cut **(S,T)** separating *s* and *t* are such that equality exists in **A-13** then both the flow and the cut are optimal, that is, φ_{sN} is maximal and κ_{ST} is minimal.

Proof: (*indirect way*) Let us denote the value of the primal objective function by P and that of the dual by D.

Let us assume that P=D, but the path P is not optimal, that is, a better primal solution P* exist. In this case P*<P=D but this is a contradiction according to the lemma.

Let us assume that P=D, but there exist a better dual solution D* where D*>D. In this case D*>D=P but this is a contradiction according to the lemma.

Now it is proved that the two models form a primal-dual pair. In the next theorem we prove that there exists a solution in which the value of the two objective functions are the same, that is, both solutions are optimal.

Theorem A-3 (*duality theorem*) In a network $[\mathbf{N},\mathbf{A},\kappa]$ there exist such a flow φ_{ij} from *s* to *t* and the cut separating nodes *s*, *t* that $\varphi_{s,N}=\kappa(\mathbf{S},\mathbf{T})$, that is, the flow is maximal and the cut is minimal.

Proof: Let us assume that there exists a flow from *s* to *t* satisfying **A-8**, **A-9**, **A-10**. At the beginning this can be the $\varphi_{ij}=0\ \forall\ (i,j)\in\mathbf{N}$ system.

Let us construct digraph **[N,B]** from the nodes and from the arcs which are not saturated, that is, $(ij)\in\mathbf{B}$ if $(ij)\in\mathbf{A}$ and $\varphi_{ij}<\kappa_{ij}$. Let us find a path from *s* to *t* in the digraph **[N,B]**. According to Theorem A-1 there exist either a path or an empty cut.

If there is an empty cut, that is, no path exists from *s* to *t*, then due to the construction of **[N,B]**, all the arcs of the cut in **[N,A]** are saturated. This means that

- according to the consequence of the lemma - the flow is maximal and the cut is minimal.

If there is a path from s to t then set **B** will be decreased in such a way that the flow has to be increased along the path until at least one arc on the path will be saturated. Of course, dual conditions **A-8**, **A-9**, **A-10** have to be satisfied at the end of the flow increment. A value δ_{ij} will be defined for all arcs (ij) in the path in the following way: $\delta_{ij}:=\{\kappa_{ij}-\varphi_{ij}\}$

Let us construct a new flow system φ^* with the help of the below defined value of δ:

$$\delta := \min_{\substack{(ij)\in P \\ (ij)\in A}} \delta_{ij}$$

Let the new flow system φ^* be constructed in the following way:

$$\varphi^*_{ij} = \begin{cases} \varphi_{ij} + \delta & \text{if } (ij) \in P \\ \varphi_{ij} - \delta & \text{if } (ji) \in P \\ \varphi_{ij} & \text{otherwise} \end{cases}$$

After this step set **B** will be decreased by, at least, one saturated arc. The new flow φ^* still satisfy **A-8, A-10**.

For the new flow system the above procedure has to be repeated. Since the maximal flow is bounded above and the flow is increased by $\delta>0$ positive integer in each step, the procedure ends in a finite number of steps.

With this the theorem is proved.

The following corollary of the theorem tell us that in case of optimal solutions the value of the two objective functions is equal.

Corollary A-4: (*strong form of equilibrium*) If the flow φ is maximal and the cut **(S,T)** separating nodes s, t is minimal then their values are the same, that is, **A-13** holds with equality.

Proof: It follows from Theorem A-3 that there exist such a system φ^* and cut (S,T)* that are optimal, that is, $\varphi_{sN}^*=\kappa_{ST}^*$. Since the flow φ and the cut **(S,T)** are also optimal, therefore $\varphi_{sN}^*=\varphi_{sN}$ and $\kappa_{ST}^*=\kappa_{ST}$, but then $\varphi_{sN}=\kappa_{ST}$.

Let us solve the following example to illustrate the procedure!

Example: Let us find the maximal flow from node '1' to node '6' and the minimal cut in the below given network. (Figure A-17 Table A-10)

	1	2	3	4	5	6
1		3	2			
2	*0*		*0*	4		
3	*0*	6		*0*	4	
4			2			1
5			*0*			4
6				*0*	*0*	

Table A-10 *Capacity matrix*

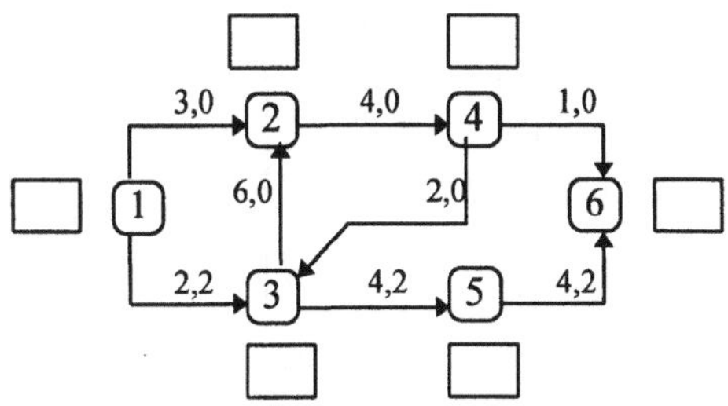

Figure A-17 *The sample digraph*

Table A-10 shows the capacity matrix with an additional row below the matrix. This row will be used for storing the labels during the path searching procedure. The cells that are written in italic format indicate the arcs with which the network is being completed according to the skew simmetricity. We can start with any flow system that satisfies **A-8**, **A-10**. We start from the flow $\varphi=2$ in the path $P=\{1\to3\to5\to6\}$, but any system can be used, which satisfies the flow conditions. Figure A-17 shows the sample network. The capacities and the flows are displayed on the arcs respectively. The backward arcs are not displayed. The boxes above or below the nodes contain the labels.

As it derives from the proof of the duality theorem, digraph **[N,B]** has to be constructed and decreased during the algorithm. If there is an empty cut exists in the digraph **[N,B]** from node '1' to node '6', then, due to the construction of **[N,B]** the solution is optimal. Remember that an arc (i,j) is an element of the **[N,B]** if $\delta_{ij}=\kappa_{ij}-\varphi_{ij}=0$. The δ_{ij} is also called the free capacity. To make the calculation easier we transform the capacity matrix to a matrix δ_{ij} . Those cells are the elements of the digraph **[N,B]** in which $\delta_{ij}>0$ values are written. The matrix δ_{ij} can be seen in Table A-11. The free capacity of the „backward" arcs are written in italic format. Searching for a path from node '1' to node '6' in the digraph **[N,B]**, $P\{1\to2\to4\to6\}$ has been found. The arcs of the path are depicted in bold both in the free capacity matrix and in the figure. The "backward" arcs of the path are indicated by stars in the δ_{ij} matrix. The value of δ is determined by the minimum of the δ_{ij} values on the path, that is, $\delta=1$.

	1	2	3	4	5	6
1		**3**	0			
2	*0**		*0*	**4**		
3	*2*	6		*0*	2	
4		*0**	2			**1**
5			*2*			2
6				*0**	*2*	
	+	+1	-4	+2		-4

Table A-11 *Step 1. The starting matrix* δ_{ij}

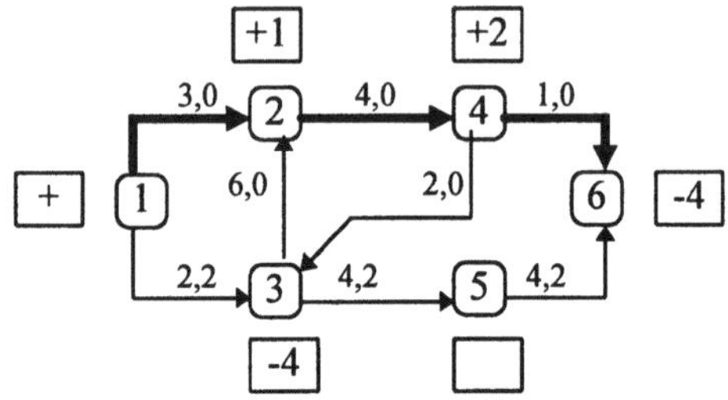

Figure A-18 *Step 1.*

The flow will be increased by this value along the path. This means that the free capacities will decrease on the path, increase in the backward arcs of the path and remain the same otherwise.The result of step 1, that is, the new matrix δ_{ij} and the flow can be seen in Table A-12 and in Figure A-19.

	1	2	3	4	5	6
1		**2**	0			
2	*1**		*0*	**3**		
3	*2*	6		*0**	**2**	
4		*1**	**2**			0
5			*2**			**2**
6				*1*	*2**	
	+	+1	+4	+2	-3	-5

Table A-12 *Step 2.*

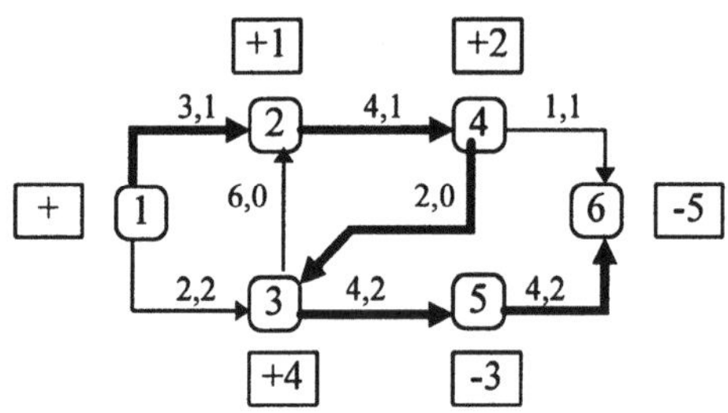

Figure A-19 *Step 2.*

In Step 2 the path P {1→2→4→3→5→6} has been found in **[N,B]** along which the flow can be increased by δ=2. The flow will be increased by this value along the path. This means that the free capacities will decrease on the path, increase in the backward arcs of the path and remain the same otherwise. The result of Step 2 can be seen in Table A-13 and in Figure A-20.

	1	2	3	4	5	6
1		0	0			
2	*3*		*0*	1		
3	*2*	6		*2*	0	
4		*3*	0			0
5			*4*			0
6				*1*	*4*	
	+					

Table A-13 *Step 3.*

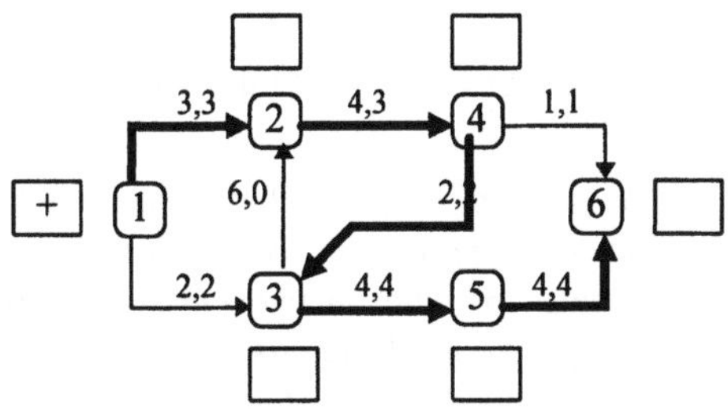

Figure A-20 *Step 3.*

In Step 3 there is no path from *s* to *t* on digraph **[N,B]**, that is, all the arcs in the cut **(S,T)** in digraph **[N,A]** are saturated, so the flow is maximal. Covering the columns of set **S** and the rows of set **T**, the arcs of the uncovered area indicate the arcs of the cut. The sum of the capacities of these arcs gives the value of the maximal flow, which is $\kappa_{ST}=\kappa_{12}+\kappa_{13}=3+2=5$. Transforming the free capacity matrix to the flow matrix by using the following formula, $\varphi_{ij}=\kappa_{ij}-\delta_{ij}$, the results can be interpreted easier. The flow matrix can be seen in Table A-14.

	1	2	3	4	5	6
1		3	2			
2				3		
3		0			4	
4			2			1
5						4
6						

Table A-14 *Flow matrix*

5. THE FIRST "K" LONGEST PATH

The problem originally has been formulated to find the first „K" *shortest* path in a network. One of the existing algorithms has been modified here to satisfy the requirements of the network-based techniques. In network-based project methodology we need the first „K" *longest* path in loopless networks, this is the reason of the modifications. Let us use the following notations in the algorithm.

$\{1,2,.....i,....j,...n\}$ for the identification of nodes
(i,j) an arbitrary arc of the digraph $i=\{1,2...n\}$, $j=\{1,2....n\}$ $i\neq j$
N the set of the nodes
A the set of the arrows
s the start node of the digraph
t the end node of the digraph
τ_{ij} the length of the directed (i,j) arc $i\neq j$, $(ij)\in \mathbf{A}$
$P^k = (s, 2^k, 3^k, ...i^k,.. t,)$ k=1,2,....K the kth longest path from s to t, where 2^k, 3^k,....i^k are the nodes of the kth longest path
P^k_i, $i=1,2....t-1$ the set of „deviations" from path P^{k-1} where the deviation starts from node i, that is, P^k_i is the same from s to t as P^k, in the $i+1$ node deviates from P^k and goes to t along the longest path.
R^k_i the root of the path P^k_i (from s to i)
S^k_i the second part of the path P^k_i (from i to t)
List #1 is the list of the longest paths
List #2 is the list of the paths candidates for getting on List #1

The algorithm can be summarized in the following manner:

Iteration 1: (determining P^1*, that is, the first longest path)*
P^1 is the longest path in the graph. Any of the existing longest path algorithms can be used. (The CPM/time forward calculations are longest path calculations from s to t.) The longest path can be placed on List #1. In case of paths with the same length, an arbitrary one has to be chosen.

Iteration k: (Determining P^k*, that is, the* k^{th} *longest path)*
For determining the P^k longest path ,P^1, P^2, P^{k-1} should be known. Knowing these paths, path P^k can be determined in the following manner:

Step 1. Check for the *i=1,2,...,t-1*, nodes for P^{k-1}

a) If the root of the path P^{k-1} from *s* to *i* is equivalent to the root from node *s* to node *k-1* of paths P^j (*j=1...k-1*), then let $\tau_{i,i+1} = -\infty$

b) Apply any longest path algorithm to detect the longest path from *i* to *t*, that is, to determine S^k_i. In case of more paths, any of them will do. The root from *s* to *i* and S^k_i make a path. If this path cannot be found either on List #1 or List #2, then this P^k_i path will be placed on List #2.

Step 2: The longest among the paths on List #2 will be moved to List #1. If the number of the path is „K" or List #2 is empty then the algorithm has been finished. Otherwise the procedure has to be continued with iteration k+1.

To illustrate the algorithm, let us find first 7 longest paths in the covered depot example. The network is shown on Figure A-21.

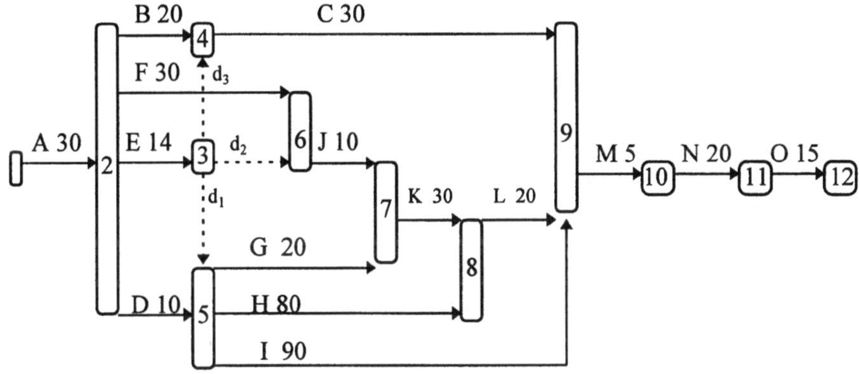

Figure A-21 *Searching for the first „K" longest path.*

Iteration 1: (Determining P^1*)*
Path A-E-H-L-M-N-O is the longest path. Its length is 184 units (days).
P^1 is placed on List #1.

Iteration II: (Determining P^2*)*

The deviation from node *1* is:

$R^2_1=0 \quad \tau_{12}=-\infty \quad S^2_1=-\infty \qquad \Rightarrow P^2_1=-\infty$

The deviation from node *2* is:

$R^2_2=30 \quad \tau_{23}=-\infty \quad S^2_2=150$ (D,H,L,M,N,O) $\Rightarrow P^2_2=180$

The deviation from node *3* is:

$R^2_3=44 \quad \tau_{35}=-\infty \quad S^2_3=100$ (d_2,J,K,L,M,N,O) $\Rightarrow P^2_3=144$

The deviation from node *5* is:
$R^2_5=44$ $\tau_{58}=-\infty$ $S^2_5=125$ (I,N,O) $\Rightarrow P^2_5=169$
The deviation from node *8* is:
$R^2_8=124$ $\tau_{89}=-\infty$ $S^2_8=-\infty$ $\Rightarrow P^2_8=-\infty$
The deviation from node *9* is:
$R^2_9=144$ $\tau_{9,10}=-\infty$ $S^2_9=-\infty$ $\Rightarrow P^2_9=-\infty$
The deviation from node *10* is:
$R^2_{10}=149$ $\tau_{10,11}=-\infty$ $S^2_{10}=-\infty$ $\Rightarrow P^2_{10}=-\infty$
The deviation from node *11* is:
$R^2_{11}=169$ $\tau_{11,12}=-\infty$ $S^2_{11}=-\infty$ $\Rightarrow P^2_{11}=-\infty$

Deviations P^2_2 (A-D-H-L-M-N-O), P^2_3 (A-E-d_2-J-K-L-M-N-O) and P^2_5 (A-E-d_1-I-N-O) have been placed on List #2. The longest is P^2_2 (180), therefore it has to move to List #1 as the second longest path, that is, P^2. Deviations P^2_3 and P^2_5 have remained on List #2.

Iteration III: (Determining P^3*)*

The deviation from node *1* is:
$R^3_1=0$ $\tau_{12}=-\infty$ $S^3_1=-\infty$ $\Rightarrow P^3_1=-\infty$
The deviation from node *2* is:
$R^3_2=30$ $\tau_{23}=-\infty$ $\tau_{25}=-\infty$ $S^3_2=130$ (F,J,K,L,M,N,O) $\Rightarrow P^3_2=160$
The deviation from node *5* is:
$R^3_5=40$ $\tau_{58}=-\infty$ $S^3_5=125$ (I,N,O) $\Rightarrow P^3_5=165$
The deviation from node *8* is:
$R^3_8=120$ $\tau_{89}=-\infty$ $S^3_8=-\infty$ $\Rightarrow P^3_8=-\infty$
The deviation from node *9* is:
$R^3_9=140$ $\tau_{9,10}=-\infty$ $S^3_9=-\infty$ $\Rightarrow P^3_9=-\infty$
The deviation from node *10* is:
$R^3_{10}=145$ $\tau_{10,11}=-\infty$ $S^3_{10}=-\infty$ $\Rightarrow P^3_{10}=-\infty$
The deviation from node *11* is:
$R^3_{11}=165$ $\tau_{11,12}=-\infty$ $S^3_{11}=-\infty$ $\Rightarrow P^3_{11}=-\infty$

Deviations P^3_2 (A-F-J-K-L-M-N-O) and P^3_5 (A,D,I,N,O) have been placed to List #2. The longest is P^2_5 (169), therefore it has to move to List #1 as the third longest path. P^3 {A-E-d_1-I-N-O}=169.

Iteration IV: (Determining P^4*)*

The deviation from node *1* is:
$R^4_1=0$ $\tau_{12}=-\infty$ $S^4_1=-\infty$ $\Rightarrow P^4_1=-\infty$
The deviation from node *2* is:

R^4_2=30 τ_{23}= -∞ τ_{25}= -∞ S^4_2=130 (F,J,K,L,M,N,O) ⇒ P^4_2=160

The deviation from node *3* is:

R^4_3=44 τ_{35}= -∞ S^4_3=100 (d2-J-K-L-M-N-O)⇒ P^4_3=144

The deviation from node *5* is:

R^4_5=44 τ_{58}= -∞ $\tau_{5,10}$= -∞ S^4_5= 110 (G-K-L-M-N-O) ⇒ P^4_5= 154

The deviation from node *10* is:

R^4_{10}=134 $\tau_{10,11}$= -∞ S^4_{10}= -∞ ⇒ P^4_{10}= -∞

The deviation from node *11* is:

R^4_{11}=154 $\tau_{11,12}$= -∞ S^4_{11}= -∞ ⇒ P^4_{11}= -∞

Deviations P^4_2 (A-F-J-K-L-M-N-O) and P^4_3 (A,D,I,N,O) coincide with P^3_2 and P^2_3, which are on List #2, therefore only P^4_5 has to be placed on the list of candidates. The longest is P^3_5 (165), therefore it has to move into List #1 as the fourth longest path. P^4 {A-D-I-N-O}=165.

Iteration V: (Determining P^5*)*

The deviation from node *1* is:

R^5_1=0 τ_{12}= -∞ S^5_1= -∞ ⇒ P^5_1= -∞

The deviation from node *2* is:

R^5_2=30 τ_{23}= -∞ τ_{25}= -∞ S^5_2=130 (F,J,K,L,M,N,O) ⇒ P^5_2=160

The deviation from node *5* is:

R^5_5=40 τ_{58}= -∞ $\tau_{5,10}$= -∞ S^5_5= 110 (G-K-L-M-N-O) ⇒ P^5_5= 150

The deviation from node *10* is:

R^5_{10}=130 $\tau_{10,11}$= -∞ S^5_{10}= -∞ ⇒ P^5_{10}= -∞

The deviation from node *11* is:

R^5_{11}=150 $\tau_{11,12}$= -∞ S^5_{11}= -∞ ⇒ P^5_{11}= -∞

Deviation P^5_2 (A-F-J-K-L-M-N-O) coincides with P^3_2, which is on List #2, therefore only P^5_5 (A-D-G-K-L-M-N-O) has to be placed on the list of candidates. The longest is P^3_2 (160), therefore it has to move to List #1 as the fifth longest path. P^5 {A-F-J-K-L-M-N-O}=160.

Iteration VI: (Determining P^6*)*

The deviation from node *1* is:

R^6_1=0 τ_{12}= -∞ S^6_1= -∞ ⇒ P^6_1= -∞

The deviation from node *2* is:

R^6_2=30 τ_{23}= τ_{25}=τ_{26}= -∞ S^6_2=90 (B,C,M,N,O) ⇒ P^6_2=120

The deviation from node *6* is:

R^6_6=60 τ_{67}= -∞ S^6_6= -∞ ⇒ P^6_6= -∞

The deviation from node *7* is:
R^6_7=70 τ_{78} = -∞ S^6_7= -∞ ⇒ P^6_7= -∞
The deviation from node *8* is:
R^6_8=100 τ_{89}= -∞ S^6_8= -∞ ⇒ P^6_8= -∞
The deviation from node *9* is:
R^6_9=120 $\tau_{9,10}$= -∞ S^6_9= -∞ ⇒ P^6_9= -∞
The deviation from node *10* is:
R^6_{10}=125 $\tau_{10,11}$= -∞ S^6_{10}= -∞ ⇒ P^6_{10}= -∞
The deviation from node *11* is:
R^6_{11}=145 $\tau_{11,12}$= -∞ S^6_{11}= -∞ ⇒ P^6_{11}= -∞

Deviation P^6_2 (A-bB-C-M-N-O) has to be placed on List #2. The longest is P^4_5 (154), therefore it has to move to List #1 as the sixth longest path. P^6 {A-E-d_1-G-K-L-M-N-O}=154.

Iteration VII: (Determining P^7*)*
The deviation from node *1* is:
R^7_1=0 τ_{12}= -∞ S^7_1= -∞ ⇒ P^7_1= -∞
The deviation from node *2* is:
R^7_2=30 τ_{23}= τ_{25}=τ_{26}= -∞ S^7_2=90 (B,C,M,N,O) ⇒ P^7_2=120
The deviation from node *3* is:
R^7_3=44 τ_{35}= -∞ S^7_3= 100 (J-K-L-M-N-O) ⇒ P^7_3= 144
The deviation from node *5* is:
R^7_5=44 τ_{57}= -∞ S^7_5= -∞ ⇒ P^7_5= -∞
The deviation from node *7* is:
R^7_8=64 τ_{89} = -∞ S^7_8= -∞ ⇒ P^7_8= -∞
The deviation from node *8* is:
R^7_8=94 τ_{89}= -∞ S^7_8= -∞ ⇒ P^7_8= -∞
The deviation from node *9* is:
R^7_9=114 $\tau_{9,10}$= -∞ S^7_9= -∞ ⇒ P^7_9= -∞
The deviation from node *10* is:
R^7_{10}=119 $\tau_{10,11}$= -∞ S^7_{10}= -∞ ⇒ P^7_{10}= -∞
The deviation from node *11* is:
R^7_{11}=139 $\tau_{11,12}$= -∞ S^7_{11}= -∞ ⇒ P^7_{11}= -∞

Deviation P^7_3 coincides with P^6_2 therefore no path has been placed on List #2. Path P^5_5 is the longest among the candidates. This is the seventh longest path. P^7 {A-D-G-K-L-M-N-O}=150. The procedure has been finished.

The first 7 longest paths are:

P^1 {A-E-H-L-M-N-O} =184
P^2 {A-D-H-L-M-N-O} =180
P^3 {A-E-d1-I-N-O} =169
P^4 {A-D-I-N-O} =165
P^5 {A-F-J-K-L-M-N-O} =160
P^6 {A-E-d_1-G-K-L-M-N-O}=154
P^7 {A-D-G-K-L-M-N-O} =150

6. LINEAR PROGRAMMING AND DUALITY

In Chapter 2:CPM time, Chapter 3: CPM Least Cost Scheduling, Chapter 4: Precedence Diagramming, Chapter 7: PDM Least Cost Scheduling, we showed that some network technique problems can be solved with operational research methods. We have solved these problems by using the primal-dual methodology of operational research. All of our models that have been solved with primal-dual methods are based on linear programming, and all of the given algorithms can be derived from linear programming. The mathematical formulation of the linear programming (LP) problem and the duality theorem is discussed in this chapter.

Let us use the following notations for the introduction of the general LP problem:

Let the following vectors be given:

$\mathbf{x}\ (\xi_1,\xi_2,\ldots\ldots\xi_j,\ldots\ldots.\xi_m)$
$\mathbf{c}\ (\gamma_1,\gamma_2,\ldots\ldots.\gamma_j,\ldots\ldots.\gamma_m)$
$\mathbf{y}\ (\eta_1,\eta_2,\ldots..\eta_i,\ldots\ldots.\eta_n)$
$\mathbf{b}\ (\beta_1,\beta_2,\ldots..\beta_i,\ldots\ldots.\beta_n)$

Let a matrix $\underline{\underline{\mathbf{A}}}$ be given. Its elements are denoted by the a_{ij} $(i=1\ldots n, j=1\ldots m)$ symbol. The rows of this matrix are denoted by vectors $\mathbf{a}^{(i)}$ the columns are denoted by $\mathbf{a}_{(j)}$. The set of indices of vectors $\mathbf{a}^{(i)}$ are denoted by I. Let I be divided into two disjoint sets of indices I_1 and I_2. The set of indices of vectors $\mathbf{a}_{(j)}$ are denoted by J. Let J be divided into two disjopint nonempty sets, J_1 and J_2.

The notations are summarized in Figure A-22.

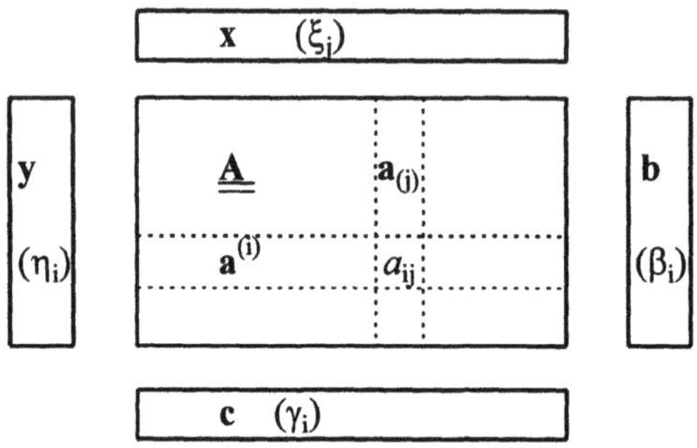

Figure A-22 *Notations of the LP*

The linear programming problem can be formulated in the following manner: The ξ_j values of vector **x** should be determined such that the conditions

$$\mathbf{a}^{(i)}\mathbf{x}=\beta_i \quad \forall\ i\in I_1$$
$$\mathbf{a}^{(i)}\mathbf{x}\geq\beta_i \quad \forall\ i\in I_2$$
$$\xi_j \geq 0 \quad \forall\ j\in J_1$$
$$\xi_j > -\infty \quad \forall\ j\in J_2$$

will be satisfied and the value of the objective function

$$\mathbf{cx}\rightarrow\mathbf{min}$$

will be the minimum.

The problem formulated by the conditions and the objective function is called the primal linear programming problem.

The dual problem can be formulated as it is shown below:

The ξ_j values of vector **x** should be determined such that the conditions

$$\mathbf{ya}_{(j)}=\gamma_j \quad \forall\ j\in J_1$$
$$\mathbf{ya}_{(j)}\geq\gamma_j \quad \forall\ j\in J_2$$
$$\eta_i \geq 0 \quad \forall\ i\in I_1$$
$$\eta i > -\infty \quad \forall\ i\in I_2$$

will be satisfied and the value of the objective function

$$\mathbf{by}\rightarrow\mathbf{max}$$

will be the maximum.

The connection between the objective functions of the primal and dual problems is given by the following lemma.

Lemma A-3 If vectors **x** and **y** are feasible, that is,the primal and the dual conditions are satisfied, then for the objective functions

$\mathbf{cx} \leq \mathbf{by}$

holds and there is an equality if and only if

$(\beta_i - \mathbf{a}^{(i)}\mathbf{x})\eta_i = 0 \quad \forall i \in I_2$

$(\gamma_j - \mathbf{y}\mathbf{a}_{(j)})\xi_j = 0 \quad \forall j \in J_1$

Proof:

$$\mathbf{cx} = \sum_{j=1}^{m} \gamma_j \xi_j \geq \sum_{j=1}^{m} \xi_j \Big(\sum_{i=1}^{n} a_{ij}\eta_j\Big) = \sum_{i=1}^{n} \eta_i \Big(\sum_{j=1}^{m} a_{ij}\xi_j\Big) \geq \sum_{i=1}^{n} \beta_i \eta_i = \mathbf{by}$$

With this the inequality is proved and equality holds if and only if:

$\beta_i = \mathbf{a}^{(i)}\mathbf{x}$ or $\eta_i = 0 \quad \forall i \in I_2$

$\gamma_j = \mathbf{y}\mathbf{a}_{(j)}$ or $\xi_j = 0 \quad \forall j \in J_1$

These are exactly the equilibrium conditions of the lemma.

An important consequence of the lemma is presented hereby.

Corollary A-5: (*weak form of equilibrium*)

If x satisfying the primal and y satisfying the dual conditions are such that the values of their objective functions are the same, then they are optimal.

Proof: (*indirect way*) Let us denote the value of the primal objective function by P and that of the dual by D.

Let us assume that P=D, but the path P is not optimal, that is, a better primal solution P* exists. In this case P*<P=D, but this is a contradiction according to the lemma.

Let us assume that P=D, but there exists a better dual solution D* where D*>D. In this case D*>D=P, but this is a contradiction according to the lemma.

We need only the above mentioned lemma from the theory of linear programming. Those wishing to get deeper knowledge of LP, important text will be cited in references.

7. PRACTICAL PROBLEMS AND SOLUTIONS

Problems

1. Find a path from node s=3 to node t=8 in the below given digraph.

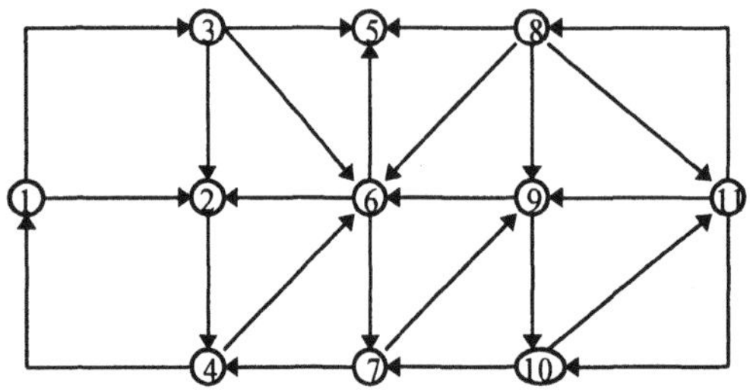

2. Find a path from s=14 to node t=11 in the below given digraph.

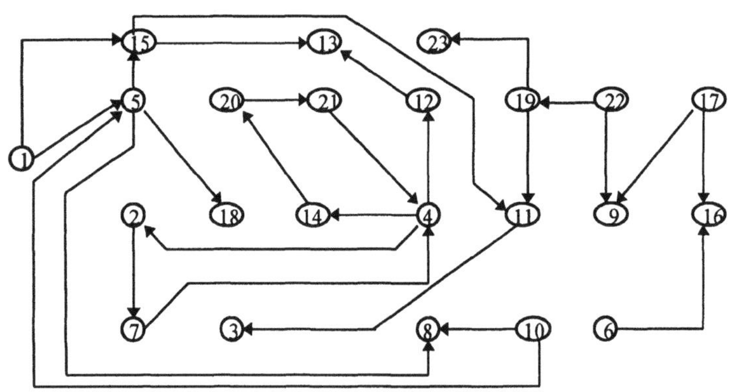

3. Find a path from s=14 to node t=17 in the below given digraph.

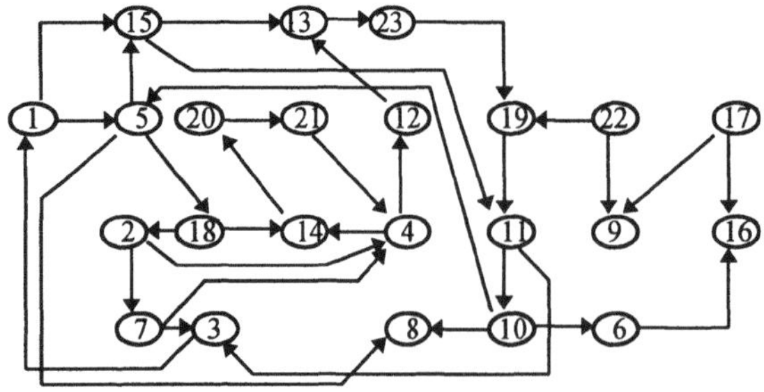

4. Find the shortest path from node s=1 to node t=6 in the below given network.

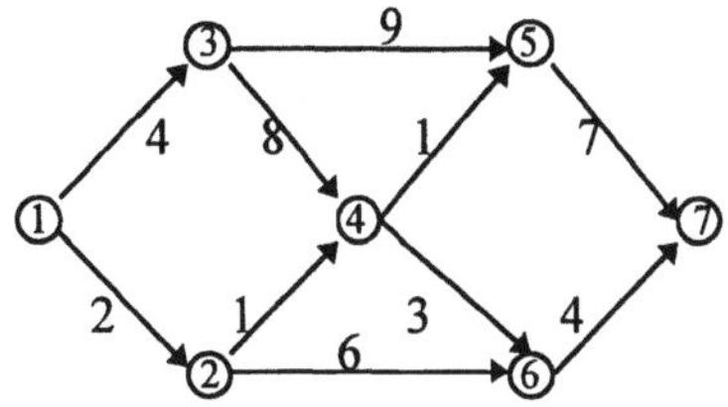

5. Find the shortest path from s=1 to t=11 in the below given network.

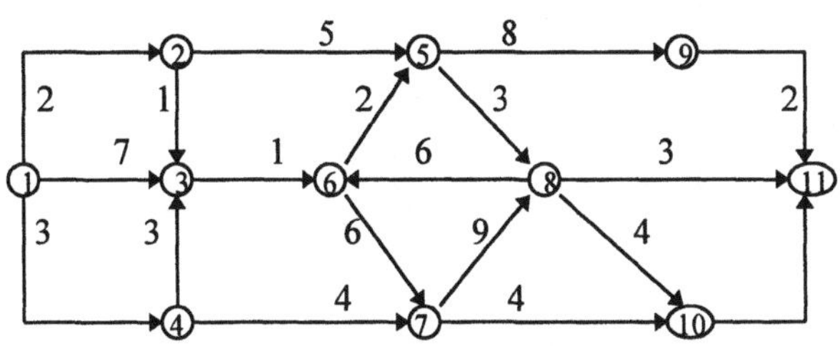

6. Find the shortest path from node s=1 to node t=11 in the below given network.

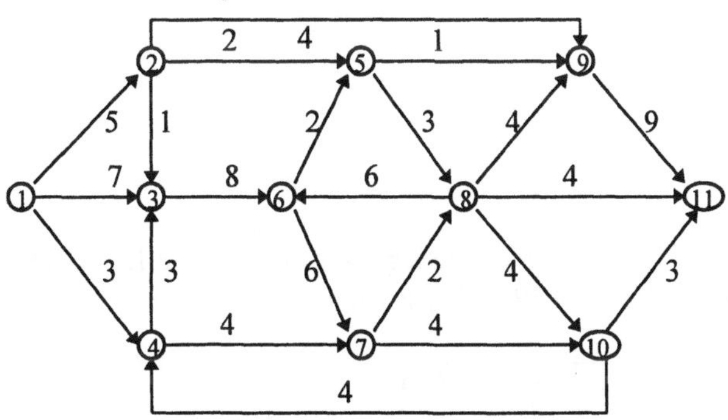

7. Find the maximal flow from node s to t. The capacities are displayed on the arcs. (s=1 t=4) Start the computation from the flow φ=2 on the path 1-3-5-2-6-4.

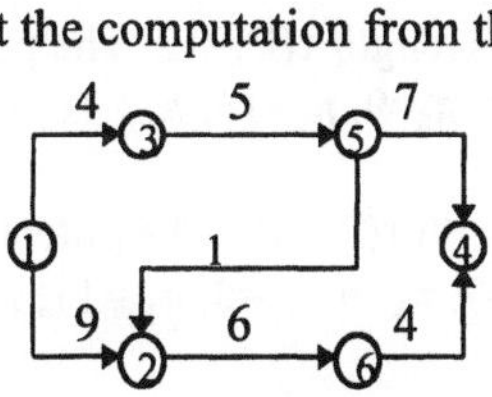

8. Find the maximal flow on the below given network from node s=1 to node t=7.

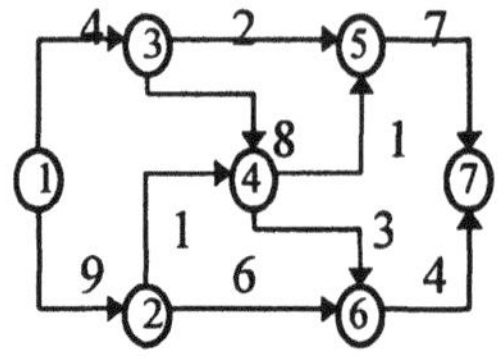

9. Find the maximal flow on the below given network from node s=1 to node t=7.

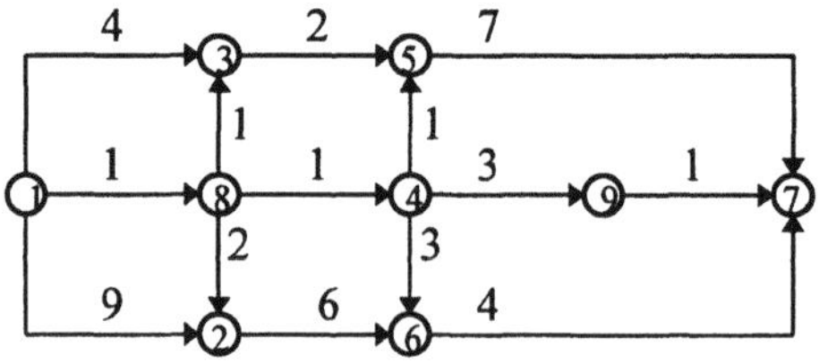

10. Look for all the paths longer than 18 units using the network of Problem 5. (Delete arc (8,6) to remove loops from the network.)

Solutions

1. P {3-6-7-9-10-11-8}

2. There is no path from s to t. The cut separating set **S** {14,2,4,7,12,13,20,21} from set **T** {1,3,5,6,8,9,10,11,15,16,17,18,19,22,23} is empty.

3. P {14-20-21-4-12-13-23-19-11-10-5-18-2-7}

4. The shortest path is P {1-2-6-7}. Its length $\tau(P)=10$. The potentials are:
$\pi_1=0,\ \pi_2=2,\ \pi_3=3,\ \pi_4=3,\ \pi_5=4,\ \pi_6=6,\ \pi_7=10$

5. The shortest path is P {1-2-3-6-5-8-11}. Its length $\tau(P)=12$. The potentials are:
$\pi_1=0,\ \pi_2=2,\ \pi_3=3,\ \pi_4=3,\ \pi_5=6,\ \pi_6=4,\ \pi_7=7,\ \pi_8=9,\ \pi_9=14,\ \pi_{10}=10,\ \pi_{11}=12.$

6. The shortest path is P {1-4-7-8-11}. Its length $\tau(P)=12$. The potentials are:
$\pi_1=0,\ \pi_2=5,\ \pi_3=6,\ \pi_4=3,\ \pi_5=7,\ \pi_6=14,\ \pi_7=7,\ \pi_8=9,\ \pi_9=8,\ \pi_{10}=11,\ \pi_{11}=12.$

7. The maximal flow is $\varphi(s \to t)=8$. The flow on the arcs is:
$\varphi_{12}=4$, $\varphi_{13}=4$, $\varphi_{26}=4$, $\varphi_{35}=4$, $\varphi_{52}=0$, $\varphi_{57}=4$, $\varphi_{64}=4$.
The arcs in the minimal cut (S,T) are: {((*1*,*3*);(*1*,*2*)}

8. The maximal flow is $\varphi(s \to t)=7$. The flow on the arcs is:
$\varphi_{12}=5$, $\varphi_{13}=2$, $\varphi_{24}=1$, $\varphi_{26}=4$, $\varphi_{34}=0$, $\varphi_{35}=2$, $\varphi_{45}=1$.
The arcs in the minimal cut (S,T) are: {(*3*,*5*);(*4*,*5*);(*6*,*7*)}

9. The maximal flow is $\varphi(s \to t)=7$. The flow on the arcs is:
$\varphi_{12}=4$, $\varphi_{13}=2$, $\varphi_{18}=1$, $\varphi_{26}=4$, $\varphi_{35}=2$, $\varphi_{45}=1$, $\varphi_{46}=0$
$\varphi_{49}=0$, $\varphi_{57}=3$, $\varphi_{67}=4$, $\varphi_{82}=0$, $\varphi_{83}=0$, $\varphi_{84}=1$, $\varphi_{97}=0$.
The arcs in the minimal cut (S,T) are: {(*3*,*5*);(*8*,*4*);(*6*,*7*)}
Arc (*4*,*6*) goes backward in the cut, that is, according the the skew simmetricity arc (*6*,*4*) is saturated and is in the minimal cut.

10. P_1 {1-3-6-7-8-11} =26 P_2 {1-4-3-6-7-8-11} =25
P_3 {1-2-3-6-7-8-11} =22 P_4 {1-3-6-7-10-11} =22
P_5 {1-4-3-6-7-11} =21 P_6 {1-3-6-5-9-11} =20
P_7 {1-4-3-6-5-9-11} =19 P_8 {1-4-7-8-11} =19

8. HISTORICAL REVIEW

The basic definitions of digraph as well as path and cut can be found in any book discussing network flow theory. Standard citation in this field is the classical work of Ford [19].

The duality theorem of path and cut has first been discussed in Minty's [49] paper.

The bibliography of the minimal path maximal potential problem is too vast even for overviewing. It is hard to decide about the priority for the loopless digraph with positive length of arcs. We followed the primal-dual methodology of Klafszky [43]. In the networks with positive arcs Dijkstra [13] has given an efficient algorithm. In general case - loops and arcs of any length are allowed - the procedure developed by Ford [19] and Bellmann [7] can be used.

Standard citations for netwok flow theory is the classical work of Ford and Fulkerson [22] and the book of Busacker and Saaty [8]. The theorem for maximal flow and minimal cut has originally been published in Ford and Fulkerson's [20],[21] papers. The algorithm discussed in this book is based on this "augmenting path" algorithm and uses the theoritical improvements of Edmonds and Karp [14].

We have to mention that another family of maximal flow algorithms exists, the family of the "preflow-push" algorithms. These algorithms are theoritically faster than the algorithms based on "augmenting paths".

The algorithm for finding the first "K" longest loopless path was originally developed by Yen [68] for finding the first "K" shortest path.

Thousands of great works can be found on the theory of linear programming. Standard citation is the book of Dantzig [12]. There are different methods for solving LP problems. A survey of pivot rules can be found in Terlaky and Zang's [63] paper. The simplest but not the fastest simplex rule has also been developed by Terlaky [62],[64].

Appendix B

COMPUTER APPLICATIONS

1. CHOOSING AMONG AVAILABLE APPLICATIONS

Hundreds of different computer software are available for any type of computers and any kind of project. Before purchasing a new software one has to carefully analyze the need and the possibilities of its own company. The nature and the size of the average project of the company defines the main requirements towards the desired project management tool. On the other hand the price of the suitable tools can greatly influence the decision. A general rule in these kinds of development is, that it is advisable to ask for help from experts in project manegement tools. These experts know the pros and cons of available software and can help to make optimal decisions. Like everywhere in life, here it is also true that the cheapest is not always the best.

Today 95 percent of the industrial applications are based on Precedence Diagramming which was comprehensively discussed through this book.

We have shown that different approaches to PDM can be found in practice. In computer software usually the activity-splitting-not-allowed approach is implemented. The activity-splitting-allowed approach takes off the greatest advantages of network techniques, that is, to get the information on carrying out individual processes automatically (see Chapter 4). Using the no-splitting-allowed approach it seems obvious that different software have to give the same result in case of same network. However, this is not always true. The algorithmical differences between different software and the different theoretical definitions of critical path, critical activity, etc., can confuse even the experienced project managers.

We are going to prove this hypothesis with the help of the following example. A pipeline has to be built which has been divided into three activities. There should be at least 2 days difference between activities at all stages of the project. The example project can be seen in Figure B-1.

A - excavation (5 days),
B - pipe laying down (5days),
C - backfilling (5 days)

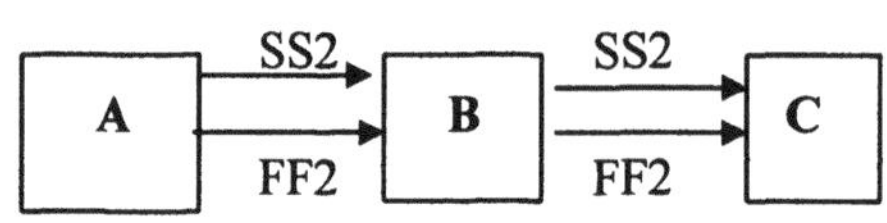

Figure B-1 *Example project*

Using the precedence diagramming calculation (no splitting allowed) the results can be followed in Figure B-2. The results are displayed in calendar format.

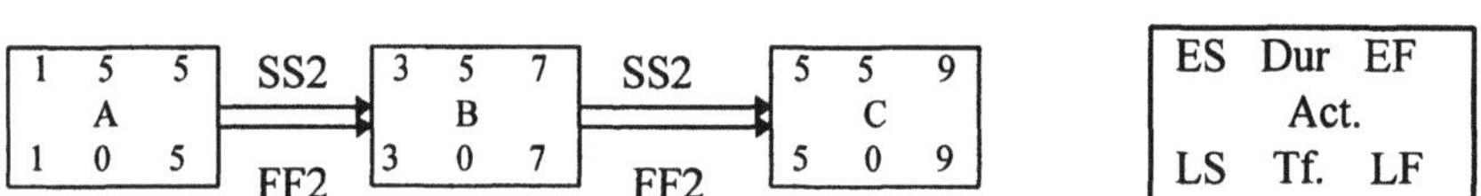

Figure B-2 *Result of the analysis*

If we are trying to analyze this network with Microsoft Project we will find that MS Project cannot analyze this network as it does not allow more than one precedence relationship between activities. As MS Project does not know the ZZ combination the planner has to analyze which relationships are the determinant ones and enter only these relationships.

In this case, as the durations are the same, either SS or FF can be used, and all the four possible combinations of them will result in the same activity dates as in Figure B-2.

However, the calculations with MS Project lead to different results in each of the cases as shown in Figure B-3 - Figure B-6. The Gantt charts are copied from MS Project.

ID	Name	Dur.	E.S	E.F	L.S	L.F	T.i
1	A	5d	1	5	3	7	2d
2	B	5d	3	7	3	9	0d
3	C	5d	5	9	5	9	0d

Figure B-3 *Gantt chart in case of A -FF2 - B - SS2 - C combination*

ID	Name	Dur.	E.S	E.F	L.S	L.F	T.i
1	A	5d	1	5	1	5	0d
2	B	5d	3	7	3	7	0d
3	C	5d	5	9	5	9	0d

Figure B-4 *Gantt chart in case of A- FF2 - B - FF2 - C combination*

ID	Name	Dur.	E.S	E.F	L.S	L.F	T.i
1	A	5d	1	5	1	9	0d
2	B	5d	3	7	3	9	0d
3	C	5d	5	9	5	9	0d

Figure B-5 *Gantt chart in case of A- SS2 - B - SS2 - C combination*

ID	Name	Dur.	E.S	E.F	L.S	L.F	T.i
1	A	5d	1	5	1	9	0d
2	B	5d	3	7	3	7	0d
3	C	5d	5	9	5	9	0d

Figure B-6 *Gantt Chart in case of A - SS2 - B - FF2 - C combination.*

We got different results in each case although we tried to describe the same project. CA Superproject or Primavera give the same solutions as shown in Figure B-2 for all the cases.

What can be behind these strange results? If the activities are assumed to be splittable we also get four different results but these results do not match the results from MS Project. The results using the splitting allowed assumption can be followed in Figure B-7. Due to the application of only one precedence relationship between finish activity times INITIAL and TERMINAL times have been used in the calculations.

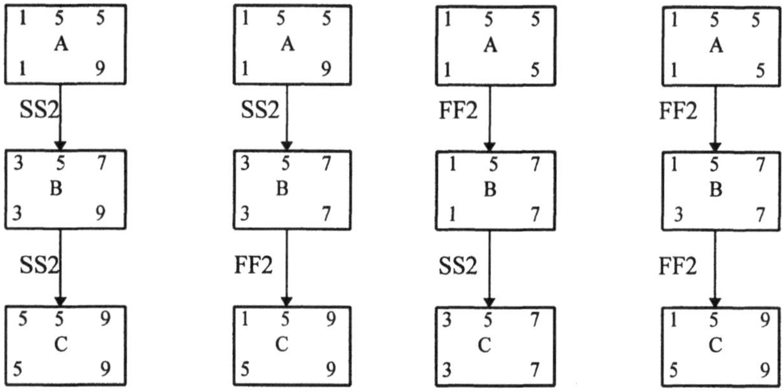

Figure B-7 *Results: splitting allowed*

In this book we have presented two approaches to PDM, one when activity splitting is allowed and one when it is not. A third approach is the mixture of the two above mentioned. In MS Project a new algorithm has been implemented. In the forward pass all the activities are assumed to be non-splittable, in the backward pass all the activities are assumed to be splittable. This approach is theoretically bad in our opinion. If an activity is assumed to be splittable then it has to be handled as splittable in both calculation phases, and if it is assumed to be non-splittable then it has to be handled so in the calculations. This calculation rule usually lead to different critical path, and activity times from the results using the normal rules. Fortunately, most of the software developers have implemented correct algorithms in time analysis but in some cases other minor differences can also arise by using different softwares. The applied algorithms and the optional differences in calculation rules have to be known prior to purchasing the product. The necessity of hiring an expert in the early phase of such kind of developments cannot be overemphasized.

2. PROJECTDIRECTOR

ProjectDirector has been developed by Planetary Systems Budapest, Hungary (http://www.plansys.hu). It contains all the theoritical improvements described in this book.

Any number of precedence relationships can be given between activities therefore the planner does not have to decide which relationship will be the determinant one. It can handle the ZZ combination, and this is the only computer tool on the market that handles maximal relationships. It shows critical path characteristics so it is easy to adjust the project duration to the desired deadline. Another standalone function in the time analysis is that activities can be made splittable if necessary. Different activity types have been implemented in the software. It can handle start milestones, finish milestones and hammock activities besides the normal activities. Constraints can be assigned to the activities on two levels. Constraints of the first level can affect the early and late activity times, second level constraints can be applied to reschedule the activities within their floats. Both resource allocation and leveling can be performed with ProjectDirector. Calendars can be assigned to individual activities and resources. Its standalone loop detecting function helps the users to make logically correct plans. The user can create WBS, RBS and OBS to better organize his or her own project. It supports the whole project management life-cycle from planning to control. Sophisticated earned value analysis and tracking functions help to keep the project on track in an optimal way.

ProjectDirector runs under Windows 3.1 or Windows 95. It is not only the arguably most sophisticated project management software from the theoritical point of view but its graphical interface is also one of the best. Different output formats like Gantt chart, PERT chart, WBS chart, and Calendar view are available.

BIBLIOGRAPHY

[1] Adamiecki, Karol, "Harmonygraph," Przeglad Organizaciji (Polish Journal on Organizational review), 1931.

[2] Ahuja N. H, Dozzi S.P, Abourizk S. M, "Project Management: Techniques in Planning and Controlling Construction Projects" John Wiley & Sons, 2nd Edition 1994

[3] Anon, "Critical Path Analysis: Eighteen Case Histories" British Productivity Council, London 1964

[4] Arisawa, S., Elmaghraby, S. E., "Optimal Time-Cost Trade-Offs in GERT Networks" Management Science, Vol. 18, No. 11, pp. 589-599, July 1972.

[5] Backer, F, "A Discussion of Problem Involved in LESS (Least Cost Estimating and Scheduling)," IBM Applied Science, Dallas, Texas, 1960.

[6] Battersby, A., "Network Analysis for Planning and Scheduling" 3rd edition Macmillan and Co. Limited, London 1970.

[7] Bellman, R,E,: "On a Routing Problem" Quart. Appl. Math. Vol 16. pp87-90, 1958

[8] Busacker, R.G. - Saaty, T.L.: "Finite Graphs and Networks" McGraw Hill, New York, 1964

[9] Clarke, E. W., "An Introduction to Critical Path Analysis," Graduate School of Business, Stanford University, March 1961.

[10] Crandall, K. K.; Hajdu, M., "A CPM költségtervezési feladat 'legrosszabb' megoldása," Közlekedéstudományi Szemle Vol. XLIV. No.5 1994 pp. 173-176

[11] Crandall. K. C., "Project Planning With Precedence Lead/Lag Factors," Project Management Quaterly, Vol. 4. No. 3. Sept. 1973, pp. 18-27

[12] Dantzig, G.B.: "Linear Programming and Extensions", Princeton University Press, Princeton, New Jersey, 1963.

[13] Dijkstra, E.W.: "A Note on Two Problems in Connection with Graphs" Numerische Mathematik, Vol.1. 1959 pp269-271.

[14] Edmonds, J - Karp, R.M.: "Theoritical Improvements in Algorithm Efficiency for Network Flow Problems" Journal of ACM. Vol 19. pp248-264, 1972

[15] Eisner, H., "A Generalized Network Approach to the Planning and Scheduling of a Research Project," Operations Research, Vol. 10, No. 1, pp. 115-125, 1962.

[16] Elmagrabhy, S.E., "Activity Networks: Project Planning and Control by Network Models" John Wiley & Sons, New York, 1977.

[17] Elmegrabhy, S.E., "An Algebra for the Analysis of generalized Activity Networks," Management Science, Vol. 10. No. 3, pp.494-514, 1964.

[18] Fondahl, J. W., "A Non-Computer Approach to the Critical Path Method for the Construction Industry," 1st ed. 1961, 2nd ed. 1962, Department of Civil Engineering, Stanford University

[19] Ford, L. R. Jr., "Network Flow Theory" The Rand Corporation., p923, August 1956

[20] Ford, L.R. Jr -Fulkerson D.R.: "A Sample Algorithm for Finding Maximal Network Flows and an Application to the Hitchcock Problem" Canadian Journal of math. Vol.9. pp210-218, 1957

[21] Ford, L.R. Jr -Fulkerson D.R.: "Maximal Flow Through a Network" Canadian Journal of math. Vol.8. pp399-404, 1956

[22] Ford,L.R.Jr -Fulkerson D.R.: "Flows in Networks" Princeton University Press, Princeton, New Jersey, 1962

[23] Freeman, R.J., "A Generalized PERT," Operations Research, Vol. 8, No. 2, p. 28, 1960.

[24] Fulkerson D. R., "" Operations Research, vol. 10, no., p. 809. 1962

[25] Fulkerson, D. R., "A Network Flow Computation for Project Cost Curves," Management Science Vol. 7 No.2. January, 1961, pp. 167-178

[26] Gondzio, J., Terlaky, T.: "Computational View of Interior Point Methods" to appear in 1996 Advances in Linear and Integer Programming" editor J. Beasley, Oxford University Press, Oxford, England

[27] Hajdu, Miklós "Precedence Diagramming Time-Cost Trade-Off: Splitting Allowed," has been sent to Management Science, 1996

[28] Hajdu, Miklós "Precedence Diagramming Time-Cost Trade-Off: Activities are Splittable or Non-Splittable" to appear in Optimization, 1996

[29] Hajdu, M,; Klafszky, E., "An Algorithm for Solving the CPM Time-Cost Trade-off problem," Periodica Polytechnika Vol. 37, No3, Budapest, 1993 pp. 231-247

[30] Hajdu, M., Klafszky, E. "Hálós tervezési technikák az építések tervezésében és irányításában" (Techniques for Construction Network Scheduling) published by Müegyetemi Kiadó, Budapest, Hungary 1994. 250pp.

[31] Hajdu, M., Mályusz L., "A beruházások gyorsításának vizsgálata matematikai módszerekkel" Közlekedésépítés- és Mélyépítéstudományi Szemle, Vol. XLV. No. 6, 1995

[32] Hajdu, Miklós "A költségtervezési feladat megoldása változó függőségi kapcsolatokat tartalmazó tevékenység csomópontú tervütemhálón," (Time Cost Trade Off in Precedence Diagramming) Ph.D. dissertation 1993, Budapest University of Technology, Hungary

[33] Hajdu, Miklós "An Algorithm for Solving the Cost Optimization Problem in Precedence Diagramming Method," Periodica Polytecthnika ser. Civil Eng. Vol. 37. No.3 pp231-247, 1993

[34] Hajdu, Miklós "Precedence Diagramming Time-Cost Trade-Off: Mathematical Basis," has been sent to Operations Research, 1995

[35] Haward and Robinson, "Preliminary Analysis of the Construction Scheduling Problem" Internal Paper, Engineering Department, DuPont Company, December 1956.

[36] Hebert, J. E. III, "Critical Path Analysis and Simulation Program for Resource Constrained Activity Scheduling in GERT Project Networks," Ph.D. dissertation, Purdue University, 1975

[37] Hebert, J. E. III, "Project Management with PROJECT/GERT," ORSA/TIMS Joint National Meeting, , Los Angeles, California, November 1978.

[38] Hindelang, T.J.; Muth J.F., "A Dynamic Programming Algorithm for Decision CPM Networks," Operations Research, Vol. 27 No.2, 1979, pp. 225-241

[39] IBM: "Users Manual for IBM 1440 Project Control System (PCS)" 1964

[40] Kas, Peter, "A Feasibility Theorem on the Earliest Starting and latest Finishing Times of Critical Path Analysis," Central European Journal of Operational Research and Economics, Vol. 2, No.4, 1993, pp. 319-324

[41] Kelley, E. James, "Critical Path Planning and Scheduling:Mathematical Basis" Operations Research, vol. 9, no. 3, pp. 296-320. 1961

[42] Kelley, E. James; Walker, E. Morgan, "Critical Path Planning and Scheduling," Proceedings of the Eastern Joint Computer Conference, pp. 160-173, Dec. 1-3, 1959.

[43] Klafszky Emil "Hálózati folyamok" (Network Flows) published by Bolyai János Mathematical Society 1969 Budapest, Hungary pp. 201-216

[44] Lockyer, Keith and Gordon, James, "Critical Path Analysis and other Project Network Techniques" Pitman Publishing, 5th Edition, 1991

[45] Malcolm, D. G.; Roseboom, J. H.; Clark, C. E.; Fazar, W., "Applications of a Technique for Research and Development Program Evaulation," Operations Research, Vol. 7, No. 5, pp 646-669, 1959.

[46] McLaren, K. G., Buesnel, E. L. "Network Analysis in Project Management" Cassell, pp11-19, 1969.

[47] Meyer, L. W.; Shaffer, R. L., " Extending CPM for Multiform Project Cost Curves" Proc. of the ASCE, Journal of the Construction Division, Vol. 91 No. Construction1, may 1965.

[48] Meyer, W.L.; Shaffer, L.R., "Extensions of the Critical Path Method Through the Application of Integer Programming," Technical Report

Dept. Of Civil Engineering, University of Illinois, Urbana Ill., July 1963
[49] Minty, G.J.: "A Comment on the Shortest Route Problem" Operations Research, Vol.5. p724 1957
[50] Moder, J. Joseph; Phillips, R. Cecil; Davis, W. Edward, "Project Management with CPM, PERT and Precedence Diagramming" 3rd ed. Van Nostrand Reinhold Company, 1984
[51] O'Brien J. J., "CPM in Construction Management" 4th ed. McGraw-Hill, Inc. 1993.
[52] Perini Corp. "Pioneers CPM, NEw Tool for Job Management," Engineering News Record, January 26, 1961.
[53] PERT, Program Evaulation Research Task, Phase I. Summary Report, Special Projects Office, Bureau of Ordnance, Department of the navy, Washington, July, 1958
[54] Pritsker, A. A. B., "The precedence GERT User's Manual" Pritsker & Associates Inc., Lafayette, IN, 1974.
[55] Pritsker, A. A..B., "GERT: Graphical Evaulation and Review Technique," The Rand Corporation, RM-4973-NASA, Santa Monica, California, 1966.
[56] Pritsker, A. A..B., and Happ, W. W., "GERT Graphical Evaulation and Review Technique, Part I. Fundamentals," Journal of Industrial Engineering, Vol. 17. No. 5, pp. 267-274, 1966.
[57] Pritsker, A. A..B., and Whitehouse, G. E., "GERT Graphical Evaulation and Review Technique, Part II. Applications," Journal of Industrial Engineering, Vol. 17. No. 5, pp. 293-301, 1966.
[58] Roy, B.,"Théorie des graphes:Contribution de la théorie des graphes á l'étude de certains problémes linéaries" Comptes rendus des Séances de l'Acedémie des Sciences, séence du Avril 1959, s2437-2449, 1959
[59] Roy,B., "Graphes et ordonnancements" Revue Francaise de Recherche Operationelle, s323-333 1962.
[60] Sayer, J. S., Kelley, E. James., Walker, E. Morgan, "Critical Path Scheduling" Factory, July 1960.
[61] Siemens, Nicolai, "A Simple CPM Time-Cost Trade-Off Algorithm," Management Science Vol. 17. No. 6. Feb. 1971 pp. 345-363
[62] Terlaky, T. : "A Finite Criss-Cross Method For Linear Programming and The Proof of Its Finiteness" (in Hungarian) Alkalmazott Matematikai Lapok, Vol. 10. pp289-296, 1984
[63] Terlaky, T., Zang, S.: "Pivot Rules for Linear Programming - A Survey" Annales of Operations Research Vol. 46, pp203-233, 1993
[64] Terlaky, T.: "A Convergent Criss-Cross Method" Optimization Vol. 16 No 5. pp683-690, 1985

[65] Weist, J. D.: "Precedence Diagramming Method: Some Unusual Characteristics and Their Implications for Project Management" Journal of Operations Management Vol. 1. No. 3. Feb. 1981 pp 121-136.

[66] Wiest, D. J., Levy, K. Ferdinand "A Management Guide to PERT/CPM," Prentice-Hall inc., Englewood Cliffs, New York, 1969

[67] Yen, J.Y., "An Algorithm for Finding Shortest Routes from all Sources Nodes to a Given Destination in General Network," Quart. Appl. Math. Vol.27, pp526-530, 1970

[68] Yen, J.Y., "Finding the K shortest Loopless Paths in a Network," management Science Vol. 17. 1971 pp. 712-716

INDEX

I

K

L

M

N

O

P

W

Y

Nonconvex Optimization and Its Applications

1. D.-Z. Du and J. Sun (eds.): *Advances in Optimization and Approximation.* 1994. ISBN 0-7923-2785-3
2. R. Horst and P.M. Pardalos (eds.): *Handbook of Global Optimization.* 1995 ISBN 0-7923-3120-6
3. R. Horst, P.M. Pardalos and N.V. Thoai: *Introduction to Global Optimization* 1995 ISBN 0-7923-3556-2; Pb 0-7923-3557-0
4. D.-Z. Du and P.M. Pardalos (eds.): *Minimax and Applications.* 1995 ISBN 0-7923-3615-1
5. P.M. Pardalos, Y. Siskos and C. Zopounidis (eds.): *Advances in Multicriteria Analysis.* 1995 ISBN 0-7923-3671-2
6. J.D. Pintér: *Global Optimization in Action.* Continuous and Lipschitz Optimization: Algorithms, Implementations and Applications. 1996 ISBN 0-7923-3757-3
7. C.A. Floudas and P.M. Pardalos (eds.): *State of the Art in Global Optimization.* Computational Methods and Applications. 1996 ISBN 0-7923-3838-3
8. J.L. Higle and S. Sen: *Stochastic Decomposition.* A Statistical Method for Large Scale Stochastic Linear Programming. 1996 ISBN 0-7923-3840-5
9. I.E. Grossmann (ed.): *Global Optimization in Engineering Design.* 1996 ISBN 0-7923-3881-2
10. V.F. Dem'yanov, G.E. Stavroulakis, L.N. Polyakova and P.D. Panagiotopoulos: *Quasidifferentiability and Nonsmooth Modelling in Mechanics, Engineering and Economics.* 1996 ISBN 0-7923-4093-0
11. B. Mirkin: *Mathematical Classification and Clustering.* 1996 ISBN 0-7923-4159-7
12. B. Roy: *Multicriteria Methodology for Decision Aiding.* 1996 ISBN 0-7923-4166-X
13. R.B. Kearfott: *Rigorous Global Search: Continuous Problems.* 1996 ISBN 0-7923-4238-0
14. P. Kouvelis and G. Yu: *Robust Discrete Optimization and Its Applications.* 1997 ISBN 0-7923-4291-7
15. H. Konno, P.T. Thach and H. Tuy: *Optimization on Low Rank Nonconvex Structures.* 1997 ISBN 0-7923-4308-5
16. M. Hajdu: *Network Scheduling Techniques for Construction Project Management.* 1997 ISBN 0-7923-4309-3
17. J. Mockus, W. Eddy, A. Mockus, L. Mockus and G. Reklaitis: *Bayesian Heuristic Approach to Discrete and Global Optimization.* Algorithms, Visualization, Software, and Applications. 1997 ISBN 0-7923-4327-1

KLUWER ACADEMIC PUBLISHERS – DORDRECHT / BOSTON / LONDON

MIX
Papier aus verantwortungsvollen Quellen
Paper from responsible sources
FSC® C105338

If you have any concerns about our products,
you can contact us on
ProductSafety@springernature.com

In case Publisher is established outside the EU,
the EU authorized representative is:
Springer Nature Customer Service Center GmbH
Europaplatz 3, 69115 Heidelberg, Germany

Printed by Libri Plureos GmbH
in Hamburg, Germany